Advances in Spatial Science

Springer

Berlin
Heidelberg
New York
Hong Kong
London
Milan
Paris
Tokyo

Titles in the Series

C. S. Bertuglia, M. M. Fischer and *G. Preto* (Eds.)
Technological Change, Economic Development and Space
XVI, 354 pages. 1995. ISBN 3-540-59288-1
(out of print)

H. Coccossis and *P. Nijkamp* (Eds.)
Overcoming Isolation
VII, 272 pages. 1995. ISBN 3-540-59423-X
(out of print)

L. Anselin and *R. J. G. M. Florax* (Eds.)
New Directions in Spatial Econometrics
XVX, 420 pages. 1995. ISBN 3-540-60020-5
(out of print)

H. Eskelinen and *F. Snickars* (Eds.)
Competitive European Peripheries
VIII, 271 pages. 1995. ISBN 3-540-60211-9

J. C. J. M. van den Bergh, P. Nijkamp and *P. Rietveld* (Eds.)
Recent Advances in Spatial Equilibrium Modelling
VIII, 392 pages. 1996. ISBN 3-540-60708-0

P. Nijkamp, G. Pepping and *D. Banister* (Eds.)
Telematics and Transport Behaviour
XII, 227 pages. 1996. ISBN 3-540-60919-9

D. F. Batten and *C. Karlsson* (Eds.)
Infrastructure and the Complexity of Economic Development
VIII, 298 pages. 1996. ISBN 3-540-61333-1
(out of print)

Y. Leung
Intelligent Spatial Decision Support Systems
XV, 470 pages. 1997. ISBN 3-540-62518-6

C. S. Bertuglia, S. Lombardo and *P. Nijkamp* (Eds.)
Innovative Behaviour in Space and Time
X, 437 pages. 1997. ISBN 3-540-62542-9

A. Nagurney and *S. Siokos*
Financial Networks
XVI, 492 pages. 1997. ISBN 3-540-63116-X

M. M. Fischer and *A. Getis* (Eds.)
Recent Developments in Spatial Analysis
X, 434 pages. 1997. ISBN 3-540-63180-1

R. H. M. Emmerink
Information and Pricing in Road Transportation
XVI, 294 pages. 1998. ISBN 3-540-64088-6

P. Rietveld and *F. Bruinsma*
Is Transport Infrastructure Effective?
XVI, 384 pages. 1998. ISBN 3-540-64542-X

P. McCann
The Economics of Industrial Location
XII, 228 pages. 1998. ISBN 3-540-64586-1

L. Lundqvist, L.-G. Mattsson and *T. J. Kim* (Eds.)
Network Infrastructure and the Urban Environment
IX, 414 pages. 1998. ISBN 3-540-64585-3

R. Capello, P. Nijkamp and *G. Pepping* (Eds.)
Sustainable Cities and Energy Policies
XI, 282 pages. 1999. ISBN 3-540-64805-4

M. M. Fischer and *P. Nijkamp* (Eds.)
Spatial Dynamics of European integration
XII, 367 pages. 1999. ISBN 3-540-65817-3

M. M. Fischer, L. Suarez-Villa and *M. Steiner* (Eds.)
Innovation, Networks and Localities
XI, 336 pages. 1999. ISBN 3-540-65853-X

J. Stillwell, S. Geertman and *S. Openshaw* (Eds.)
Geographical Information and Planning
X, 454 pages. 1999. ISBN 3-540-65902-1

G. J. D. Hewings, M. Sonis, M. Madden and *Y. Kimura* (Eds.)
Understanding and Interpreting Economic Structure
X, 365 pages. 1999. ISBN 3-540-66045-3

A. Reggiani (Ed.)
Spatial Economic Science
XII, 457 pages. 2000. ISBN 3-540-67493-4

D. G. Janelle and *D. C. Hodge* (Eds.)
Information, Place, and Cyberspace
XII, 381 pages. 2000. ISBN 3-540-67492-6

P. W. J. Batey and *P. Friedrich* (Eds.)
Regional Competition
VIII, 290 pages. 2000. ISBN 3-540-67548-5

B. Johansson, Ch. Karlsson and *R. R. Stough* (Eds.)
Theories of Endogenous Regional Growth
IX, 428 pages. 2001. ISBN 3-540-67988-X
(out of print)

G. Clarke and *M. Madden* (Eds.)
Regional Science in Business
VIII, 363 pages. 2001. ISBN 3-540-41780-X

M. M. Fischer and *Y. Leung* (Eds.)
GeoComputational Modelling
XII, 279 pages. 2001. ISBN 3-540-41968-3

For information about further volumes please see end of the book

Michel Beuthe · Veli Himanen
Aura Reggiani · Luca Zamparini
Editors

Transport Developments and Innovations in an Evolving World

With 69 Figures
and 60 Tables

Springer

Professor Michel Beuthe
Groupe Transport et Mobilité (GTM)
Facultés Universitaires
Catholiques de Mons (FUCAM)
151, Ch. de Binche
7000 Mons
Belgium
michel.beuthe@fucam.ac.be

Professor Aura Reggiani
Department of Economics
Faculty of Statistics
University of Bologna
Piazza Scaravilli 2
40126 Bologna (BO)
Italy
reggiani@economia.unibo.it

Dr. Veli Himanen
JP-Transplan Ltd
Jaakonkatu 3
01621 Vantaa
Finland
veli.himanen@poyry.fi

Dr. Luca Zamparini
Department of Law
University of Lecce
Via Monteroni, snc
73100 Lecce (LE)
Italy
luca.zamparini@sesia.unile.it

ISBN 3-540-00961-2 Springer-Verlag Berlin Heidelberg New York

Cataloging-in-Publication Data applied for

A catalog record for this book is available from the Library of Congress.

Bibliographic information published by Die Deutsche Bibliothek
Die Deutsche Bibliothek lists this publication in the Deutsche Nationalbibliografie; detailed bibliographic data available in the internet at *http.//dnb.ddb.de*

Springer-Verlag Berlin Heidelberg New York
is a part of Springer Science+Business Media
springeronline.com

Cover design: Erich Kirchner, Heidelberg

SPIN 10923474 42/3130 – 5 4 3 2 1 0 – Printed on acid-free paper

"Before new knowledge changes knowledge structures and systems of activity within an organization, knowledge has to be accessed, understood and accepted."

Ilkka Tuomi

Preface

Our world is continuously changing as new technologies are developed, demographic patterns evolve and new social and political configurations emerge. However, most people would agree that, not unlike the time of the Industrial Revolution, the rate of change has accelerated over the last few decades. It is certainly the case of many economic activities, where the trend towards globalization, supported by a faster diffusion of knowledge and information, economies of scale, the opening of borders and improved transport networks, has induced a spatial redistribution of production and ownership. Until recently, the concomitant economic growth eased the way towards the necessary reorganization, but, at the same time, created additional environmental and transport problems.

This phenomenon is pervasive all around the globe, but it particularly affects the European continent, which is still fragmented by administrative, cultural and linguistic differences, national boundaries and rivalries. In this context, the slow, but persistent drive towards market integration and deregulation makes Europe an extraordinary laboratory for politicians, economic researchers and especially for transport analysts.

For all these reasons, dynamic research programs have been financed by European authorities, who rightly view an efficient international transport network as a necessary condition for developing an integrated European market. These programs have played an important role in supporting many research activities in universities, and furthering analytical advances in transportation. Part of that work in Europe has taken place within the framework of the Network on European Communications and Transport Activity Research (NECTAR). This scientific network organizes specialized workshops and network conferences on a regular basis all over Europe, with the objective of the development and dissemination of new knowledge.

The present volume is the outcome of one of these NECTAR activities, viz. a research conference held at Espoo (Helsinki) in May 2001, and organized with the help of VTT Building and Transport, the Government Institute for Economic Research (VATT), the Ministry of Transport and Communications, Finland,

ETLA – The Research Institute of the Finnish Economy, Helsinki University of Technology and the City of Espoo. The overall theme was European strategies in the global markets with a focus on transport innovations, competitiveness and sustainability in the Information Age. More than 60 papers were presented and discussed by NECTAR members during the conference. From among the various presentations, the editors made a selection of new contributions that analyzed contemporary transport developments and innovations in our evolving world. They were reviewed thoroughly and set in their final form as they appear in this volume.

The organization of a conference and the making of a book require many contributions beyond the papers concerned. The following persons in VTT Building and Transport are to be thanked for their support of the Helsinki research conference: Marko Nokkala, Jukka Rasanen, Raija Sahlstedt, and Anu Tuominen contributed to the perfect organization of the conference and the social events that made it a convivial as well as a productive event.

The editors wish to thank the Editorial Board of the series *Advances in Spatial Sciences*, David F. Batten, Manfred M. Fisher, Geoffrey J.D. Hewings, Peter Nijkamp and Folke Snickars, as well as Marianne Bopp (Economics Editor, Springer-Verlag) for giving their kind support to the idea of a volume on these topics. Last but not least, we thank Patricia Ellman (Perfect Copy, Amsterdam) for her careful checking of the English text, tables and figures, and Dianne Biederberg (Contact Europe, Amsterdam) whose help in producing the final text was indispensable for carrying out the editing of the book.

We were happy to benefit from the collaboration of all those people, who, like most of the authors, are citizens of many different European countries. As such, this book is a small example of what European cooperation can achieve.

Michel Beuthe
Veli Himanen
Aura Reggiani
Luca Zamparini

Mons/ Helsinki/ Bologna/ Lecce
Autumn 2003

Contents

1 Introduction

Michel Beuthe*, Veli Himanen**, Aura Reggiani***
and Luca Zamparini****

*Facultés Universitaires Catholiques de Mons
**JP-Transplan Ltd
***University of Bologna
**** University of Leece

We are living in the Information Age with ever-more complex societies acting in global networks. The impacts of ICTs (information and communication technologies) embrace not only faster and denser communication but also a reorganization of value chains, firms, labour relations and management structures, and consequently of spatial forms and developments. Our daily activities, and therefore travel demand, are influenced directly by our own use of new ICT devices and indirectly by the above-mentioned changes in our society. Also the transport supply side is influenced directly through the application of ICTs and indirectly through demands arising from changing society. We have to remember, however, that: i) transport still is for improving accessibility, i.e. overcoming distances, and ii) because of the current huge volumes of motorized transport, serious safety and environmental problems have emerged together with local or even regional congestion.

Among the characteristics of the Information Age, there are three – economic growth, complexity and globalization – which have a distinct influence on travelling. Economic growth increases travelling in general. We know that affluent people travel more than poor ones and employed people travel more than the unemployed. Growing complexity in various issues of society increases the need for information and thereby also travelling. All information cannot be obtained from the Internet, and this holds true even more for knowledge and understanding. Globalization increases the length of trips and especially favours air transport. On the other hand, globalization means that negative consequences are also spreading rapidly all over the world. The misery of the airline industry in 2003 was produced by a chain of successive phenomena: economic recession, the 11/9 attack, the Iraq war, and SARS.

ICTs would seem to provide many new options that are not actually adopted, due to the fixed routines of individual persons and firms. Inertia produced by these routines is the major reason why our daily living – including travelling – has not changed that much. The conflict between transport innovations, often triggered by ICTs, and the various kinds of resistance to their adoption is a basic theme in this book. Both sides of this discrepancy are highlighted in the following chapters through methodological discussions and practical case studies. Some chapters present possible transport developments and innovations and their expected influences, some discuss the resistance and inertia for adoption, and some embrace both aspects. We have not advanced far enough into the Information Age to be able to thoroughly understand the implications that ICTs have in our transport system and daily travelling. However, the following chapters give interesting glimpses of the possible directions, as well as highlighting the existing inertia in adopting new ways in passenger and freight transport.

This book is divided into four parts, whose structure is illustrated in Figure 1.1. In brief, Part A includes four chapters, which provide a broad base for the whole book, by touching on most of the major topics mentioned above. The first of these chapters provides a model for analyzing the adoption of transport innovations, and discusses the reasons for rapid and slow adoption of different innovations in various transport sectors (Chapter 2). Accessibility changes stimulated by the applications of ICT are then considered and conclusions concerning their influence on our daily action spaces are drawn (Chapter 3). The reasons for accepting various ICT applications in the global shipping industry are also presented (Chapter 4). In the longer run, the use of ICTs, together with structural changes in economy, will have an impact on urban functions and spatial development (Chapter 5).

Looking in more detail at Part A, first, in Chapter 2, Eran Feitelson and Ilan Salomon focus on the issue of why some innovations are adopted and some are not. To answer this they develop a political economy model of transport innovations and analyze factors that affect the adoption of innovations. They also apply the model in practical cases and give answers as to why light rail is often adopted, road pricing schemes are not realized, and telecommuting centres survive only as far as the funding of public agencies continues.

Our daily action space – the area where we undertake activities – is already influenced by the application of ICTs, both through our own use of new communication facilities and through the changes happening in our society. In Chapter 3, Martin Dijst analyzes the possible future changes in our action spaces through the potential impacts of ICTs on accessibility. He comes to the conclusion that our daily action space will expand and the actual action space will then come nearer to the perceived one, which will in turn better fit the potential action space.

Globalization of economies means, in principle, ever-longer transport distances. The container shipping industry, in particular, benefited from this development. Shipping companies are widely using some ICT applications – Electronic Data Interchange (EDI) and Electronic Data Processing systems – but they are far from adopting fully integrated e-business systems. In Chapter 4, Hans van Ham and Bart Kuipers describe the container shipping market, which is dominated by some 30 carrier companies. They also clarify the basic features of e-business and apply these in container shipping. Their conclusion is that the reluctance to give up earlier investments in EDI and move to the Internet is the main reason for the current situation, but also that the prevailing old-fashioned culture in carrier companies has some influence.

The wide use of ICTs leads to structural changes in our economy and society. In Chapter 5, Marina van Geenhuizen and Peter Nijkamp deal with both the actual and the potential impacts of these structural changes on urban functions. They study various new developments, such as customer driven production, customer services, desintermediation and reintermediation, e-firms and e-webs of firms, and flexibility of work places. They also discuss the possible constraints that different forms of knowledge impose on electronic communication. They found a few obvious trends but many uncertainties in the impacts that the e-economy has on cities.

Part B deepens the understanding of the conflict between options provided by transport innovations and the resistance to their adoption. All the chapters in this part are based on primary data obtained through questioning people about their travel habits and preferences – two in local and two in European-wide studies. It is found that individuals do not freely consider all possible options for travelling but are constrained by their current habits (Chapter 6). As is also pointed out, methodologies can, and have to, internalize the inertia coming from current habits, in order to provide correct policy guidance (Chapter 7). The more traditional measures – parking charges and the improvement of public transport – for influencing modal split are analyzed (Chapter 8), followed by a rigorous examination of the acceptability of road pricing – one of the most promising and, at the same time, most debated measures (Chapter 9).

Stated preference analyses are commonly used in order to forecast changes in travel behaviour before implementing a new transport measure. In Chapter 6, Job van Exel and Piet Rietveld challenge the standard behavioural model used and prove – by analyzing the answers of Dutch commuters – that our preferences over transport modes as well as preferences towards choice determinants are influenced by our current commuting habits. This bias has impacts not only on the use of stated preference analyses but also on policy implementation. Individuals respond from their current situation and may picture themselves as captive users of their current mode.

In theory, individuals have to decide their next activity, its place and mode of transport before going anywhere. In Chapter 7, Martin Lanzendorf has found that routines also exist for leisure travel, which in principle can be more freely executed than our daily obligatory trips like commuting. He proves, on the basis of interviews in various European countries, that travel mode choices in leisure travel are, in most cases, routines – only 16 percent of them are not made as routines. This again has serious implications in transport policy implementation and has to be considered in transport surveys when forecasting the impacts of any transport development or policy action.

There are various ways to improve the information provided by stated preference analyses. In Chapter 8, Jari Kurri presents an exercise where the preference survey was arranged in two parts, the first dealing with choice of parking location, and the second with choice of mode of travel. Data from the survey were then used to estimate logit models of transport mode and parking choices. The models showed that parkers are not very willing to walk even a couple of hundred metres extra, but that the effects of habit and imperfect information can also have a significant effect on choice. Furthermore, it seems that parking policy measures, such as doubling parking costs, are more efficient as ways to influence modal share than those measures that make public transport more attractive.

Economists have for many decades promoted road pricing as the most efficient measure to reduce traffic congestion and pollution – at least inside a limited area. Yet, its breakthrough is still to be seen. In Chapter 9, Jens Schade presents a European-wide study on the acceptability of road pricing. Already in Chapter 2, Freitelson and Salomon argued that the lack of acceptability is the main obstacle for the implementation of road pricing. The motorists surveyed in Schade's study confirm this. The stronger – i.e. more effective – the strategy, the less they liked it. This result is, of course, in conflict with their general agreement that traffic ought to be limited in order to reduce congestion and air pollution.

Part C highlights freight transport developments and innovations. It starts with network competencies, focusing on the organizational aspects (Chapter 10), and then continues with intermodal terminals (Chapter 11). The findings suggest that geographical proximity – both among the local firms and between these and local transport firms – represents an important condition for generating external economies in terms of a regional transport and logistics system. In other words, even in the Information Age, local accessibility – together with shared knowledge networks – has a strong impact on the success of firms. Intermodal transport has the potential to advance sustainability – especially when most of the route is travelled by rail, an inland waterway system, or via an ocean-going vessel, with the shortest possible initial and final journeys by road. An intermodal route includes, by definition, more than one transport mode. Therefore, the location of a terminal where unit loads are moved from one mode to another has a great impact for the success of that route. At the terminal itself, both efficiency and environmental quality can be improved by the use of ICTs in its operations (Chapter12).

Analyses of the spatial organization of industries and networks of firms often focus on production and distribution systems. The significance of transport operations and logistical activities are often neglected, or usually referred to as an activity which is solely a consequence of the particular industrial organization. In Chapter 10, Leif Hansen suggests that transport is more than infrastructure and transport capacity supplied to industry. It is an integrated part of the competitive advantage for single firms' production systems, as well as for localized networks of industrial organization. Relatively small transport operators have gained specialized competencies in handling goods and organizing transport chains as direct door-to-door delivery from factory to store or final consumer. The development of such localized competence could prove to be more important for the industrial activity of a region than traditional improvements of its physical transport infrastructure. Hansen draws his conclusions from a case study based on a questionnaire and in-depth interviews in Denmark.

In Chapter 11, Cathy Macharis proposes a methodological framework for determining the optimal location for an intermodal terminal. She presents a network model in order to identify the market areas of the existing and possible new intermodal terminals. The effect of a new terminal on the existing ones and on the environment is assessed. Then, a discrete number of potential locations are evaluated with the aid of a multi-criteria decision support model. Here, the goals of the various parties economically affected by the location of a new terminal (the users, the terminal operators, the investors and the society as a whole) are taken into consideration.

Modern terminal operations strongly rely on the advanced possibilities of ICTs. Due to ICTs, more complex and more simultaneous operations can be carried out. In Chapter 12, Yvonne Bontekoning and Jan Jacob Tripp address the problem of shunting in hub-and-spoke networks. The disadvantages of shunting are that it is time- and land-consuming and noisy. Recently, alternative solutions for shunting in the form of hub-terminals have been proposed and elaborated into detailed designs. The in-depth qualitative comparative case study on the operational details and structure indicate that new-generation hub-terminals could be a feasible alternative for shunting.

Part D embraces many issues discussed above – acceptability, accessibility, sustainability and transport demand – and provides methodologies and insights for the assessment of transport developments in relation to future cities and regions. European urban regions are more and more polynuclear. Promoting the economic vitality of the city centres has an increasing political priority, and new solutions for public transport – and also restrictions on car traffic – are being sought.

The transformation from traditional industries to a service economy, stimulated by ICTs, gives new opportunities for urban development and urban strength. There is, however, an increasing danger that the development potential of cities will be frustrated through shortcomings in the accessibility of the cities. The opportunities to travel to a city by car and to find a parking place on arrival are bound by physical limitations. Light rail networks – as a potential solution for improved accessibility in urban regions – are discussed in Chapter 13 by Hugo Priemus (for the acceptability of light rail, see also Chapter 2). According to his findings, light rail seems to be an effective instrument to bring about a synergy between public transport, real estate development and urban vitality.

In many cities, current severe problems caused by transportation, such as air pollution, traffic congestion and open land depletion, are being tackled by various transport developments. In Chapter 14, Yoram Shiftan, Sigal Kaplan and Shalom Hakkert examine potential policy measures (for the acceptability of policy measures, see also Chapters 2, 8 and 9) which may contribute to a sustainable transportation system in Tel Aviv. The implementation probabilities and the desirability of the policy measures were evaluated by means of a Delphi expert-based survey among leading Israeli experts in the fields of transportation and regional planning. From the opinions of the experts, an expected scenario and a desired scenario were formulated for the year 2030. According to the results of the first round of this survey, the key elements in the desired scenario are: a highly developed public transportation system, better coordination between the spatial development and the transportation system, traffic restraint policy measures (high parking fees, congestion pricing, restricted car zones), and maintaining the functional role of the Central Business District area.

Alain L'Hostis, Philippe Menerault and Christophe Decoupigny have studied – in Chapter 15 – accessibility provided by rail services into Lille city centre from the polynuclear region of the Nord-Pas-de-Calais. They have developed an accessibility indicator for rail services, also including the impacts of fixed timetables and feeder buses on travel time. They have applied the indicator in four different scenarios relating to the region.

Transport developments and innovations are implemented in an ever-more complex world. This complexity makes policy makers ask for more comprehensive assessments about the benefits provided by new transport developments. In the final Chapter 16, John Nellthorp highlights four areas of multi-modal assessment in which progress has been made: firstly, attempts to bring spatial employment and output effects into the analysis; secondly, refinements in the theory and practice of multi-modal appraisal; thirdly, the rise of participative evaluation; and fourthly, a substantial shift in 'official' evaluation methods, from an efficiency perspective to a strategy-led perspective, with implications for the breadth and scope of the assessment in many countries. In summary, he identifies as the main driving forces behind recent appraisal developments: i) transport policy itself, acting through the objectives-led appraisal approach which is dominant at present; ii) multi-modality, which has helped to rejuvenate transport cost-benefit analysis; and iii) changes in political structures, particularly increased public-private partnership and the role of interest groups as 'stakeholders'.

From this review of the chapters, the following general conclusions can be drawn:

1 In the current Information Age, the network economy, globalization, complexity and sustainability influence transport demand.
2 ICTs are influencing transport demand directly by the use of new devices, and indirectly by changing the society in which we live, thereby inducing a change in urban structures.
3 The above influences have not diminished the value of geographical proximity and accessibility.
4 New transport developments and innovations have manifold objectives: i) to improve accessibility; ii) to restrict car use; iii) to diminish environmental impacts; iv) to increase the use of public transport; v) to increase the efficiency of transport firms.
5 Transport policy measures have to gain both social and political acceptability before they can be implemented.
6 There is considerable resistance and inertia in passenger and freight sectors for new forms of travel behaviour and business practices.

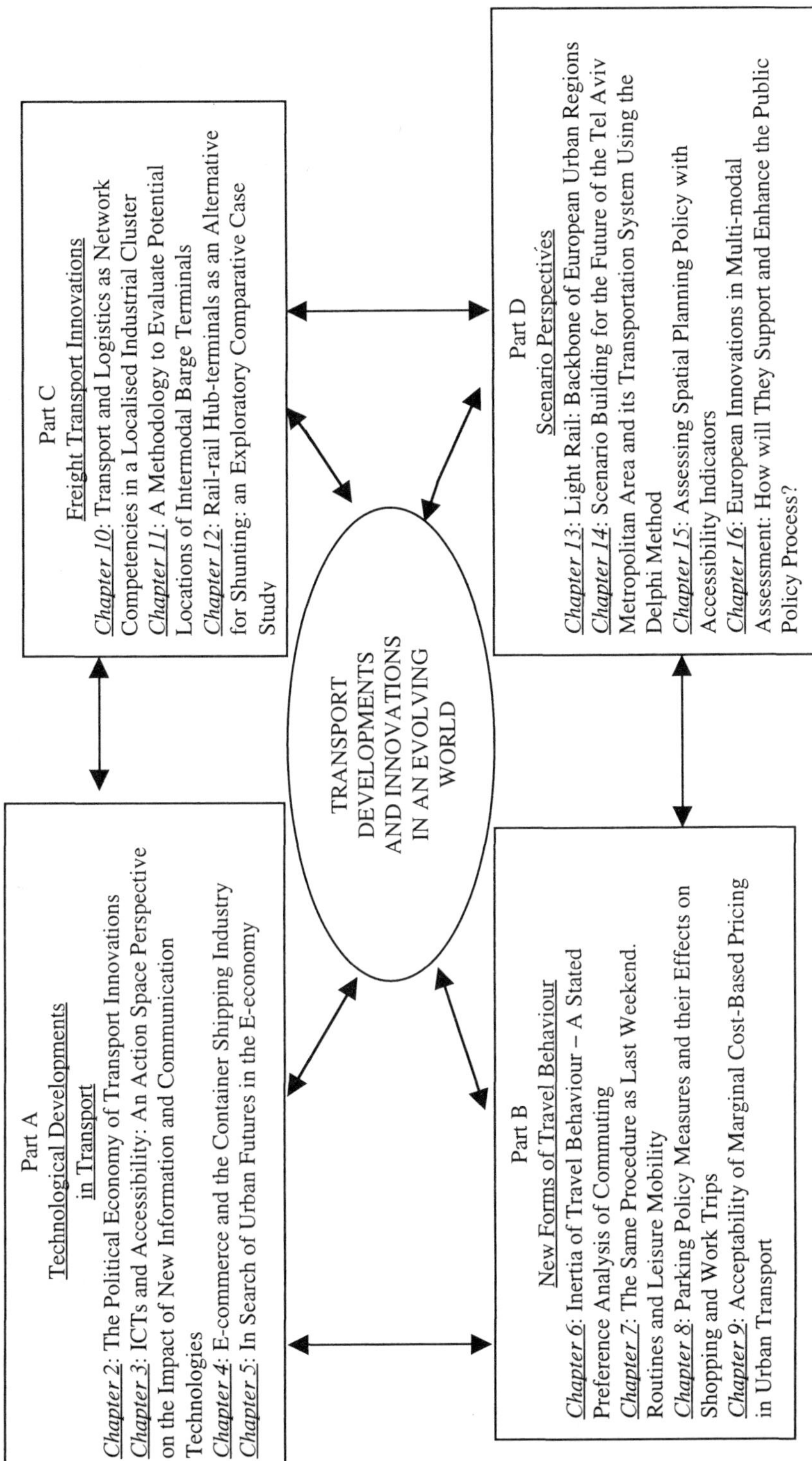

Figure 1.1 Structure of the book

PART A: Technological Developments in Transport

2 The Political Economy of Transport Innovations

Eran Feitelson and Ilan Salomon

The Hebrew University of Jerusalem, Jerusalem, Israel

2.1 Introduction

Many transportation and telecommunication innovations were proposed over the course of the last century. They include new technologies, new ways of organization and new ways to manage these systems. Overall, the innovations that were adopted have changed the way we conduct our affairs, as the cost and effort associated with overcoming distance plummeted. Yet, many additional options were proposed, but never adopted, or were adopted only in a very limited and modified form. In some cases, innovations were first adopted and later abandoned. These observations raise the question: Why have certain innovations been adopted while others have not?

Clearly, new ideas will be adopted only if they are deemed useful, and thus people are willing to use them. Much of the research effort on transport and telecommunication innovations has therefore focused on the potential of the innovations and the way they are perceived by consumers. Yet, the implementation of most transport innovations requires large-scale infrastructure investments, or changes in rules and statutes, and thus it is not really something that can be analyzed purely as an outcome of atomistic decision making processes. Rather, it should be viewed as an outcome of societal processes, often with significant government involvement. In this chapter, we analyze the factors that affect the adoption of innovations at the societal level.

In the next section, the factors that affect the adoption of transportation innovations are reviewed. Then, a political economy framework for analyzing the adoption of transport innovations is advanced in Section 2.3. This framework argues that the adoption of innovations is predicated on their economic, technical, social and political feasibility. Thus, it is insufficient that an innovation meets a strict benefit-cost criterion, or that there is a majority of voters supporting it. Rather, it suggests that the coalitions behind such innovations should be analyzed, and the incentives for decision makers need to be reckoned with.

To test the generality of this framework, several different proposed innovations are analyzed in the subsequent section, thus widening the scope of analysis from analyses of single policy elements (such as road pricing) or technological options (such as rail) to the adoption of transport innovations as such. The chapter concludes with a discussion of the main factors that affect the adoption of different innovations.

2.2 Factors That Affect Adopting of Innovations: A Brief Review

The term 'transport innovation' includes both new ways to manage transport systems (mostly by use of various policy tools) and new technologies. The adoption of any such innovation implies that it comes into use. Thus, it is necessary to distinguish between penetration and adoption of innovations. The former means that an innovation has become available, while the latter refers to the actual use of an innovation that has already penetrated the market. Yet, such use is predicated on a series of factors.

Perhaps the most fundamental factor is that the innovation will be seen as technically feasible – that it can be used. The first problem that most innovators face, therefore, is to convince others that the innovation can work. This is true for both technology-driven innovations, and for policy innovations. Road pricing, for example, was proposed over 80 years ago, and toll roads have been known for over 200 years (Levinson 1998). But the modern concepts were given impetus only after the early experiments in Singapore and Hong Kong, which duly received wide attention. The importance of the Hong Kong electronic road pricing scheme was that it proved that such a system is capable of working technically. Yet, the same experiment also showed that technical success may be insufficient, as the scheme was scrapped due to the privacy concerns it raised, which led to its failure in the political test (Borins 1988).

When the innovation is not perceived as technically feasible, it is likely to fail. This in essence was the story of ARAMIS – the personal rapid transit technology developed and studied in France for over 17 years (Latour 1996). Yet, purely technical feasibility is a necessary but not sufficient condition for an innovation to be adopted. There has to be a demand for it. Indeed the question of whether there is sufficient demand for cars powered by alternative fuels has received considerable attention in recent years (for a recent example, see Brownstone et al. 2000), as this is seen as a crucial issue in the attempts to curb CO_2 emissions.

The technical and behavioural analyses which attempt to assess the technical feasibility (which may often be a function of cost) and the latent demand for the innovation are inputs that are necessary for mainstream cost-benefit analyses of such innovations. Essentially, the logic of such studies is that, if an innovation can be shown to be cost effective and to provide benefits above costs, it will eventually be adopted. This has been the underlying rational behind many economic analyses of potential innovations.

Yet, innovations that have been shown to be cost effective have not necessarily been implemented. This is the case of road pricing. Despite the multiple studies showing that such schemes are efficient and technologically feasible, they have not been widely implemented. This failure has drawn considerable attention in recent years (Giuliano 1992; Jones 1998; Small and Gomez-Ibanez 1998).

The studies analyzing the impediments to the implementation of road pricing have identified multiple factors that seem to impede the adoption of such schemes. Jones (1998), drawing upon UK case studies, has identified a series of arguments that have been raised against road pricing, and suggested improvements in the way they are advanced so as to allay these concerns. These include: the need to convince the public and decision makers that action is indeed necessary; the need to allay public qualms about the effectiveness of road pricing; the need to address equity concerns; and the need to explain the use of revenues and relate it to drivers' concerns.

Other studies tried to identify whether there are inherent limitations to the adoption of such instruments. Lave (1994), for example, shows, in a theoretical model, that it is unclear whether road pricing does indeed improve the welfare of drivers. Small and Gomez-Ibanez (1998) looking at empirical evidence suggest that the main impediment to the use of such taxes is indeed that most motorists stand to lose, and do not perceive what they stand to gain from the revenues they generate. Rietveld and Verhoef (1998) note that the main beneficiaries are likely to be those suffering from the roads' environmental externalities, and the citizens at large enjoying the additional funds in the general coffer. They propose, therefore, that part of the revenues be directed back toward transport issues, so the benefits will be more obvious.

The discussion of road pricing is, however, only illustrative of the more general concern over the distribution of benefits and costs of different innovations. Rienstra et al. (1999) have analyzed the factors that determine the support for different policy options in the Netherlands. They found that perceptions of both the salience of the problem and of the effectiveness of the proposed policy tool in addressing the problem affect the readiness to support the use of the specific measure. The perception of the salience of the problem is, however, not limited to the individual concern over the problem, but is extended to also include issues seen as socially problematic.

The conceptual approach behind these studies, stated explicitly by Rietveld (1997), is of a referendum or voting context. That is, it is assumed that policies are determined by popular support. Rietveld (1997) thus adds to the transportation factors and distribution implications also the voters' ideologies, or attitudes toward environmental equity and development issues. Yet, transport policies, including decisions regarding investments necessary to implement new technologies, are most commonly made by existing institutions, and are not a focal point of referendums or elections. That is, they are not subject to direct democracy procedures. In discussing such routine decision-making contexts, the power of interest groups needs to be taken into account (Becker 1983).

The power of interest groups stems from their relatively small size. This allows them to organize themselves better than larger more diffuse groups, as their interests are more focused and the transaction cost of organizing small groups are lower (Olson 1965; Becker 1983). This focus and superior organization allows interest groups to devote their often-considerable resources (whether in terms of time, money or personnel) to very concrete issues. In this way, they often succeed in swaying policies their way, regardless of their actual size.

As there are multiple interest groups, whose interests may or may not coincide in any particular case, the outcome according to this perspective is a function of the political power they wield. This can be enhanced by coalitions with other interest groups.[1] Whitt and Yago (1985) analyze the evolution of urban transportation systems in the US as a function of the changing political and market power of different industrial interest groups and the changing relations among social classes. They show how the shift from transit to car-based systems was an outcome of the rise in market and political power of the oil-rubber-car oligopolies at the expense of the local transit systems and the industries that supplied transit vehicles, which were developed when the coal rail and iron industries were supreme. They go on to argue that the recent transit systems built in the US are essentially political tools for urban redevelopment and pacification, wielded by the growth coalitions that have interests in the urban land markets.[2]

Yet, policies are not determined solely by interest groups. Politicians are sensitive both to values, as advanced by different advocacies, and to national and international interests. Grossman and Helpman (1994) suggest that decisions are an average between the general good and the weighted power of interest groups, where this power is determined by contributions to election campaigns.

From this brief review, we can see that there is no agreement on the context within which the adoption of transportation innovations should be analyzed. Yet, it may be possible to reconcile these various strands into a single comprehensive framework.

2.3 A Political Economy Framework for Analyzing Transport Innovation Adoption

Innovations are not developed in response to need, but rather as a result of entrepreneurship. Essentially, two groups advance transportation innovations. The first is the industry, generally defined, that is driven by a profit motive and hence develops products to sell, or advances measures that will increase the productivity or profitability of its business. Thus, industry interests are not limited to producers of transport products, but also to firms advancing measures that will allow them to transport their goods or structure their business more effectively, or limit the ability of competitors to do so.

The second group that advances innovations, primarily policy innovations, is composed of experts and professionals. These are termed by Kingdon (1984) 'policy entrepreneurs', i.e. those who advance various policy suggestions in which they believe. The promotion of economic instruments by economists is an example of this type of entrepreneurship.

Although the ideas are promoted constantly, they become concrete policy proposals or investment priorities at specific points in time, termed by Kingdon

(1984) 'policy windows'. These are the politically propitious moments when ideas can be put on the policy agenda. These policy windows open as a result of a perceived 'crisis', or political changes. When a situation is deemed to be a crisis, different policy solutions are proposed as answers to the perceived problem, each competing with the others for the attention of decision makers. The media, experts and members of various epistemic communities and different industry interests, formulate and mediate such solutions. The question this chapter asks is: Why are some of the innovations that are put on the agenda adopted, while others are not?

In Figure 2.1 a political economy framework for addressing this question is presented. On the basis of the previous review, it is suggested that an innovation will be adopted only if it is seen as technically, economically, socially and politically feasible. Unless an innovation meets this minimal (essentially satisficing) criteria, it will not be adopted.

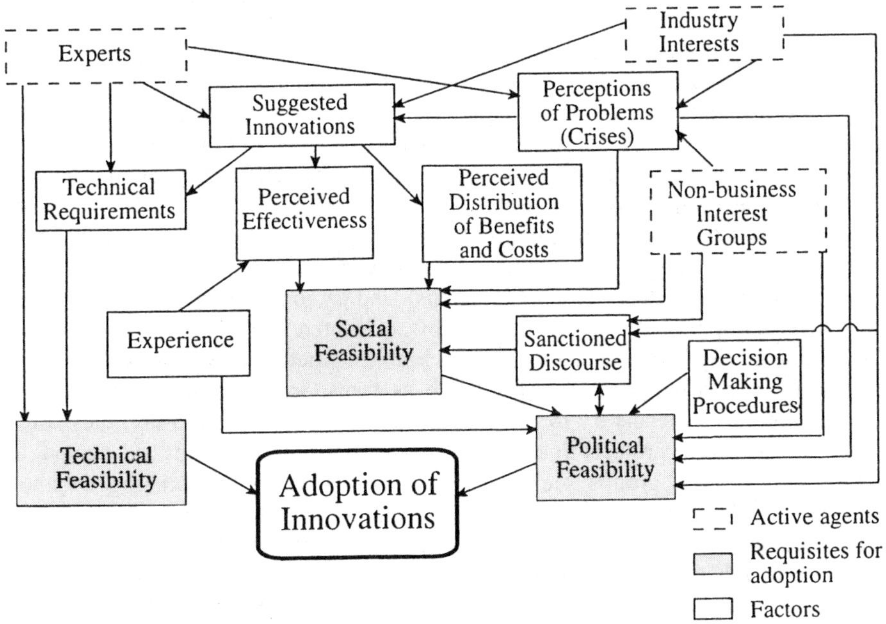

Figure 2.1 A political economy model of transport innovations

Techno-economic feasibility is composed of two elements. One, the innovation needs to be technically feasible (i.e. it can work technically). However, an innovation is not likely to be seen as feasible unless it can also pass a benefit-cost analysis (i.e. that it is also economically viable). Given the wide latitude that analysts have in order to make various assumptions in any benefit-cost analysis, this second criterion is realistically a very minimal requirement. Thus, if an innovation cannot pass a rudimentary benefit-cost criterion it should be viewed, from a societal perspective, as being unrealistic and hence socially infeasible, even if it can work technically.

From the social feasibility perspective, a project can be seen as feasible if a majority of voters are likely to support it. Following Rienstra et al. (1999), social feasibility is suggested in Figure 2.1 to be a function of the public perception of problems and the perceptions of the effectiveness of the proposed innovation in addressing these problems. These perceptions are affected by experience with similar policies or systems elsewhere. Policy entrepreneurs can affect these perceptions by publications and media appearances that support or criticize the proposed innovations. Environmental interests or transport lobbies can do likewise. By doing so they try, essentially, to affect the 'sanctioned discourse', i.e. the discourse sanctioned by decision makers as being politically feasible. This sanctioning is a function of the dominant ideologies, and what the media and elites see as publicly acceptable. Thus, it is highly unlikely today that a new highway through historic neighborhoods will be proposed, as such proposals are largely seen as wrongheaded, and are thus largely outside the sanctioned discourse.

Social acceptability is one factor that determines political acceptability, as politicians do take account of voter preferences. However, the general public does not vote upon most transport investment and policy decisions. These decisions are taken as part of more routine procedures. Under these circumstances, politicians, assuming they want to maximize the likelihood that they are re-elected to office, and knowing that these decisions will not be the crucial issues in most future elections, have additional incentives. As they do need the support of interest groups (not least to finance their campaigns), and try to prevent being the object of negative publicity that can be generated by such interest groups, politicians try to accommodate specific demands of active lobbies, whether they represent industry interests or non-business interests (such as transit or environmental lobbies). Moreover, as politicians try to receive the maximal positive exposure, they often try to direct visible projects to areas where potential support groups are located, or areas that support groups see as preferential. By using the existing decision making structures, politicians can target certain groups (which can be voting areas, jurisdictions or neighborhoods) so they receive the perceived benefit at a minimal cost (although often at the expense of the general coffer). Essentially, in this framework, decision makers try to provide answers to the perceived problems, so that they can maximize the support they receive from the various interest groups, in a way that is socially acceptable and that can be rationalized by arguments that are within the currently dominant sanctioned discourse.

If this is the case, some innovations will be seen as politically acceptable. These will be ones that are supported by a wide coalition of specific interests, and whose cost will be borne by the wider public, but is still low enough not to generate opposition, and that can be rationalized as meeting the public good and addressing an urgent problem.

However, not all politically acceptable innovations will ultimately be adopted. It is possible that a politically acceptable innovation may not ultimately be technically feasible in practice (in contrast to earlier perceptions). This will happen when the actual implementation costs become apparent, forcing difficult choices, often after the 'crisis' recedes from the public eye. If the cost proves to be high, it is possible that innovations that were accepted formally by decision makers will not be implemented ultimately, or will be implemented in a minor or highly modified way, as both decision makers and the technocracy will try to limit their cost. This additional criterion will primarily affect very expensive innovations.

2.4 The Experience of Three Innovations: Road Pricing, Rail and Telecommuting Centres

To test the usefulness of the analytical framework, the adoption of three innovations is reviewed. On the basis of the literature and reports, the factors that seem to affect the adoption of road pricing, rail systems (mainly light rail) and telecommuting centres are identified. These three innovations differ in terms of the capital cost required for their adoption, the degree to which they are technologically innovative, and hence the degree to which the technology on which they rely is known, and in their distributional implications. Thus, it can be presumed that, if the framework is indeed useful for such different innovations, it may indeed be useful for analyses of a wide variety of innovations.

2.4.1 Road Pricing

Road pricing has been widely promoted as a measure to reduce traffic congestion and pollution for over 80 years (Button and Verhoef 1998). The most active promoters of this idea are economists, who have argued time and again that it is the most efficient or cost-effective way to tackle these problems (for a review of the economic theory behind this statement, see Hau 1998). Yet, as recent analysts increasingly recognize, it has not been widely implemented.

Road pricing faces multiple obstacles, such as the need to account for variation in traffic flows over space and time (Mohring 1976), and the possible spillover effects to untolled roads (Lave 1994). Many of these can be overcome through electronic road pricing. Such pricing has been proved to be technically feasible. Thus, the question remains: Why has it not been more widely implemented?

Studies analyzing this question suggest several reasons. Most suggest that such pricing may have regressive distribution effects (Richardson 1974; Layard 1977). Recent analyses note that distribution should not be analyzed only from an income perspective, but rather according to user and non-user groups (Rietveld and Verhoef 1998). Small and Gomez-Ibanez (1998) show, on the basis of empirical evidence, that most current road users (motorists) stand to lose from road pricing. While non-users may gain (Rietveld and Verhoef 1998), they remain usually unorganized, and hence are only rarely heard in discussions about road pricing.

Giuliano (1992) suggests that while such concerns are indeed real, they are not the main reason for non-implementation. She suggests a list of issues including the distribution of tolls, skepticism regarding the outcome (effectiveness), concerns over impacts on business and the objections to pricing what is free (prior experience). Other analysts voice similar concerns. Thus, Rietveld and Verhoef (1998) note doubts over effectiveness, and Jones (1998) suggests that road pricing will be viewed simply as another form of taxation. He also notes the concerns over privacy issues in the case of electronic road pricing.

There have been many attempts to address these issues. Richardson and Bae (1998), for example, discuss the ability of FASTRACK lanes[3] to address some of these issues, particularly the equity and privacy concerns. Others have proposed that the revenues be redistributed so as to address such concerns.

These studies focus almost exclusively on the question of social and technical feasibility. They do not discuss political feasibility. By extending Dijkstra's (1999) work to this issue, we can suggest that the bureaucracies involved have only little incentive to introduce such measures, especially as they are likely to face opposition by the users who will be adversely affected, and only little support from those non-users who may benefit from the implementation of such measures.[4]

In a recent contribution, Levinson (2000) suggests that the only case where there will be political support for such taxes is when most of the traffic that is taxed is through traffic. This will occur when the authority that imposes the taxes is small, as in this case the local voters-drivers are hardly affected. Yet in modern times road pricing schemes usually pertain to relatively large areas, and are not in the purview of local government. Hence, in most cases many of the voters will be adversely affected in practice, and are thus unlikely to support such measures.

2.4.2 (Light) Rail

Rail systems, and particularly light rail, seem to show something of a renaissance. After being phased out around the middle of the 20th century, they seem to be making a comeback in many European and North American cities.

In contrast to the case of road pricing, analysts often criticize such systems, pointing out that the projections on which they are based are only seldom met. In several well-known analyses in the US, it has been shown that in most cases where light rail has been developed in the last few decades, the number of passengers has been lower and the costs higher than originally projected (Pickrell 1992; Gomez-Ibanez 1985; Soberman 1997). This does not seem to be an accident, as there is evidence in some cases that the biases are intended to promote rail (Kain 1990).

Due to current market trends, particularly the transition to post-industrial economies and the advent of polycentric metropolitan areas (where most of the employment is located outside the traditional central business district (CBD)), and the existence of a latent demand for road travel, the potential share of light rail in total trips within the metropolitan areas is limited, and the likelihood that it will significantly reduce road travel is remote (Feitelson 1994; Salomon et al. 1997). Hence, the likelihood that such systems will meet their stated goals, such as alleviation of congestion and reduction of atmospheric pollution, is low. Moreover, also the purported land-use effects of rail, and particularly light rail have been questioned (Mackett and Edwards 1998).

Therefore, the popularity of rail solutions despite these misgivings, and the willingness to use overly optimistic projections in order to advance such systems, requires an explanation. Some analysts suggest that this can be partially explained as the outcome of financing procedures. Edwards and Mackett (1996) argue that in the UK the deregulation of buses and the greater readiness to finance discrete projects that are highly visible give an advantage to light rail projects over improvements in bus systems, even in cases where improving the bus systems may be more effective. Similarly, in the US the availability of federal funds gives local jurisdictions the incentive to advance such projects, as the benefits accrue locally, while most of the cost is borne by the US taxpayers. From a local perspective it is sufficient that the local benefits exceed local cost for a (light rail) project to be supported. This can also happen when the benefits do not exceed total cost.

Mackett and Edwards (1998) also note that rail is highly visible, and viewed as more effective than bus by the public. As public transport is largely viewed upon favorably from a distributional perspective, and rail is seen as an effective means of public transport, it is not surprising that rail in general, and light rail in particular, are usually viewed as socially desirable. Rail is also seen as environmentally-friendly, as the pollution it generates is not perceived directly because it is part of the general electricity production, and because of the reported low energy use per passenger. Thus, the qualms of Feitelson (1994) not withstanding, environmentalists tend to support such systems.

In Israel, for example, environmentalists have upheld the development of rail systems as a one of the major environmental causes in recent years, and demand that they be developed rather than roads. They have strenuously supported increases in rail budgets. In this effort, environmentalists are in coalition with industry interests, especially those who stand to benefit from rail development. Given the multiple interest groups that support rail, it is not surprising that rail projects, and the highly visible light rail projects in particular, enjoy widespread political support.

2.4.3 Telecommuting Centres

Telecommuting centres do not enjoy the same level of attention as rail or road pricing. They are a relatively new idea, stemming from the rapid advances in telecommunications. Essentially, they attempt to reduce the problematic aspects of work from home, such as loneliness, lack of infrastructure, and difficulties in support, by concentrating teleworkers in centres where they have all the facilities. Nilles et al. (1976) suggest that such centres may become a new and important element in business structures and location. They envision that such centres will become the base for independent contractors, who can then provide services to multiple corporate headquarters or to multiple small employers.

In a report on telecommuting centres Bagley et al. (1994) review the experience of eight such centres in North America, and bring some data on 19 additional centres in other parts of the world. The main finding of the report is that the usage of such centres is extremely low. In North America, the average level of utilization was around 34 percent. The number of employees working there is minuscule, and thus has no discernible effect on the transport situation. They go on to note several concerns that employers have with such centres, such as about the security of proprietary information, and a discomfort with remote supervision.

These centres receive considerable funding and support by public agencies. However, employers have shown a marked disinclination to pay market rate rents for workplaces in such centres. Consequently, most centres have operated as pilot or demonstration projects. The report suggests that the utility of such centres may be limited to relatively remote rural areas, where there is an ex-urban population with very long commute time.

On the basis of this report, it seems that such centres seem to be receiving more political support than is warranted by their actual usage. This can be explained by the fact that the logic underlying them is very attractive. They hold promise of a way to do things differently, at very low cost from a political perspective. There are no reports of controversies around them, and they seem to benefit employers, workers and the transport system. They seem thus like a win-win-win proposition. In practice, however, employers do not seem to support them, thus limiting their usefulness. It can be suggested, therefore, that telecommuting centres have been found to be technically infeasible, in a wide sense, though they are clearly feasible in a narrow sense. While they can work, they are not utilized to a sufficient degree to make any difference.

2.5 Discussion: The Sources of Support or Opposition

The three innovations discussed above share similar goals. All three are intended to reduce congestion, and vehicle-miles traveled (VMT), thus also providing environmental benefits. Yet, only light rail is widely adopted. Many analysts, however, do not support this innovation. It is thus obvious that the adoption of innovations is not an outcome of some rational process based solely on the best advice possible. Rather, it is indeed a political process.

Rail enjoys several advantages over the other innovations. There is widespread experience with rail systems, and they are highly visible. Both users and non-users perceive rail as being efficient and effective. In addition to its direct benefits to users, it can support landed interests, primarily in or near the CBD area, and is a major public works project and thus supported by developers. As shown in Table 2.1, the costs of rail projects are mostly borne by the taxpayers at large. The groups that may be adversely affected directly are those living near the rail lines, bus riders, if their level of service falls due to a greater need for transfers or decline in budgets for buses, and bus companies. As bus riders are often from the weaker strata of society, and are not well-organized, their voice may not be heard.

If the bus sector is fragmented, or the bus companies are party to the light rail project, they may either be disorganized or acquiescent too. Thus, the only real opposition may be from residents, and this can be mitigated by ameliorative measures (such as going underground in particularly sensitive spots). Moreover, as residents are also likely to enjoy better accessibility and perhaps a rise in their property values, due to enhanced accessibility, their opposition may not be too fierce.

Rail also enjoys the support of industry interests, and of environmental and transit lobbies. The skeptical experts are thus left in a small minority, which has only limited political clout. Another factor that enhances the attractiveness of rail is the way it is funded. If much of the cost is borne at the national level, while benefits accrue locally, the local authorities have an incentive to promote such systems. Rail systems thus enjoy widespread social acceptability. Moreover, the promotion of rail conforms to the current sanctioned discourse regarding the need to reduce traffic externalities, and these systems are promoted by business and lobby groups. It is not surprising, therefore, that rail systems are seen as politically desirable. Moreover, as there is ample experience with rail systems, they are seen as technically viable, and thus seem to be a low risk venture. The widespread adoption of such systems, despite their high costs, and low performance in practice, is therefore very logical.

Road pricing, in contrast, aggravates a large number of groups, while providing benefits only to those drivers remaining on the road with a high value of time. The other beneficiaries, mostly non-users, would only rarely relate road pricing to any perceived utility they may gain. Moreover, as these non-users are widely spread, and hence disorganized, they are not likely to be active in the political arena. Consequently, road pricing is likely to remain an innovation whose political acceptability is questionable. To introduce it, experts have to conduct an uphill struggle, and introduce many compromises in terms of the features of the scheme. As Rietveld (1997) suggests, the outcome is likely to be a third-best scheme, a far cry from the type of schemes suggested in the more theoretical presentations. Road pricing also lacks the backing of any substantial industry interests, and usually enjoys only lukewarm support from other lobbies. On the other side, businesses that may fear that they will lose clientele may actively oppose any scheme that raises the cost of motoring to them.

New electronic road pricing schemes, which overcome some of the previously raised objections, may not, however, be socially feasible. While the Hong Kong experience does prove that the systems can work, it also demonstrates the concerns such systems raise. It may prove difficult to convince the public that such a scheme will be effective, and be able to protect privacy. While, again, it may be possible to introduce elements that will mitigate these concerns, they are unlikely to disappear.

Telecommuting centres do not generate the same intensity of debate that the previous innovations do. This is partly due to the lack of experience with them, and the small number of people who may object to them. At the same time, they provide very small benefits. While they are very attractive, socially, politically and technically, at the first stage, later on they have not proved to be financially viable.

Companies do not seem inclined to pay market rents for them, due to the concerns such centres raise in company headquarters. Thus, there are many pilot and demonstration projects of this sort, supported by public funds. But these may close if public funds do not continue to support them. Given the very small number of beneficiaries, and subsequently very limited benefits from them, which almost all accrue only to the direct users, it is doubtful whether continued public support can be justified. From a political perspective, the benefit of such projects is at the time they are launched, not in their continual usage. Thus, it is unlikely that they will receive substantial public subsidies over time. It seems, therefore, this innovation will be adopted, only on a limited and temporary level.

Table 2.1 Potential support and opposition to transport innovations

Innovation	Perceived Benefits	Potential Beneficiaries	Cost Burden	Potential Supporters
Rail Light	Accessibility to CBD; reduced emissions and energy use per passenger-mile travelled; safety	New users; developers; CBD landed interests;	General tax payers, nearby residents; bus riders(?)	Transit industry and producers; drivers; environment-alists; passengers
Road User Charges	Reduced congestion; reduced emissions;	Drivers with high time value; non-users(?)	Drivers; switchers to transit; transit users;	Economists; environmental ists; budget departments
Tele-commuting Centres	Reduced congestion and emissions; time savings	Telecommu-ters drivers remaining on the roads	Tax payers initially, later companies	Environmental ists telecommuting industry; telecommu-ters

The analysis of the three cases based on the concepts advanced in the theoretical framework suggests that the adoption of innovations is a function of the coalitions that adopt or oppose the innovations. Moreover, the adoption is a function of the degree to which members in those coalitions are ready to act for or against the adoption of the innovations. It seems many interests are ready to act for light rail. Only a few are ready to act for road pricing. This finding corroborates the approach advanced by Olson (1965) and Becker (1983), who suggest that it is the degree to which interests, or coalitions of interests, are likely to organize which will determine the adoption of policies (and innovations). The transaction cost involved in organizing a pro-rail coalition are simply lower than those of a coalition for road pricing, and the scope of such a coalition is likely to be wider.

2.6 Conclusions

This chapter argues that the adoption of transportation innovations, widely defined to include policy innovations, is the outcome of a political process, and not merely of innovation diffusion. Subsequently, we advance a political economy framework to analyze the likelihood that different innovations will be adopted.

Discussion of three innovations suggests that the range of support each innovation is likely to garner in the political field, and the range and intensity of opposition, will determine the likelihood that it is adopted. The support and opposition are a function of the scope of proponents and opponents, their influence and the transaction costs involved in organizing the pro- or anti-coalition. Thus, an analysis of the likelihood of the adoption of different innovations cannot be limited to their technical feasibility or to their social acceptability. While these two factors are important, it is crucial that the political dimension be addressed as well. This discussion also shows that, while experts play an important role in formulating innovations and advancing them, and in their technical evaluation (including economic viability), their influence on political feasibility is limited.

Thus, it seems that future studies of innovation adoption in the transport field should pay greater attention to interest group politics, and to the implications of decision making structures for the incentives to decision makers with regard to the different innovations. The model advanced in this chapter notes these factors. But it does not indicate the relative importance of such factors. It also does not account for the variation over time in the importance of the different factors, something that can be affected by the accumulating experience with the innovations adopted. Hence, there is place for the development of dynamic models of innovation adoption over time, something that is beyond the scope of this chapter.

References

Bagley M., Mannering J. and Mokhtarian P. 1994: Telecommuting Centers and related Concepts: A Review of Practice, University of California Davis Institute of Transportation Studies, Davis

Becker G. 1983: A Theory of Competition among Pressure Groups for Political Influence, Quarterly Journal of Economics, Vol. 98, 371–400

Borins S. 1988: Electronic Road Pricing: An Idea whose Time May Never Come, Transportation Research A, Vol. 22, 37–44

Brownstone D., Bunch D. and Train K. 2000: Joint Mixed Logit Models of Stated and Revealed Preferences for Alternative-fuel Vehicles, Transportation Research B, Vol. 34, 315–338

Button K. and Verhoef E. 1998: Introduction, in K. Button and E. Verhoef (eds.), Road Pricing, Traffic Congestion and the Environment, Edward Elgar, Cheltenham, UK and Northamption, MA, 3–13

Dijkstra B.R. 1999: The Political Economy of Environmental Policy, Edward Elgar, Cheltenham, UK and Northamption, MA

Edwards M. and Mackett R. 1996: Developing Urban Public Transport Systems, Transportation Policy, Vol. 3, 225-239

Feitelson E. 1994: The Potential of Rail as an Environmental Solution: Setting the Agenda, Transportation Research A, Vol. 28, 209–221

Giuliano G. 1992: An Assessment of Political Acceptability of Congestion Pricing, Transportation, Vol. 19, 335–358

Gomez-Ibanez J. 1985: The Dark Side of Light Rail: the Experience of Three New Transit Systems, Journal of the American Planning Association, Vol. 51, 337-351

Grossman G. and Helpman E. 1994: Protection for Sale, American Economic Review, Vol. 84, 833–850

Hau T. 1998: Congestion Pricing and Road Investment, in Button, K. and Verhoef, E. (eds), Road Pricing, Traffic Congestion and the Environment, Edward Elgar, Cheltenham, UK and Northampton, MA, 39–78

Jones P. 1998: Urban Road Pricing: Public Acceptability and Barriers to Implementation, in Button, K. and Verhoef, E. (eds.), Road Pricing, Traffic Congestion and the Environment, Edward Elgar, Cheltenham, UK, and Northampton, MA, 263–284

Kain J. 1990: Deception in Dallas: Strategic Misrepresentation in Rail Transit Promotion and Evaluation, Journal of the American Planning Association, Vol. 56, 184–196

Kingdon J. 1984: Agendas, Alternatives and Public Policies, Harpers Collins, New York

Lave C. 1994: The Demand Curve under Road Pricing and the Problem of Political Feasibility, Transportation Research A, Vol. 28, 83–91

Latour B. 1996: ARAMIS or the Love of Technology, Harvard University Press, Cambridge, MA

Layard R. 1977: The Distributional Effects of Congestion Pricing, Economica, Vol. 44, 297–304

Levinson D. 1998: Road Pricing in Practice, in Button K. and Verhoef E. (eds), Road Pricing, Traffic Congestion and the Environment, Edward Elgar, Cheltenham, UK and Northampton, MA, 14–38.

Levinson D. 2000: Revenue Choice on a Serial Network, Journal of Transport Economics and Policy, Vol. 34, 69–98

Logan J. and Molotch H. 1987: Urban Fortunes: The Political Economy of Place, University of California Press, Berkeley

Machiavelli N. 1532: The Prince, Translated by Luigi Ricci, The New American Library, N.Y. (1952)

Mackett R. and Edwards M. 1998: The Impact of New Urban Public Transport Systems: Will the Expectations be Met? Transportation Research A, Vol. 32, 231–245

Mohring H. 1976: Transportation Economics, Ballinger, Cambridge, MA

Nilles J., Carlson F.R., Gray P. and Hanneman G. 1976: The Telecommunications – Transportation Trade-off: Options for Tomorrow, Wiley and Sons, New York

Olson M. 1965: The Logic of Collective Action: Public Goods and the Theory of Groups, Schocken Books, New York

Pickrell D. 1992: A Desire Named Streetcar: Fantasy and Fact in Rail Transit Planning, Journal of the American Planning Association, Vol. 58, 158-176

Richardson H.W. 1974: A Note on the Distributional Effects of Road Pricing, Journal of Transport Economics and Policy, Vol. 8, 82–85

Richardson H.W. and Bae C-H.C. 1998: The Equity Impacts of Road Congestion Pricing, in Button K. and Verhoef E. (eds), Road Pricing, Traffic Congestion and the Environment, Edward Elgar, Cheltenham, UK, and Northampton, MA, 247–262

Rienstra S., Rietveld P. and Verhoef E. 1999: The Social Support for Policy Measures in Passenger Transport: a Statistical Analysis for the Netherlands, Transportation Research D, Vol. 4, 181–200

Rietveld P. 1997: Political Economy Issues of Environmentally Friendly Transport Policies, International Journal of Environment and Pollution, Vol. 7, 398–416

Rietveld P. and Verhoef E. 1998: Social Feasibility of Policies to Reduce Externalities in Transport, in Button K. and Verhoef E. (eds), Road Pricing, Traffic Congestion and the Environment, Edward Elgar, Cheltenham, UK and Northampton, MA, 285–307

Salomon I., Feitelson E. and Cohen G. 1997: The Rail Solution: Promises and Limitations (in Hewbrew), Riveon Lekalkala (Economic Quarterly), Vol. 44, 663–699

Small K. and Gomez-Ibanez J. 1998: Road Pricing for Congestion Management: the Transition from Theory to Practice, in Button K. and Verhoef E. (eds.), Road Pricing, Traffic Congestion and the Environment, Edward Elgar, Cheltenham, UK, and Northampton, MA, 213–246

Soberman R. 1997: Rethinking Urban Transportation: Lessons from Toronto, TRR, Vol. 1606, 33–39

Whitt J.A. and Yago G. 1995: Corporate Strategies and the Decline of Transit in US Cities, Urban Affairs Quarterly, Vol. 21, 37–65

Endnotes

[1] For a recent analysis of the how rent seeking interests that work cooperatively can prevent the application of economic instruments in environmental policies, see Dijkstra (1999).

[2] On the definitions, functions and workings of growth coalitions, see Logan and Molotch (1987).

[3] FASTRACK is a privately financed 10-mile toll road in Orange County, California, where tolls are collected automatically, and thus can change as frequently as desired to limit congestion. It also allows for a free carpool lane for vehicles with three or more occupants (Richardson and Bae 1998).

[4] This conforms to the general observation made by Machiavelli (1532), albeit in a different context: "For the reformer has enemies in all those who profit by the old order, and only lukewarm defenders in all those who would profit by the new. This lukewarmness arising … partly from the incredulity of mankind, who do not truly believe in anything new until they had actually experience in it" (p. 55).

3 ICTs and Accessibility: An Action Space Perspective on the Impact of New Information and Communication Technologies

Martin Dijst

Utrecht University, Utrecht, The Netherlands

3.1 Introduction

Personal computers, mobile phones and the Internet are conquering the world apace. A few figures will illustrate this. Mokhtarian (1998) estimates that, at present, about 6 percent of the American workforce is telecommuting, while Nilles (1995) forecasts that, in the year 2020, 60 million American full-time employees will telecommute full-time or part-time. In 2001, the revenues of worldwide e-commerce were US$ 400 billions, probably 20 percent of which can be classified as business-to-consumer traffic. The revenue in the year 2003 is estimated at US$ 1.3 trillion (Golob and Regan 2001). Viswanathan and Goulias (2001) expect 15 percent to 20 percent of all shopping in 2010 in the USA to be labelled as e-commerce. These developments are being equipped technologically by faster and cheaper microchips, increased transmission speeds on the Internet, growth in the number of web pages on the Internet, small and portable wireless equipment, and so forth, (Golob and Regan 2001).

Of course, we have to remember that almost anything can be put down on paper. This is not the first time that advocates of new information technologies and online services have argued that they are on the point of a breakthrough into rapid growth. According to Gillespie (in Hall 1998), these breakthrough points have repeatedly been postponed, but only until tomorrow. The impact of new ICTs is difficult to isolate from other factors, such as sociodemographic and economic developments in society, which have complicated relationships with the new technologies. We agree with Giuliano (1998) that new ICTs form necessary, but not sufficient, conditions for change.

Notwithstanding these uncertainties with respect to the use of new ICTs, we can speculate about the impact they will have on the spatial configuration of opportunities. These spatial opportunities refer to the most central concept for understanding spatial and social impacts of telecommunications and transportation: *accessibility* (Couclelis 2000). Couclelis (1998) asserts that human activity in post-industrial society is becoming increasingly person-based as opposed to place-based; growing numbers of economic and other activities are no longer firmly linked to fixed geographic locations. This development will have the effect that accessibility will no longer be measurable solely in terms of travel time, distance, or generalized cost. 'Virtual accessibility' will also be important (Golob and Regan 2001).

It is our purpose in this chapter to clarify the potential impact of new ICTs on accessibility in the physical, or 'real' world. We have not elaborated on the issue of accessibility in cyberspace or virtual space (see Kwan 2001). For our purpose, we have distinguished a spatiotemporal and a perceptual dimension of accessibility. It is shown that new ICTs may impact on the spatial, temporal, and interpretable characteristics of three fundamental elements of accessibility: the reference location from which access to one or more other locations is determined; the set and attractiveness of opportunities; and the travel impedance. The study has been placed within an action space framework. The concept of *action space* refers to the area in which people undertake activities, or could do so if they wished.

As the literature makes only too clear, our current understanding of the impact of new ICTs is limited. Unavoidably, the discussion in this chapter is qualitative and may at times seem ambiguous. With these reservations, some hypotheses have been developed which could serve as starting points for future research. Next, in Section 3.2 of this chapter, the general action space framework is presented. Section 3.3 contains an assessment of the potential impact of new ICTs on action spaces from a spatiotemporal perspective. In Section 3.4, the procedure is repeated from a perceptual perspective. In the conclusion, Section 3.5, some directions for future research are indicated.

3.2 Accessibility Dimensions and Action Spaces

According to Couclelis (2000), accessibility is the geographic definitions of opportunity. Which *opportunity set* is relevant for individuals seeking to participate in activities depends on their *ability* to reach these opportunities. This, in turn, depends on the time, costs, and effort an individual is prepared to spend on travelling (Janelle 1969). We argue that the new ICTs will have an impact on the ability to overcome distance and on the opportunities which individuals can reach from two not mutually exclusive perspectives: a spatiotemporal and a perceptual perspective. To discuss these impacts from both perspectives we use an action space framework (Figure 3.1).

In general terms, the action space refers to the area in which a set of opportunities is located which people could use for their activities. As discussed elsewhere (Dijst 1995; 1999), we can find in the literature a large number of related and sometimes confusing terms for the concept *action space*. For that reason we developed a new framework for the study of action spaces (op. cit.). In this chapter, we use this framework to elaborate the two accessibility perspectives shown in Figure 3.1.

At the core of the *spatiotemporal perspective* is the notion that individual behaviour has both spatial and temporal attributes. Individual behaviour is determined by three types of space-time constraints (Hägerstrand 1970):

a. *Capability constraints* refer to physiological and biological limitations, abilities, and access to tools. People need to spend some time of the day sleeping and eating. In addition, the maximum speeds of some transport modes limit the opportunity to visit remote activity places.
b. *Coupling constraints* define where, when, and for how long an individual has to associate with other individuals, tools and materials in order to produce, consume and transact.
c. *Authority or regulatory constraints,* like the capacity of activity places and rules and laws, prohibit the use of facilities at certain times.

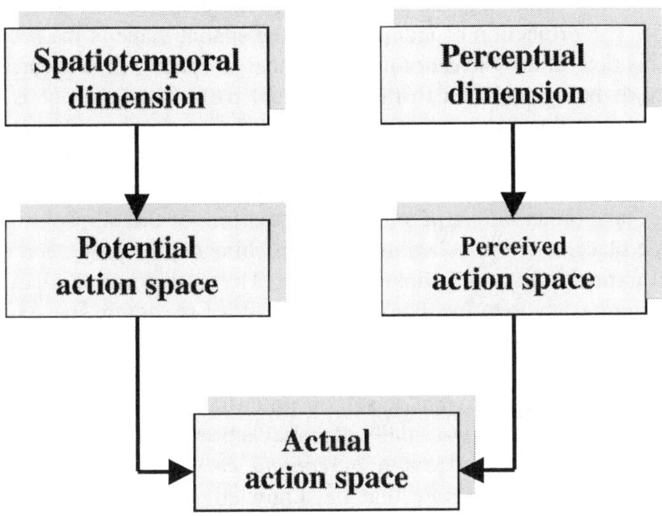

Figure 3.1 Two dimensions of accessibility experienced by individuals

These constraints impact on four basic determinants of space-time accessibility (Lenntorp 1976; Dijst 1999; Dijst et al. 2002):

- *Location of bases*: working, eating, and sleeping, for instance, are relatively fixed in certain places for a long period.
- *Number and length of time intervals*: the time arrangements at each base break up the day into intervals, or time windows (Schwanen and Dijst 2003). These arrangements concern working hours and mealtimes, for example. The uninterrupted stretch of time that can be devoted to activities outside the home for shopping, school, or leisure between leaving a base and arriving at the same or a different base is called the *available time interval.* These intervals are important, because of their length. The longer an available

interval, the more choices there may be in ways in which that time can be spent.

- *Travel time ratio*: given the available time interval, an individual has the opportunity to trade-off travel time to one or more activity places and the amount of time spent there (Dijst and Vidakovic 2000; Schwanen and Dijst 2002). For example, a person may choose to stay for a long time at a nearby location. In that event, the actual action space would be relatively small and would limit the range of options to visit other activity places. The alternative is to opt for a short stay at a distant location; then, the actual action space would be relatively large.
- *Travel speed*: determines the travel time in physical space.

All four determinants restrict what an individual can physically reach from specified locations and within a given period. The spatiotemporal unit is the *prism* (Lenntorp 1976). The projection of this prism on the spatial plane is the *potential action space*. This determines the set of activity places in space with a greater than zero probability of being included in the individual path. This concept is most closely associated with the *accessibility* concept, or such concepts as *activity area* Lenntorp (1976), or *travel probability field* (Zahavi 1979; Beckmann et al. 1983).

Three ideal forms of action space can be distinguished. The projections on the spatial plane can take on an *elliptical shape*. The two foci of the ellipse might be the home and the place of work. When the two foci coincide, the projection of the prism takes on a special form of ellipse: a *circle*. The third form of a potential action space is the *line*. The two foci constitute the ends of the action space, within which the opportunities to visit other activity places are exhausted and no other activity places are visited. Individuals can have several time intervals in one day (Schwanen and Dijst 2002), so that a set of potential action spaces can be delimited. They can be aggregated to a daily potential action space.

Accessibility can be studied from a *perceptual perspective* as well as a spatiotemporal perspective. We know that the knowledge people have of their choice sets is incomplete and subjectively filtered (Kitamura et al. 1997). People's limitations in information processing lead them to strive for satisfactory rather than optimal solutions (Simon 1975). Bias and stereotypes play a substantial role in this process selection (Horton and Reynolds 1971; Klingbeil 1978). The set of opportunities known by the individual is called the *perceived action space*. Related concepts include *awareness space* (Brown and Moore 1970), *action space* (Wolpert 1965), and *mental map* (Lynch 1984). Circular and elliptical action spaces can be distinguished on the basis of the number of base locations.

The area containing all the activity places a person (or group) has visited during a given period is defined as the *actual action space*. This concept is approximately coterminous with *activity space* (Brown and Moore 1970), *travel field* (Zahavi 1979), *spatial usage fields* (Potter 1979) and *daily contact space* (Kolars and Nystuen 1974). Three basic forms of actual action spaces which can be compared with potential action spaces can be distinguished: circle, ellipse, and line.

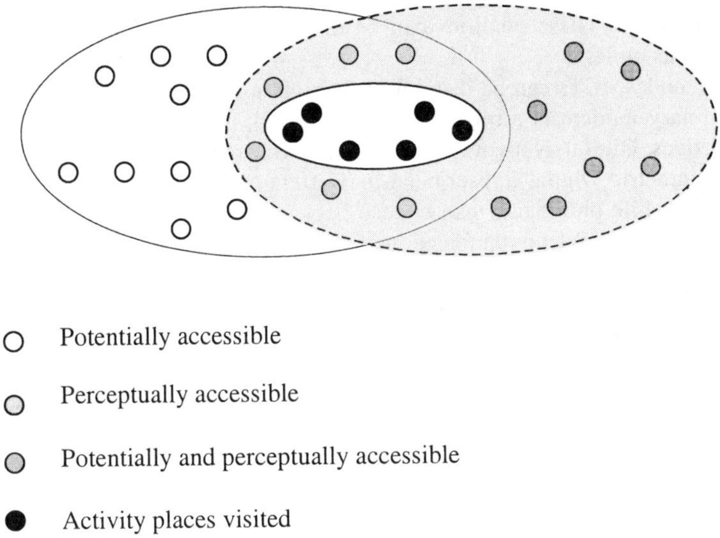

Figure 3.2 Relationship between potential, perceived, and actual action spaces

As shown in Figure 3.2, these three types of action spaces are interrelated.[1] In theory, the actual action space lies entirely within the potential action space. An individual's perceived action space also contains the actual action space together with part or all of the potential action space. Figure 3.2 also shows that three *accessibility situations* can be distinguished:

- Potentially accessible: opportunities which are exclusively accessible from a spatiotemporal perspective; and a perceptual perspective;
- Perceptually accessible: opportunities which are exclusively accessible from a perceptual perspective;
- Potentially and perceptually accessible: opportunities which are accessible from both a spatiotemporal.

On the basis of this conceptual framework, the impact of new ICTs is elaborated from a spatiotemporal and a perceptual perspective, respectively, in the next two sections.

3.3 Spatio-temporal Perspective on New ICTs

3.3.1 Constraints on ICTs

Certain qualities ascribed to the new ICTs lead to situations which allow us to participate in certain activities *at any time, at any place*. Is it indeed the case that

ICTs are not bound by time and space? Are Hägerstrands' space-time constraints not at work here at all? These qualities can be disputed. Figure 3.3 lists some aspects of constraints on ICTs.

Like transport modes, ICTs can be distinguished by transmission speed, and so by cost. An ordinary modem is a relatively slow network interface. With ISDN (Integrated Services Digital Network), the bits are transmitted faster. ADSL technology (Asymmetric Digital Subscriber Line) offers electronic highways for information. The mobile phone user can choose between different travel speeds; UMTS (Universal Mobile Telecommunications Systems), the new world standard for mobile communications, offers transmission speeds six times faster than the top speed of ISDN; an increase in transmission speed is accompanied by higher costs. We can draw a distinction between *captives*, with no access to ICTs at all or only via very slow information modes, and *non-captives* who can choose between all the different types of ICTs. Under the condition that new transport and communication technologies facilitate the spatial separation of activities, Scott (2000) asserts that the captives will be faced with a diminishing intra-metropolitan accessibility.

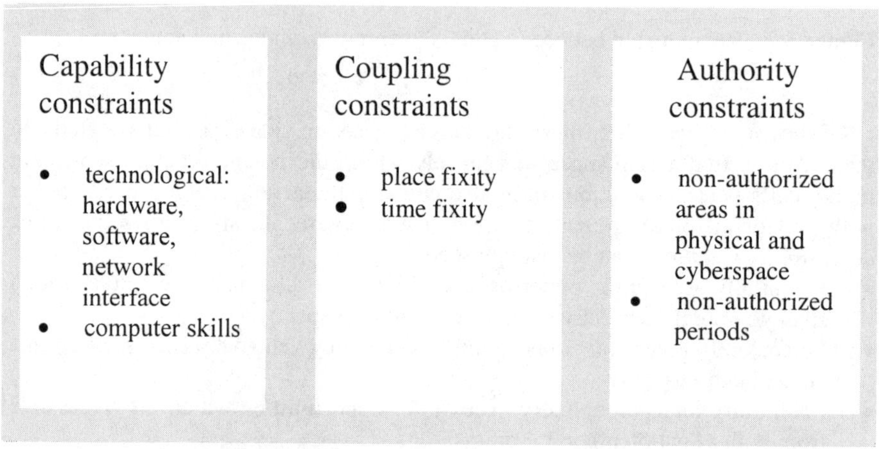

Figure 3.3 Constraints on the use of ICTs

In addition to the technological capabilities of ICTs, use depends on the ability of individuals to operate the equipment and their software (Harvey and Macnab 2000; Golob and Regan 2001). Hanson (1998) sees the dominance of English thought structures embedded in cyberspace as important barriers for some people. She asserts that ICTs, like telecommuting, can have a negative impact on the contacts and friendships people develop through face-to-face interactions and thereby on their opportunities, such as job search (op. cit.).

Although some activities needing a commitment to time and place might be satisfied by replacing personal contact by telecommunications (Golob and Regan 2001), we consider that coupling constraints are still at work. Usually, the user of a personal computer has to be located at home, at work, in an Internet café, or at a public terminal. Laptops and mobile phone users accessing relevant information are freed from fixed locations, although the technological capabilities of mobile phones are not (yet) comparable with the personal computer in either quantitative or qualitative respects. Focusing on interactive communication, Harvey and Macnab (2000) point out the necessity of the temporal coincidence of individuals who are free to communicate. This condition depends on people's activity patterns. For example, when one has made arrangements to eat with other people, enjoy face-to-face social contacts, or give a lecture, using ICTs for one's own purposes would not be possible.

Authority constraints in ICTs are expressed in domains in which the use of mobile phones is not appreciated by the public at large, or is forbidden for safety reasons. Mobile phones can be annoying in a cinema, a theatre, or on public transport, and could be dangerous in airplanes or hospitals. Some employers try to ban the use of the Internet for personal communication during working hours. Harvey and Macnab (2000) see the use of filters to limit exposure to some types of undesirable information (to protect children from information on pornography or violence, for example) as good examples of authority constraints.

We hypothesize that the need for face-to-face interactions stimulates the development of new authority constraints on social occasions. This need could limit Locke's (1999) 'de-voicing' of society. In his opinion, our society is characterized by a replacement of intimate personal conversation by detached, efficient exchange of information alone. In our private lives, the extent to which we want to be accessible is debatable. We might like to have some off-line periods in which we can relax undisturbed. Such regulation rules could become more pressing when people use mobile phones or other ICTs more intensively.

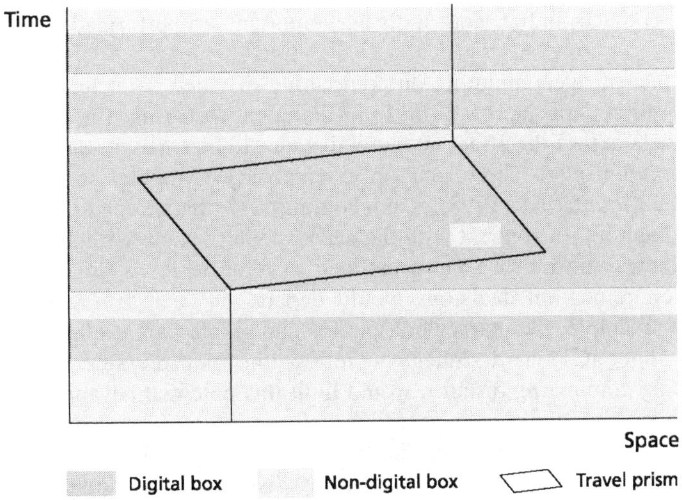

Figure 3.4 Travel prism, digital boxes, and non-digital boxes (non-authorized areas for the use of new ICTs)

We can show the impact that ICTs could have on the opportunities offered to individuals from a spatiotemporal perspective. In Figure 3.4, besides the familiar travel prism, we also see *digital boxes* and *non-digital boxes*. Within the digital boxes, individuals can use ICTs to gain access to relevant information sources. In principle, these boxes stretch out over the whole physical area. Within the non-digital boxes, represented by the white layers and white cells between the digital boxes, activity places (or transport modes) are situated where the use of ICTs would either be not authorized, or impossible.

3.3.2 Determinants of Potential Action Spaces

In Section 3.2 we referred to several determinants of potential action spaces. What impact will the new ICTs have on these determinants? Our aim is not to predict a particular outcome, but rather to raise some issues likely to be associated with increasing use of the new technologies.

Base Locations

The industrial society was a society of sedentary workers; buildings, machines, and face-to-face contacts tied people to fixed workplaces (Couclelis 1998). ICTs offer people the opportunity to perform work activities in a variety of temporary workplaces, such as at (a second) home, on the road, or in telecommuting centers. At present, most telecommuting is home-based with an average frequency of one or two days a week (Mokhtarian 1998). The fact that telecommuting is only of importance once or twice a week means that in a full-time workweek, work is still fixed in time and place in the traditional workplace on 2 or 3 days. Activities for which face-to-face contacts and/or work tools are important are performed in this location.

The probable effect of telecommuting on commuting distance is unclear. Van Reissen (1997) predicts an increase in long-distance traveling for Dutch telecommuters. Others expect the effect on travel distance to be small, because the adoption of telecommuting is slow and only temporary (Mokhtarian 1997; Mokhtarian, 2000). Mokhtarian (1998) is noncommittal with respect to future adopters of telecommuting. In contrast with the early adopters, they will probably have shorter commutes and may be more inclined to relocate for them. In our view, this effect on relocation decisions would depend on sociodemographic characteristics. For example, two-earner households and single-parent employed households could apply ICTs as a strategy to relieve the time pressure on the household. Increasing commuting distance would limit this potential advantage of ICTs.

As a consequence of telecommuting, participation in types of action space may change. If telecommuters increased their commuting distance on working days, the size of their line action spaces would probably increase; however, more elliptical action spaces could be created on commuting days. Empirical research has revealed (Pendyala et al. 1991; Saxena and Mokhtarian 1997; Van Reisen 1997) that telecommuters choose non-work destinations that were closer to home. This inclination means that more smaller action spaces would be created and that circle or elliptical action spaces would be substituted for line action spaces.

Time Intervals

On commuting days, four time intervals are usually relevant: before work, lunchtime, after work, and after dinner. The limits of the time intervals are not so sharply defined on telecommuting days as on commuting days. There are fewer coupling constraints, so that individuals working at home have more flexibility in spending time on activities. Couclelis goes further in her estimation of the effects of ICTs. She states (2000, p. 346): *"...the proverbial nine-to-five weekday job gradually gets fragmented into chunks spread out over arbitrary hours of the day (and many of the nights), interspersed with tasks from other activities occurring at equally odd – by traditional standards – times as well as places."*

We endorse Couclelis' argument that the use of ICTs increases the ability to move work activities around on a day, or in the week, or to combine work with travel. We also feel that this fragmentation and flexibility has limits. We expect large differences between population categories with respect to the opportunities they have to reschedule their day. Two-earner families and single parent households are dependent on their children's agendas. Parents take their children to school and pick them up on a daily basis. As a consequence, they show little flexibility in rescheduling their activities compared with childless households.

Secondly, two-earner families and couples follow different strategies, such as monetarization (contracting out house maintenance, providing care, laundry, buying microwaves, going out for dinner, and so forth) or flexibilization (use of flexible working schemes, for example) to save time for leisure activities (Vijgen and van Engelsdorp Gastelaars 1991; Beckers 1995; Mingioni 1983). They prefer a few larger time intervals to many smaller intervals. Large intervals give them the opportunity to spend time on outside-the-home leisure activities (Vijgen and Van Engelsdorp Gastelaars 1991; Dijst and Vidakovic 2000).

A third comment on Couclelis' statement is that not all activities can be easily combined because of the temporal and spatial separation of activity places. Some activity places are exclusively open in the evening (theaters and concert halls), or by day (schools and many shops). Sometimes the stay time is long and prescribed (as in many cultural and sports facilities). We have to remember that we have inherited a spatial separation of activity places which is often based on the use of the car and which put some constraints on the flexibilization offered by new ICTs (Dijst and Vidakovic 2000).

Notwithstanding these reservations, we hypothesize a fragmentation of time intervals on working days. This fragmentation increases the number of relatively small, home-based action spaces. At weekends, we hypothesize that the number of time intervals changes very little.

Travel Time

There are two effects of the use of ICTs which are relevant to travel time: the size of the travel budget and travel speed. Use of new ICTs could bring about a substitution of travel by electronics. Telecommuting and teleshopping would then lead to travel time savings. The use of Intelligent Transportation Systems (ITS) in public and car transport could have similar results. These systems increase an individual's travel speed by improving the efficiency of transit operations and traffic flows (Suen and Mitchell 2000). Of course, this new capacity may stimulate new travel demand and this could reduce the advantages of the increased efficiency (Lanzendorf 1999).

In spite of this potential for saving travel time, we think this saved time would be used for other trips. Although critiques have been formulated (see Supernak 1982), several authors have found stability in travel time budgets. Hupkes (1982) formulated the Law of Constant Travel Time and Trips (*BREVER-wet*), which implies that, notwithstanding changes in modal split, an individual's total time spent on transport remains (and will remain) unchanged. The arguments on which Hupkes based his 'law' need considerably more elaboration. For empirical evidence, he used highly aggregated data from various years and countries without the necessary statistical elaboration. However, his hypothesis of a constant travel-time budget of 70 minutes per person per day in all societies was supported by the research of Schafer and Victor (1997). We do not think a 'no-travel' option offered by ICTs would reduce the amount of travel time. For several reasons, people have an instinctive drive for mobility (Michon 1980). Salomon and Mokhtarian (1998) give some of these reasons: adventure-seeking, variety-seeking, independence, and so forth. If there is such a thing as a stable travel time budget, this drive for mobility would lead to an expansion of the size of action spaces for those activities which cannot readily be substituted for by ICTs.

We hypothesize that all these changes in saving travel time will have an increasing effect on the size of potential action spaces.

Opportunities

At the macro-level, as with any other innovation in transportation and communication technologies, new ICTs will bring a new time-space convergence (Gould and Golob 1997). In general, time-space convergence offers opportunities for specialization, centralization, suburbanization, or some other form of the spatiotemporal reorganization of activities (Janelle 1969). As a consequence, the use of new ICTs could lead to an increased choice of opportunities. On the other hand, another possible outcome is a reduction in accessibility: time-space convergence facilitates a dispersion of activities (Scott 2000). Overall, we hypothesize that new ICTs will increase the range of opportunities within reach.

At the micro-level, we have previously noted that people are inclined to protect their free time by using such strategies as monetarization and flexibilization. The use of new ICTs could offer another strategy to meet this need for protecting free time. Handy and Mokhtarian (1996, p. 235) remark: *"A desire for more free time may be a strong motivator for telecommuting."* Gould and Golob (1997) suggest that electronic shoppers will convert a portion of travel time savings from shopping trips to trips for personal and household business. These findings indicate that both telecommuting and teleshopping will at least change the type of opportunities users will want to visit. We hypothesize that the value will decrease for accessible maintenance activity places and increase for discretionary activity places.

3.4 Perceptual Perspective on New ICTs

Within the perceptual perspective, a large range of issues can be discussed. We restrict ourselves here to two issues: the revaluation of opportunities, and the acquisition of information about opportunities.

Revaluation of Opportunities

ICTs can have an impact on the type of activity places people visit and the types of activities they perform in these places. To appreciate this point we must first understand what activities are.

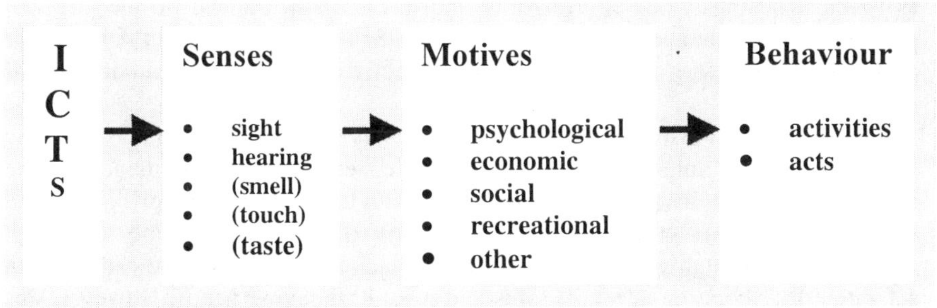

Figure 3.5 Impact of ICTs on senses, motives, and behavioural components

In transport planning, human *behaviour* consists traditionally of clearly defined elements called *activities*, characterized by a start time, duration, and purpose (Axhausen 1998). Normally, we distinguish very broadly defined categories, such as working, shopping, leisure, and so forth. These activity types constitute a stream of several smaller activities called *acts* (op. cit.) or *events* (Gärling and Garvill 1993). For example, an activity such as shopping is composed of such acts as the acquisition of information, examination of goods, purchase transaction, and so forth. These compositional characteristics of activities clearly differ from day to day and from person to person. Each act is connected with a specific *goal* or *motive/purpose* (Gärling and Garvill 1993; Axhausen 1998). Different motives can be distinguished: psychological, economic, social, recreational, and so forth, (see Figure 3.5).

The use of new ICTs could have an impact on the motives for participating in acts or activities (see Salomon and Koppelman 1992; Gould and Golob 1997; Mokhtarian 1998):

- psychological: gratification, reduction of uncertainty, reduction of stress, exposure to information/exposure to stimuli (meaning and variety), privacy;
- economic: savings in time and monetary costs (see also the spatio-temporal perspective in this chapter);
- social: encounters with sales personnel, watching and meeting people, social encounters;
- recreational: physical effort, entertainment, dining out.

We hypothesize that the use of ICTs will only influence some motives, and some acts of all other motives and acts which are connected to one activity. For this statement to be meaningful, we must first discuss the role of the senses.

New ICTs are, in quantitative as well as in qualitative terms, limited in their stimulation of the *senses*. They send out information to the visual and audio senses in a narrow range of size, colors, and sounds (Salomon and Koppelman 1992). Obtaining more qualitative stimuli and stimuli through the other senses (smell, taste, and touch) which do not (as yet) receive any signals from ICTs, face-to-face contacts and physical travel are necessary and still very important. Activities such as finance, business, power, creativity and tourism still depend on the physical proximity of people and facilities (see Hall 1998, 1999). In general, the relevance of these stimuli will vary across types of activities, products and individual preferences.

New ICTs only have an impact on some motives for activities and acts, because of their narrow range of stimuli of the senses. As a consequence, some activities/acts may disappear from an individual's activity patterns, or be replaced by others. Consider for example the activity of electronic home shopping. Ordering on-line highly standardized products such as coffee, beer, or cat food can substitute for shopping at retail stores and meet cost saving needs. However, this substitution process could lead to a revaluation of other motives for shopping, such as the need for social encounters with other customers, drinking coffee, smelling freshly-baked bread, and so forth, which shopping in a store could fulfil. As a consequence, this change in the values of motives and acts could lead to visiting retail shops for leisure purposes as well as for maintenance shopping (see also Gould 1998). This could reinforce the long-existing trend for the integration of leisure elements in shopping malls (Van der Toorn Vrijthoff 1998). As Gould and Golob (1997) showed for the USA, working at home in relative isolation from others may also make the need for social contacts or relaxation more pressing. This need could account for their finding that full-time teleworkers spent a larger amount of time on shopping or discretionary activities than away-from-home workers.

The use of new ICTs could also change the value of travel time. Instead of just driving a car or sitting in public transport, the use of ICTs enables more activities to be performed during travelling, such as communicating with others or searching the Internet. The disutility of traveling could be reduced. Psychological studies (Hoorens 1998) on the perception of time confirm the aphorism: *Time flies when you are having fun*.

These changes in the valuation of motives, acts and activities will bring about a change in the desired opportunities for activities. Quantitative as well as qualitative aspects of opportunities are concerned. Since electronic means can be substituted for some maintenance trips, we presume that accessibility will become more important for discretionary activities than for maintenance activities. Of course, the characteristics of the relevant sets of opportunities are highly dependent on individual attributes.

Acquisition of Information on Opportunities

Normally, consumers gather information on retail activity places from their trips in a city (Potter 1979). This information is supplemented by advertisements and suggestions from other people. The spatial search-and-learn process is linked with people's activity patterns. A person's home, place of work, other activity places visited, and the paths between them all, bias the composition of the perceived action space (Golledge et al. 1985). It is conceivable that sociodemographics and the characteristics of activity pattern generate large differences between individuals in perceived action spaces.

Use of ICTs may change the characteristics of perceived action spaces dramatically. Search engines on the Internet can take care of a consumer's lack of knowledge of products, facilities and locations. The knowledge gained of previously unknown products and places can expand the perceived action spaces. The information density of action spaces can also change. Normally, a perceived action space is composed of a number of holes with activity places of which an individual has no knowledge. These holes can be filled through using new ICTs. Isolated activity places lying outside the concentrations of activity places, such as shopping centers, can improve their competitiveness. We can hypothesize that using ICTs reduces the differences between population categories with respect to the characteristics of their perceived action spaces, e.g. size and information density. Of course, the ability of an individual to use these technologies is also important.

New ICTs could change the travel behaviour of individuals not only at home or in some other fixed place, but also on the road. Access to the Internet on mobile phones offers individuals the opportunity to ask for information about activity places such as local restaurants, museums, public parks, police stations, hospitals, and 'the best' walking and driving routes through an area. Mobile phones can increase familiarity with certain areas and reduce feelings of uncertainty. People usually prefer behavioural options providing certain outcomes to options providing uncertain outcomes (Vugt et al. 1996; Hall 1983). Uncontrollable events may elicit feelings of anger, frustration and stress and may lead individuals to opt for behavioural alternatives providing more controllable outcomes.

These changes in the acquisition of information could lead us to expect a spread of outside-the-home activities. Isolated areas and lower order segments of transport networks in particular could see an increase in number of visitors. Because of their less routine nature, the impact of these new ICTs applies more to discretionary than to maintenance activities.

3.5 Conclusions and Future Research

The history of the new ICTs is relatively short, so most discussions on the effect of ICTs on the spatial configuration of opportunities are highly speculative. This chapter is no exception. On the basis of a review of the relevant literature, we described in a systematic fashion the possible effects of ICTs on *physical accessibility* from a spatiotemporal and a perceptual perspective. In this section we summarize our findings in the form of some hypotheses:

a. *More authority constraints (based on the discussion in Section 3.3.1)*
   - With the growth in the use of new ICTs, authority constraints will become more desirable and urgent. More non-authorized areas (non-digital boxes) will be the result.

b. *Expansion of potential action spaces (based on the discussion in Section 3.3.2)*
   - When telecommuting is adopted on a more permanent basis and by employees with short commutes, commuting distance and line and elliptical action spaces will expand.
   - The substitution of electronic shopping for trips to the shops will cause an expansion of (circle) action spaces.
   - The use of new ICTs, particularly ITS, will increase travel speed, which will result in larger action spaces. This time-space convergence will increase the range of opportunities within reach.

c. *Change in the composition of daily action spaces (based on the discussion in Section 3.3.2)*
   - Increasing popularity of telecommuting will lead to fragmentation in time intervals and thereby to an increase in the number of action spaces. More circle (home based) action spaces will also be created.

d. *Revaluation of accessible opportunities (based on the discussion in Sections 3.3.2 and 3.4)*
   - The use of new ICTs will cause a revaluation in favor of discretionary activities and acts. As a consequence, the importance of accessible opportunities will decrease for maintenance activities and increase for discretionary activities.
   - The use of new ICTs will cause a revaluation in favour of discretionary acts of travelling.

e. *Spread of visited activity places (based on the discussion in Section 3.4)*
- The use of new ICTs will increase the knowledge of formerly unknown activity places and routes. This increased knowledge will bring about a better fit of perceived action spaces on potential action spaces. A spread of visited (discretionary) activity places in actual action spaces would be the result.

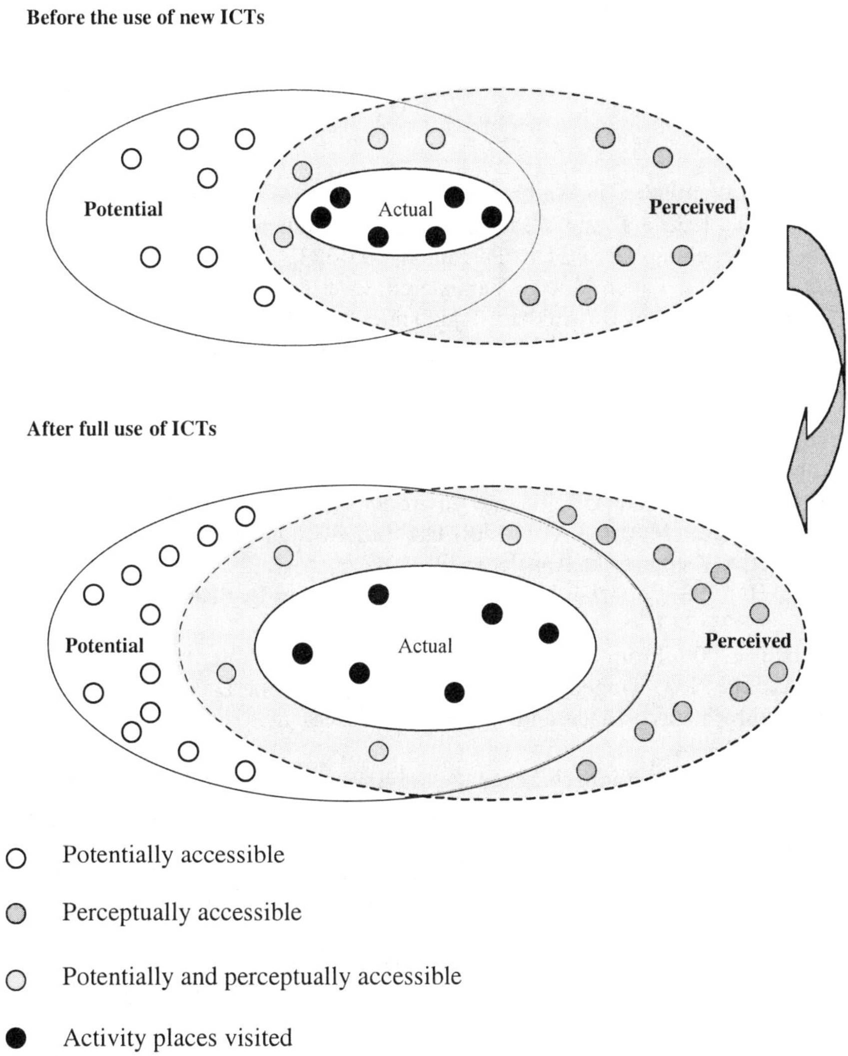

Figure 3.6 Potential impact of ICTs on action spaces

In brief: the use of new ICTs will cause a time-space convergence, a revaluation of travel time, activities and acts, and thereby of opportunities together with a greater knowledge of previously unknown opportunities. The ultimate anticipated result is shown in Figure 3.6.[2] All types of action spaces will expand, the overlap between types of action spaces will become larger, and known and visited activity places will be spread more widely within action spaces. Of course, these outcomes of the use of new ICTs are highly dependent on the psychological, sociodemographic and spatial attributes of individuals.

References

Axhausen K.W. 1998: Can We Ever Obtain the Data we Would Like to Have? In Gärling T., Laitila T. and Westin K. (eds.), Theoretical Foundations of Travel Choice Modelling, Oxford UK, Pergamon, 305–323.

Beckers Th. 1995: Tijden Moeten Veranderen: Over de Noodzaak tot een Zuinige Omgang met de ons Beschikbare Tijd (Times must change: On the necessity of an efficient use of our available time), in Van Dijk G. and Potman G (eds.), Verslavend lekker en nooit genoeg: overconsumptie motieven in deze tijd (Addictively nice and never enough: over-consumption motives at the present time), Kampen, Kok Agora, 88–98

Beckmann M.J., Golob T.F. and Zahavi Y. 1983: Travel Probability Fields and Urban Spatial Structure: 1, Theory. Environment and Planning 15A, 593–606

Brown L.A. and Moore E.G. 1970: The Intra-Urban Migration Process: A Perspective, Geografiska Annaler (52B), 1–13

Couclelis H. 1998: Editorial, Environment and Planning B, Planning and Design, Vol. 25, 321–323

Couclelis H. 2000: From Sustainable Transport to Sustainable Accessibility: Can we Avoid a New 'Tragedy of the Commons'?, in Janelle D.G. and Hodge D.C. (eds), Information, Place and Cyberspace: Issues in Accessibility, Springer, Berlin, 341–356

Dijst M.J. 1995: Het Elliptisch Leven: Actieruimte als Integrale Maat voor Bereik en Mobiliteit – Modelontwikkeling met als Voorbeeld Tweeverdieners Met Kinderen in Houten en Utrecht, (The Elliptic Life: Action Space as Integral Measure for Access and Mobility; Model Development with Two-earner Families with Children in Houten and Utrecht), (in Dutch with extensive summary in English), PhD Thesis, Koninklijk Nederlands Aardrijkskundig Genootschap/ Faculteit Bouwkunde, TU-Delft, Utrecht/Delft

Dijst M. 1999: Action Space as Planning Concept in Spatial Planning, Netherlands Journal of Housing and the Built Environment, Vol. 14, 2, 163–182

Dijst M. and Vidakovic V. 2000: Travel Time Ratio: The Key Factor of Spatial Reach, Transportation, Vol. 27, 179–199

Dijst M., de Jong T. and Ritsema van Eck J. 2002: Opportunities For Transport Mode Change: An Exploration of a Disaggregated Approach, Environment and Planning B, Vol. 29, 413–430

Gärling T. and Garvill J. 1993: Psychological Explanations of Participation in Everyday Activities, in Gärling T. and Golledge R.G. (eds.), Behavior and Environment; Psychological and Geographical Approaches, Amsterdam, Elsevier Science Publishers BV, 270–297

Giuliano G. 1998: Information Technology, Work Patterns and Intra-metropolitan Location: A Case Study. in Urban studies, Vol. 35, 7, 1077–1095

Golledge R.G., Smith T.R., Pellegrino J.W., Doherty S. and Marshall S.P. 1985: A Conceptual Model and Empirical Analysis of Children's Acquisition of Spatial Knowledge, Journal of Environment Psychology, Vol. 5, 125–152

Golob T.F. and Regan A.C. 2001: Impacts of Information Technology on Personal Travel and Commercial Vehicle Operations: Research Challenges and Opportunities, Transportation Research C, Vol. 9, 87–121

Gould J. 1998: Driven to Shop? Role of Transportation in Future Home Shopping, Transportation Research Record, Vol. 1617, 149–156

Gould J. and Golob T.F. 1997: Shopping without Travel or Travel without Shopping? An Investigation of Electronic Home Shopping, Transport Reviews, Vol. 17, 4, 355–376

Hägerstrand T. 1970: What about People in Regional Science?, Paper presented at the Ninth European Congress of the Regional Science Association

Handy S.L. and Mokhtarian P.L. 1996: The Future of Telecommuting, Futures, Vol. 28,3, 227–240

Hall D.R. 1999: Conceptualising Tourism Transport: Inequality and Externality Issues, Journal of Transport Geography, Vol. 7, 181–188

Hall P. 1998: Information Technology, Globalization and Regional Development, in Gravensteijn S.G.E., Griensven S. van and Smidt M.C. de (eds.), Timing Global Cities, Utrecht, 101–115

Hall R.W. 1983: Travel Outcome and Performance: The Effect of Uncertainty on Accessibility, Transportation Research B, Vol. 17B, 4, 275–290

Koninklijk Nederlands Aardrijkskundig Genootschap/ Vereniging van Utrechtse Geografie Studenten VUGS

Hanson S. 1998: Off the road? Reflections on Transportation Geography in the Information Age, Journal of Transport Geography, Vol. 6 4, 241–249

Harvey A.S. and Macnab P.A. 2000: Who's Up? Global Interpersonal Temporal Accessibility, in Janelle D.G. and Hodge D.C. (eds), Information, Place and Cyberspace; Issues in Accessibility, Springer, Berlin, 147–170

Hoorens V. 1998: Tijd is Geld – maar Hoeveel? (Time is Money – but How Much?), Nederlands Tijdschrift voor de Psychologie, Vol. 53, 30–40

Horton F.E. and Reynolds D.R. 1971: Effects of Urban Spatial Structure on Individual Behavior, Economic Geography, Vol. 47, 36–48

Hupkes G. 1982: The Law of Constant Travel Time and Trip-rates, Futures, Vol. 14, 38–46

Janelle D.G. 1969: Spatial Reorganization: A Model and Concept, Annals of the Association of American Geographers, Vol. 59, 348–364

Kitamura R., Chen C. and Pendyala R.M. 1997: Generation of Synthetic Daily Activity-travel Patterns, Transportation Research Record, Vol. 1607, 154–162

Klingbeil D. 1978: Aktionsräume im Verdichtungsraum: Zeitpotentiale und ihre räumliche Nutzung. Kallmünz über Regensburg: Lassleben (Münchener Geographische Hefte 41)

Kolars J.F. and Nystuen J.D. 1974: Human Geography: Spatial Design in World Society, McGraw-Hill, New York

Kwan M-P. 2001: Cyberspatial Cognition and Individual Access to Information: The Behavioural Foundation of Cybergeography, Environment and Planning B, Vol. 28, 21–37

Lanzendorf M. 1999: Technology, Travel Speed and the Environment. How to Assess The Environmental Impacts of New Transport Technologies?, Paper presented at 5th NECTAR Conference, Delft, the Netherlands

Lenntorp B. 1976: Paths in Space-Time Environment: A Time Geographic Study of Possibilities of Individuals, Lund Studies in Geography, Ser. B. Human Geography, no. 44, The Royal University of Lund, Department of Geography., Lund

Locke J.L. 1999: Why Don't We Talk to Each Other Anymore: The De-Voicing of Society, Simon and Schuster, New York

Lynch K. 1984: Good City Form. Cambridge: MIT Press

Michon J.A. 1980: Psychology: Aid or Guide for Travel-Demand Analysis?, in, Polak J.B and Van der Kamp J.B. (eds.), Changes in the Field of Transport Studies, Nijhoff, The Hague, 160–174

Mingioni E. 1983: Informalization, Restructuring and the Survival Strategies of the Working Class, International Journal of Urban and Regional Research, Vol. 7, 311–339

Mokhtarian P.L. 1997: The Transportation Impacts of Tele-Commuting: Recent Empirical Findings, in Stopher P. and Lee-Gosselin M. (eds.), Understanding Travel Behaviour in an Era of Change, 90–106

Mokhtarian P.L. 1998: A Synthetic Approach to Estimating the Impacts of Telecommuting on Travel, Urban Studies, Vol. 35, 2, 215–241

Mokhtarian P.L. 2000: Telecommunications and Travel, Transportation Research Board, Washington

Nilles J.M. 1995: Scenarios for the Development of Telwork, in, Reisen F. van and Tacken M. (eds.), A Future of Telework : Towards a New Urban Planning Concept?, KNAG/Faculteit Bouwkunde, TU-Delft, Utrecht /Delft

Pendyala R., Goulias K.G. and Kitamura R. 1991: Impact of Telecommuting on Spatial and Temporal Patterns of Households, Transportation, Vol. 18, 383–409

Potter R.B. 1979: Perception of Urban Retailing Facilities: An Analysis of Consumer Information Fields, Geografiska Annaler, Vol. 61B, 19–29

Reisen F. van 1997: Ruim Baan voor Telewerken? Effecten van Flexibele Werkvormen op Ruimtelijke Ordening en Mobiliteit als Gevolg van Veranderend Tijd-ruimtegedrag, (A Clear Path for Teleworking? The Effects of Flexible Work Forms on Spatial Planning and Mobility as a Consequence of

Changed Time-Space Behavior), KNAG/ Faculteit Bouwkunde TU Delft, Utrecht/Delft

Salomon I. and Koppelman F.S. 1992: Teleshopping or Going Shopping? An Information Acquisition Perspective, Behaviour and Information Technology, Vol. 11, 189–198

Salomon I. and Mokhtarian P.L. 1998: What Happens When Mobility-inclined Market Segments Face Accessibility-enhancing Policies, Transportation Research D, Vol. 3, 129–140

Saxena S. and Mokhtarian P.L. 1997: The Impact of Telecommuting on the Activity Spaces of Participants, Geographical Analysis, Vol. 29, 2, 124–144

Schafer A. and Victor D. 1997: The Past and Future of Global Mobility, Scientific American, Vol. 277, 36–39

Schwanen T. and Dijst M. 2002: Empirical Analysis of Travel-Time Ratios: The Association between Commuting Time and Work Duration, Transportation Research A, Vol. 36, 573–592

Schwanen T. and Dijst M. 2003: Time Windows in Worker's Activity Patterns: Empirical Evidence from the Netherlands, Transportation, Vol. 30, 261–283

Scott L.M. 2000: Evaluating Intra-metropolitan Accessibility in the Information Age: Operational Issues, Objectives, and Implementation, in Janelle D.G. and Hodge D.C. (eds), Information, Place and Cyberspace: Issues in Accessibility, Springer, Berlin, 21–46

Simon H.A. 1957: A Behavioural Model of Rational Choice, in Simon H.A., Models of Man: Social and Rational, John Wiley and Sons, London, 241–260

Suen S.L. and Mitchell C.G.B. 2000: Accessible Transportation and Mobility, Transportation Research Board, Washington

Supernak J. 1982: Transportation Modelling: Lessons from the Past and Tasks for the Future. Paper presented at 10th Planning and Transport Research and Computation Company Ltd., Summer Annual Meeting, University of Warwick, England

Toorn Vrijthoff W. van der, Jonge H. de, Draijer M., Delft A. van and Guyt P. 1998: Werk aan de Winkel: De Toekomst van de Winkelmarkt 1995–2015 (Shop at Work: The Future of the Shopping Market 1995–2015), TU-Delft, Delft

Vijgen J. and Engelsdorp Gastelaars R. van 1991: Een gevarieerd bestaan. Het gebruik van tijd en ruimte in het dagelijks leven van enkele 'oude' en 'nieuwe' groepen binnen de Nederlandse bevolking, (A Varied Existence. The Use of Time and Space in the Day-to-day Life of Some 'Old' and 'New' Groups in the Netherlands Population), Centrum voor Grootstedelijk Onderzoek, Universiteit van Amsterdam, Amsterdam

Viswanathan K. and Goulias K.G. 2001: Travel Behavior Implications of Information and Communications Technologies (ICT) in the Puget Sound Region. Paper presented at 80th Transportation Research Board, Annual Meeting, Washington DC

Vugt M. van, Lange P.A.M. van and Meertens R.M. 1996: Commuting by Car or Public Transportation? A Social Dilemma Analysis of Travel Mode Judgements, European Journal of Social Psychology, Vol. 26, 373–395

Wolpert J. 1965: Behavioral Aspects of the Decision to Migrate, Papers and Proceedings of the Regional Science Association, Vol. 15, 159–169

Zahavi Y. 1979: The UMOT Project, US Department of Transportation, Ministry of Transport, FRG, Washington DC/ Bonn

---

Endnote

[1] In reality, the form and size of the three types of action spaces can differ from each other. This figure just gives a hypothetical impression of the relationship between the three types of action spaces.

[2] See endnote 1.

4 E-commerce and the Container Shipping Industry

Hans van Ham* and Bart Kuipers**

*Delft University of Technology, Department of Transport Policy and Logistics' Organization, Delft, The Netherlands
**TNO Inro, Section Logistics and Transport, Delft, The Netherlands

4.1 Introduction

The globalization of economies is generating new maritime transport flows. In particular, the container shipping industry has benefited from this development. However, shipping a container from shipper/consignor to consignee sometimes involves over 60 documents. Hence, Information and Communication Technology (ICT) has always been important in this sector. Shipping companies have embraced Electronic Data Interchange (EDI), and Electronic Data Processing (EDP-)systems have been widely implemented. However, e-commerce, i.e. the marketing, selling and buying of products and services via the Internet, has been far less applied.

This chapter deals with the question of why the container shipping industry is relatively slow in implementing e-commerce and which way it is heading in this respect. The starting point in Section 4.2, is the description of the container shipping market. Section 4.3 deals with e-commerce and e-business from a theoretical point of view. Three types of models are discussed, i.e. sell-side business models (buyers-to-one-seller relationships); buy-side business models (one-buyer-to-many-sellers relationships); and electronic marketplaces allowing buyers and sellers to exchange information. Also, the organizational implications of e-business are considered. The empirical findings are presented in Section 4.4. Extensive research into e-commerce and container shipping have proven the existence of both auction and non-auction marketplaces, whereas only little evidence concerning the existence of buy-side business models was found. The most striking results concern sell-side models, and scrutinizing the websites of 66 major ocean carriers in the year 2000 showed various stages of development. Finally, in Section 4.5, some explanations are given and conclusions are drawn.

4.2 The Container Shipping Industry

4.2.1 Transportation

Shipping a container from origin to destination involves a large number of parties. These players can be categorized in various ways. In the first place, a distinction can be drawn between the actual shipping process and external parties. The shipping process can be further broken down into parties that deal with the physical aspects of shipping (e.g. those companies that have physical contact with the containers, i.e. transport requesters and transport suppliers), and logistic service providers that focus on managing and organizing the supply chain. External parties can be regarded as companies that do not directly take part in the shipping process but offer supporting and facilitating services. Figure 4.1 visualizes these parties and their mutual relations. Two parties play a pivotal role i.e. the carrier and the client or shipper (consignor) and consignee, respectively. The carrier, in this case the container shipping line, takes care of the actual transportation of the containers. They usually own a large number of vessels to carry out the transport.

The transportation of cargo can be organized in various ways. A shipper can decide to do it all by himself, or he can decide to have it outsourced to a third party or the carrier. Small customers often make use of intermediates, whereas large clients turn to the carrier for their shipments. In the case of complex inland transport: for example, because of transhipment, outsourcing the chain management to shipping lines, forwarders or inland carriers becomes more attractive.

The aim of many ocean carriers is to act as transport chain managers, who organize the complete transport process from door (origin) to door (destination). The term 'carrier haulage' is used when the shipping line which carries out the marine transport of the container, or its agent, also arranges the inland transport, whereas 'merchant haulage' is used when the shipper, freight forwarder or consignee arranges the inland carriage of goods (to port and/or to door). Inland carriers and freight forwarders can arrange the hinterland transport on behalf of shippers or consignees and thus play the role of chain manager in the case of merchant haulage.

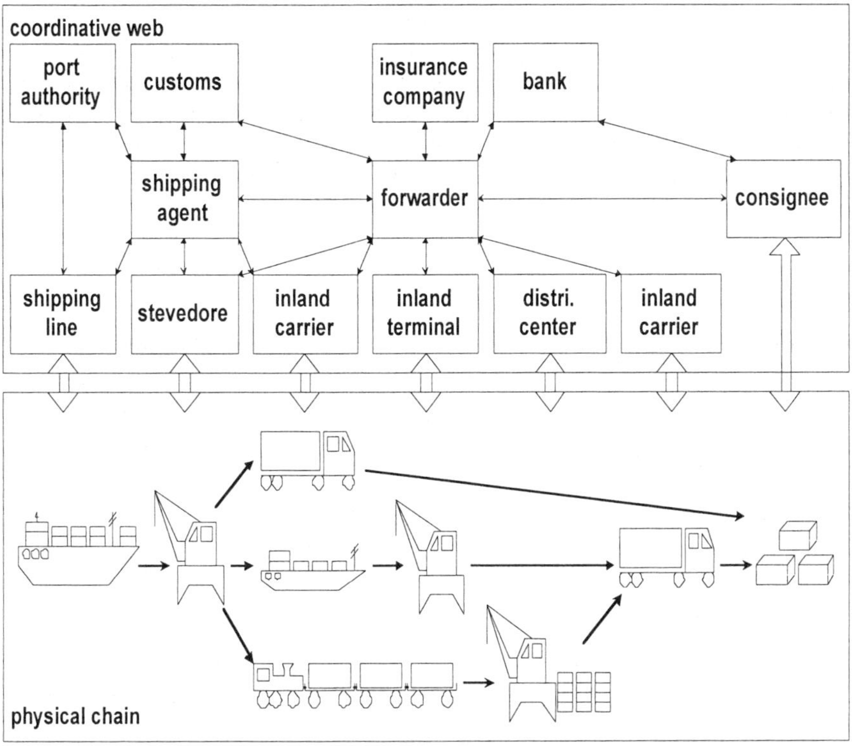

Figure 4.1 The shipping network
Source: M. den Hengst-Bruggeling 1999 (adapted).

4.2.2 Liner Shipping Sector

Industry Structure

There are currently less than 100 carrier companies with lines serving multiple continents. Only some 30 companies make up the elite, whereas others serve in niche markets or on specific routes. Like other mature industries, the liner trade is currently scaling up by means of mergers, takeovers, etc. This trend can be made clear in relation to the control of slot capacity in container traffic. In 1995, the 15 largest shipping companies controlled 62.5 percent of the slot capacity, and by 1999 their share had increased by nearly 4 percent to 66.2 percent (Heaver et al. 2001). Table 4.1 shows the top 15 container shipping companies by TEU[1] capacity.

The largest companies offer door-to-door transport services, have an extensive network of agents and focus on customer relations. Maersk-Sealand, Evergreen, P&O Nedlloyd, MSC, Hanjin, APL, NYK, Mitsui, CP Ships, OOCL, CMA-CGM, all claim to offer 'complete', 'door-to-door', 'end-to-end', 'multi-modal' solutions to their customers and to have their own agent network. Typically, smaller companies (and some relatively large companies such as Cosco, Yangming, UASC and CSAV) do not provide complete logistic services and do not have their own global agent network. In the intercontinental container shipping industry potential competitive advantages depend on size, which in turn can depend on consolidation (Maersk-Sealand, P&O Nedlloyd) or on location (Evergreen, Hanjin, Cosco).

Table 4.1 Top-15 ocean carriers (September 1999)

Rank	Carrier	Fleet Size (TEU)	Ships	Alliance
1	Maersk-Sealand	544,558	228	-
2	Evergreen/Uniglory Marine Corp	311,951	132	-
3	P&O Nedlloyd	268,625	117	Grand
4	Mediterranean Shipping Co	225,636	124	-
5	Hanjin Shipping Co	217,804	70	United
6	NOL/APL	199,881	74	New World
7	Cosco Container Lines	189,016	121	-
8	NYK Line/TSK	156,821	76	Grand
9	Mitsui OSK Lines	146,026	71	New World
10	Zim Israel Navigation	144,751	82	-
11	CP Ships	133,006	77	-
12	CMA-CGM	127,147	76	-
13	Hyundai Merchant Marine	109,105	33	New World
14	Yangming Marine Transport Corp.	101,445	40	-
15	OOCL	94,967	34	Grand

Source: Boyes 1999.

Cooperation Between Carriers

Since the advent of steam vessels in the 19th century, liner shipping has increased rapidly. On the one hand, steamers introduced a more reliable service, and on the other hand, the investments were huge compared with sailing vessels. The fluctuation in supply being the highest risk for liner carriers, shipping companies have started to collaborate in conferences. The objective of a conference is to fix agreements about rates (price cartel) and/or shipping capacity (quantity cartel) per trade. The shape of a quantity cartel may vary. If, for example, the loading and unloading rights of shipping companies are restricted to specific ports, and members are allotted a certain volume of cargo, it is called a pool. In a pool, shipping companies turn their cargo and/or income over to a common fund that is divided among the participants. Usually, individual sailing schedules and agent networks are maintained.

Bearing in mind that clients at least want a weekly sailing, it is almost impossible for a single company to independently maintain a regular service. In a consortium, the combined vessels of the member shipping companies make up a service with a single sailing schedule. In this way consignors/consignees are assured of regular services and the pooling of capacity enables shipping companies to fill their vessels. Consortia can be classified according to their level of cooperation. In contrast with 'separate marketing', where the participants retain their own marketing organization and agent network, 'joint marketing' uses a single agent for the entire consortium. Since containerization has increased the commercial and financial risks of shipping companies, the latter type of cooperation has become very popular, especially within the Far East Freight Conference (FEFC). These consortia included Ace, ScanDutch and Trio. During the first half of the 1990s, the agreements of the existing consortia expired and the reshuffling among carriers began. The new consortia are now known as alliances. The first such alliance was the Global Alliance and was set up by APL, OOCL, MOL and Nedlloyd in 1994. The objective was to establish an integrated Europe-Far East service. This initiative was soon to be followed by the creation of the Grand Alliance (1995) and United Alliance (1997) (GHR/CBS 1996).

At the moment, well-known alliances are the Grand Alliance (Hapag Lloyd, P&O Nedlloyd, NYK, OOCL, MISC), the New World Alliance (Hyunday, NOL/APL, MOL) and the United Alliance (Hanjin/ DSR-Senator, Cho Yang). Table 4.1 shows clearly that the Grand Alliance and the New World are much more powerful now than the United Alliance. Moreover, Maersk concluded a pact with Sea-Land and operates as a 'soloist'. This strategy is also pursued by MSC, Evergreen and Cosco and the China Shipping Group. Finally, there are a number of cooperation agreements regarding slot exchange (e.g. CMA and GCM) (Heaver et al. 2001).

Nowadays, the conference system is still very popular. Until recently, the liner shipping market was a classic example of an oligopoly: a limited number of large shipping companies that cooperate closely, and many (inadequately informed) shippers having hardly any influence on tariffs. However, cartels have always aroused the suspicion of the public authorities. A typical example in this respect is the Trans-Atlantic Agreement (TAA), which became operational in 1993. In order to gain tighter control of seriously loss-making shipping activities on the North Atlantic, the major shipping companies agreed upon rates, capacity supply and freight conditions. Because of allegations of rate manipulation, the European Commission banned the TAA in 1994. In this very same year, fines were imposed on a group of 14 European and Asian liner shipping companies, all members of the Far East Freight Conference, for illegitimate price fixing and discriminatory practices (Heaver et al. 2000).

Operational Management

As in other commercial sectors, revenues and costs determine the operational result of a liner shipping company. However, revenues are difficult to raise since tariffs are, to a certain extent, given in a conference system. The opposite is often the case, since (cyclical) overcapacity puts prices under pressure. Therefore, companies can only influence their revenues by such factors as vessel capacity and utilization rate.

In order to stay profitable companies must be cost-effective because competition is fierce. In this situation, low or high fuel prices can make the difference between profit and loss. Hence, carriers are constantly looking for ways to reduce costs and raise profits. Figure 4.2 shows a further decomposition of costs and revenues. The costs relating to intercontinental container shipping consist of six components, i.e. administration costs, voyage costs, maintenance and repair, container costs, port costs and capital and insurance costs. Costs relating to the hinterland transport are not taken into account as they do not form an integral part of the shipping process. However, the significance of these costs must not be underestimated, since they form a substantial part of the total transport costs.

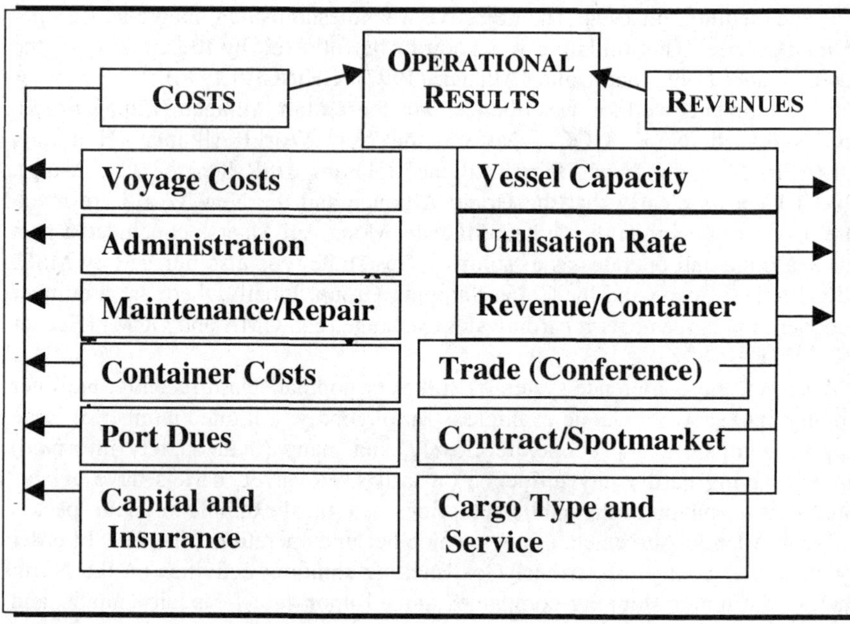

Figure 4.2 Cost-revenue breakdown of a typical ocean carrier
Source: Stopford 1997.

In relation to e-commerce, the focus is on administration costs. These costs (mostly) encompass wages for booking, documentation, planning, information technology, human resource management and other staff, but also costs related to office equipment, communications, software, etc. Dependent on the other cost categories, administration costs can amount to 20 percent of the total shipping costs (Stopford 1997). In pursuing cost reductions, administration probably is the most promising field to concentrate on.

Document management in the container shipping industry is a very difficult, time-consuming and costly business. There are three major reasons for this:

- Quantity
  It is a popular saying that 'every container filled with goods is accompanied by a container filled with paper'. Research in the port of Rotterdam resulted in more than 60 different documents in relation to transporting a container from shipper to consignee (Centrum Transport Technologie 1996). If this number is multiplied by the millions of transactions a large carrier handles, the enormous amount of paper is evident.
- Lack of Standardization
  Each carrier has its own booking system and uses different formats for documents including the Bill of Lading that is the contract of carriage between shipper and carrier. This single most important transportation document provides all the information the carrier needs to accomplish the move, stipulates the contract terms, acts as a receipt for the goods that the shipper tenders to the carrier, and, in some cases, shows certificate of title to the goods.

  Similarly, other parties in the supply chain use different systems and formats as well. This makes seamless coordination nearly impossible, and results in a lot of retyping and checking.
- Errors
  Documents being sent from one party to another can contain errors such as wrong numbers, conflicting data or incomplete information. In addition, documents arrive too late or get lost. Moreover, the purpose of a specific document can be unclear or irrelevant.

Traditionally, order processing and documentation were carried out manually; several persons had to key in the essential data. Afterwards, the information was sent by telephone and faxes (which, of course, still happens). This means that a large number of staff is required. Costs related to booking and documentation have fallen because of Electronic Data Interchange, internal computer networks and Value Added Networks, but can still mount up to $135 per shipment according to Manalytics, a group of transportation industry experts (Ryan 1999). Therefore, it is interesting to see what opportunities the Internet can offer to further reduce costs.

4.3 E-commerce and E-business

4.3.1 Introduction

Various people and organizations have given definitions of e-commerce and e-business. Even though most of these definitions do not differ widely, it is still important to define e-commerce and e-business clearly.

Here, e-commerce is the buying, selling and marketing of products and services on the Internet. It comprises the use of a digital network (the Internet) to support business transactions. A very important factor is that at least one of the participants should use the Internet for this purpose without human intervention. Therefore, e-mail cannot be used for e-commerce, but the World Wide Web can. E-commerce basically takes place at the front end of a company.

E-business is a more comprehensive concept and includes more than the on-line marketing, selling and buying of products on the Internet. It affects all aspects of an organization, varying from strategy, process, organization and systems to its trading partners and the ultimate consumer. Enterprises will examine, and probably change the way they perform basic business functions, i.e. sales, marketing, customer service, purchasing, operations, overseas initiatives, human resources, finance, and, of course, Information Technology.

The following section specifies some relevant aspects of e-business. Firstly, it deals with the issues (trends, success factors and imperatives) of e-business and, secondly, various models of e-commerce implementation are outlined. Finally, the organizational implications of e-business are considered.

4.3.2 E-business Issues

Obviously the Internet will dramatically change the way of doing business. But how will the Internet change the landscape? The answer is simple, yet profound. The Internet enables business ubiquity, allowing a company to do business everywhere, at all times. The Internet eliminates the constraints of time and distance in operating a business (Fingar et al. 2000). Seven aspects of e-business that give an overview of how the Internet may impact on business, are listed below and see Figure 4.3. Occasionally examples from the container shipping industry will be mentioned.

- Ubiquity
  The Internet offers companies and customers opportunities for doing business anywhere at any time. Even the smallest businesses can establish a global presence.
- Market entry
  Basically anybody can start up an Internet company. New entrants can be successful if they discover and deliver unique value to customers. With the elimination of the traditional middleman, new Internet intermediaries will emerge. These 'infomediaries', as they are often called, are independent third parties that offer aggregated information from various relevant sources. Infomediaries in the ocean shipping industry can compare various carriers on different lines on the basis of several criteria (time, price, security, conditions, etc.) and possibly also host insurance companies. Since market entry is relatively easy, and business processes become more transparent, other companies can easily copy these concepts.

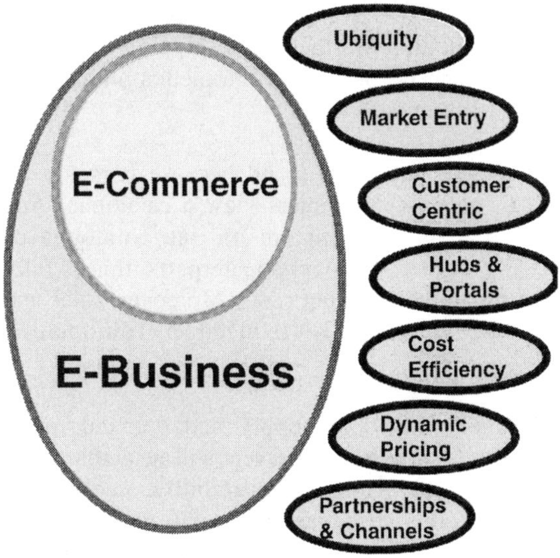

Figure 4.3 Aspects of e-business
Source: Bakker 2000.

- Power shift to the customer
  Based on the ubiquity of the Internet, the cost and effort for customers to find other suppliers is almost zero. Instead of producers 'pushing', customers can now 'pull' the information they want and reach out to all sources instantly. Therefore, businesses will have to invest in their customer relationships. By means of Web site customization, a company can achieve one-to-one marketing. By analysing buying behaviour and giving customers the opportunity to add, change and remove information on their Web site, this one-to-one marketing can be reached. With respect to the ocean shipping industry this implies that, on the one hand, carrier companies should extend their customer relationships by offering on-line services such as tracking and tracing, Bill of Lading-print-outs and tailor-made information on frequently used lines and service areas. On the other hand, by engaging in e-procurement, ocean carriers can save money on buying fuel, spare parts, paper, computer accessories, and more.
- Hubs and portals
  Simultaneously, the Internet can reduce the effort of a company to be involved in customer relationships. By means of customer service automation through a single enterprise access point, customers can seek the information they want instead of communicating via call centres, salesmen and the like. A hub or portal is an access point for Internet users. If successful, other companies will want to advertise and offer their products and services, thus further strengthening the hub's position.

In new volatile industries, a few 'first movers' emerge from a group of innovators to set the terms of competition. In the ocean shipping industry a company can become a hub by being the winner in infomediation and/or a marketplace for on-line shipping transactions.

- Cost reduction
  Variable costs for on-line catalogues are close to nil because it makes no difference if one or millions of potential customers view a catalogue. And even though designing, creating and building an on-line catalogue or marketplace may be more costly than its physical counterparts, this is fully outweighed by the lower printing and distributing costs. Moreover, sales and order-processing costs can be significantly reduced by automating transactions.
- Dynamic pricing
  When the element of time is eliminated in supply and demand, price mechanisms will change radically. Most Internet markets will resemble stock markets and fixed prices will be rare. Although availability, quality and service still count, 'dynamic pricing strategies are vital to successful marketing plans' (Fingar et al. 2000). Ocean carriers should use dynamic pricing not only on infomediary, marketplace or auction sites (e.g. for last-minute cargo space), but also on their own on-line buying/selling systems.
- Partnerships and channels
  E-business does not replace traditional ways of doing business; it just adds new possibilities. The Internet offers businesses excellent opportunities to broaden relationships with customers and suppliers. A solid network where companies can offer integrated solutions to their customers can be established by broadening one's channels and fine-tuning them. According to Keen (1999) in *Competing in Chapter 2 of Internet business* (page 23): 'Whereas on-line trading companies offered very low transaction fees and minimal service and Merrill Lynch offered outstanding service at high commission rates, Schwab (www.schwab.com) offered the best of both. It cut its commission drastically, but even so, at $29.95, it remains far higher than the $6 fee charged by 'pure' online players. It is Schwab's superb harmonization of all its channels – phones, branches, web and personal contacts – that makes it the leader of the industry.' For ocean carriers, this implies that they need to offer harmonized on-/off-line services and have these fully integrated with their partners in the supply chain and supporting services.

4.3.3 E-commerce Models

In order to categorize e-business/e-commerce models, it is important to identify the parties involved: Business-to-Consumer (B2C) and Business-to-Business (B2B). This report focuses on the latter category, since ocean carriers deal with companies and not with consumers. Moreover, the nature of transactions (selling or buying) needs to be taken into account.

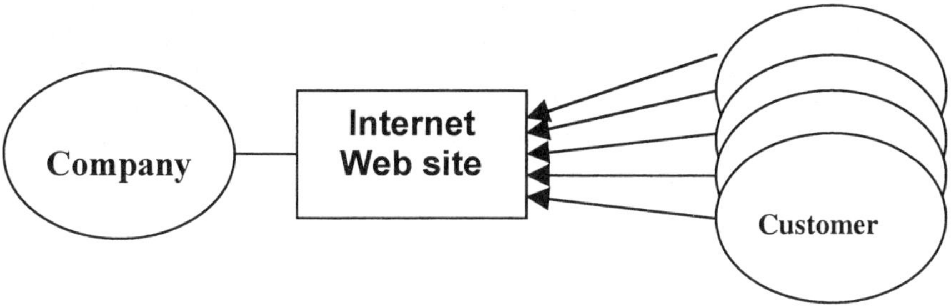

Figure 4.4 Sell-side e-commerce

Sell-side business models are the most familiar kind of e-commerce (see Figure 4.4). Here a many-to-one relationship exists between buyers and sellers. Huge numbers of customers visit a single company's Web site where they can buy products with fixed prices. Usually the products sold are highly standardized, high in supply and demand or products of strong brands. A good example is Amazon's books. In B2B, sell-side models should work with standardized services such as Web space advertising and financial transactions. Sell-side e-commerce is a good opportunity for ocean carriers. It could be argued that container space on liners is a very standardized good and that sell-side e-commerce would work very well there. In the case of regular bookings this might be true, but in other cases (e.g. special containers, peak demands, last-minute bookings) it could be false.

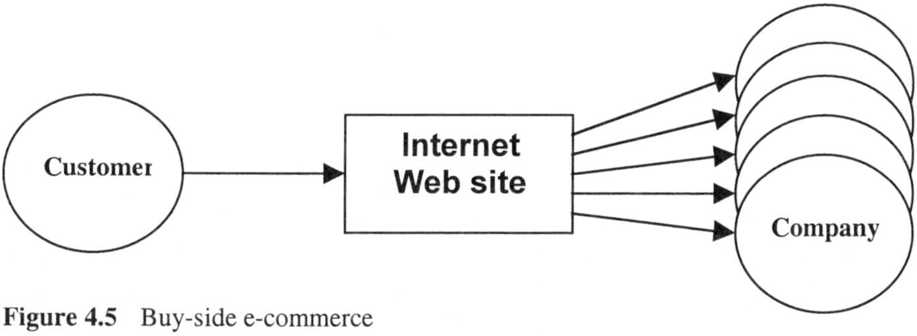

Figure 4.5 Buy-side e-commerce

Buy-side servers provide the capabilities for purchasing orders to be entered and fulfilled from the buyer's perspective. In this case a one-to-many relationship exists between buyers and sellers (see Figure 4.5). A single customer can enter the specifications of the product or service (s)he wishes to obtain. In response to this request, sellers can make an offer directly, or they can bid against each other for the best offer. The order is generally rather specific, while a large number of parties can handle the order. Furthermore, products and services with a relatively high value can best be sold in this way. From a customer's perspective it might be beneficial to make the extra effort, and from a supplier's perspective it will be an opportunity to obtain large orders. In a way, this type of e-commerce is the result of the shift from product-orientation to customer-orientation. Shippers and freight forwarders can adapt this strategy; since there are quite a few sellers in the market, the product can be rather specific (time, volume, weight, hazards, etc.), and has a high price.

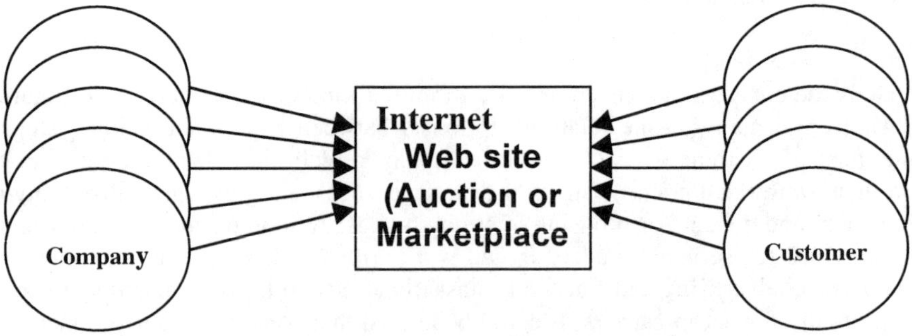

Figure 4.6 Marketplace e-commerce

An electronic marketplace (EMP) can be defined as an interorganizational information system that allows the participating buyers and sellers to exchange information about prices and product offerings (Bakos 1991). Generally, EMPs relate to both selling and buying. Within the value chain two parties are of primary importance, i.e. sellers and buyers. A third party is the intermediary, who provides the operation of the system.

A marketplace could work quite well in the ocean shipping industry. Supply and demand of cargo space can meet via the Internet. A distinction must be made between non-auction and auction marketplaces. In the latter case, prices are not fixed but depend on the interaction of supply and demand, thus creating a transparent, efficient market. It is obvious that the benefits of an EMP generally end up at the buyers and that, in the container shipping industry, carriers are not happy with a more transparent market. For an individual seller, it could be beneficial to introduce an EMP. The seller can bill buyers for using the system. The revenues would probably be higher than the decrease in profit caused by the increase in competition. However, as a group they are not better off by introducing such a system.

4.3.4 Organizational Implications

Sooner or later, the adoption of e-commerce will have its impact on the organization. In order to exploit new opportunities, companies may seek cooperation with other companies that have other fields of expertise. Companies can form a virtual web by selecting partners from the, already existing, business environment. Figure 4.7 shows such a web. According to Goldman et al. (1995), the virtual web is an open-ended collection of pre-qualified companies. They share an organizational culture and one or more areas of expertise. Moreover, they agree to form a pool of potential members of a virtual organization and to cooperate in a virtual organization if the occasion arises. Thanks to several process agreements being arranged in the virtual web beforehand, there is clarity among the partners in the virtual web about the way in which all decisions (e.g. in the formation process of a virtual corporation, are made).

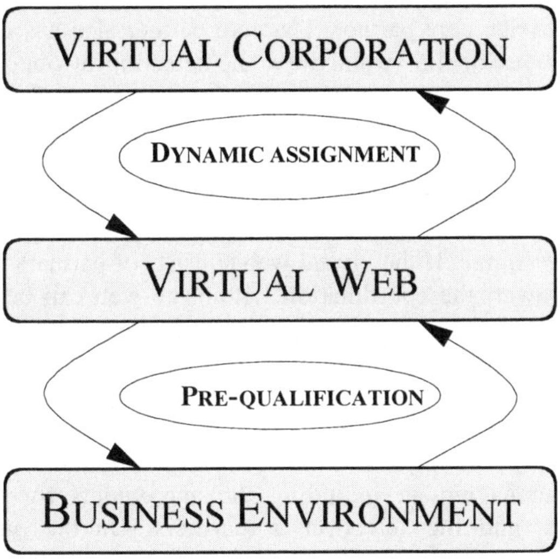

Figure 4.7 Virtual organization as the basis of network activation
Source: Becker 2001.

Within the virtual web the participating actors do not yet cooperate, but they have already made the necessary preparations to rapidly combine the expertise required. In this way the participating companies are capable of agilely responding to an emerging opportunity. The moment such an opportunity occurs, some of the companies present in the virtual Web, are selected, and form a virtual corporation. In short, a virtual corporation can be defined as a 'temporary network of independent companies, which is flexible in adapting to different conditions and meets the customer demand by sharing information, using the available knowledge and services (from the selected companies) and, if desired, by sharing costs and assets'. In the literature, several views are to be found on the virtual corporation. Most authors consider the virtual corporation a type of cooperation between companies, groups, or individuals. The essence of the virtual corporation is that each composition is based on the characteristics of one (kind of) opportunity (Grenier et al. 1995; Mertens et al. 1998), and that it accounts for the aspect of change (Katzy 1998). The set of competencies and resources can be adjusted quickly and frequently in response to changing market conditions and opportunities. When the opportunity is over, the virtual corporation is disbanded and the selected companies go back into the virtual web.

In the literature, several organizational variants are indicated in which the role of process designer and manager are allocated differently to the parties involved. Franke et al. (1999) distinguish three types of control:

1. *A core partner*. The core partner is the leading organization in both the virtual web and the virtual organization. All major activities and communication are coordinated by the core partner. The core partner also has a major influence on the objectives of the virtual web, the selection of opportunities and the prospective partners in a specific virtual cooperation. Furthermore, a core partner may take on a significant part of the operational activities necessary to support the web and the virtual corporation. The selection of the core partner can be based on market/technological power or on organization size.

2. *A steering committee*. If the virtual web consists of partners of equal size and with equal power, the coordination within the web can be performed by a steering committee: a governing body of representatives of the network partners. According to Van Aken (1998), strategic and operational management should be separated into virtual organizations. The steering committee addresses strategic issues such as the objectives of the network, the contribution by each partner, and the structure and authorities of the operational management, including the agreements for the adoption of opportunities, and the selection of partners. On the operational level, managers of the partners in the virtual corporation should be able to manage their allocated resources as if these were the resources of a typical organization.

3. *A broker*. The last major model of coordination of the virtual web is the 'net-broker concept' or broker. The broker acts as facilitator and catalyst. Brokers help companies to form strategic partnerships, organize network activities, and identify new business opportunities. Their task is to spread the network concepts, promote cooperation, organize groups or firms, and have them connected to the product designers, marketing specialists, and training providers that they need to compete successfully. However, the broker does not take part in the (productive) activities of the virtual corporation.

In conclusion, the allocation of tasks can be arranged in three different ways, i.e. a core partner; a steering committee; and a broker. Becker (2001) shows, for such varied industries as automotive (Smart), computer (Dell) and fashion (Benneton), that networks with the essential characteristics of virtual organizations do exist.

4.4 Container Shipping and E-commerce

This section combines the previous two chapters on an empirical level. It covers the current state of e-commerce in the ocean shipping industry alongside the three models described in the previous section.

The findings presented in this section are based on extensive research in the container shipping industry carried out by Bakker (2000) during the first half of the year 2000. First of all, a Web survey was conducted, including the Web sites of the major ocean carriers (in addition, the Web sites of some 30 short-sea carriers were viewed but not incorporated in this section), and some 20 Internet initiatives such as Bolero, GoCargo, Celarix, Interbox and Arena. Apart from this, interviews were held at six of the top-20 carriers, two smaller carriers were interviewed over the phone, and some 20 carriers and Internet companies were questioned via e-mail. Various experts in this field such as academics, port representatives and consultants were visited as well.

4.4.1 Current Use of E-commerce

The Internet hosts a large number of Web sites that relate to the ocean shipping industry. Most of these Web sites contain company, business or academic information and links, but some are part of an e-business initiative. There are company Web sites that provide tools for on-line bookings and additional information and services, and there are third-party infomediaries that host buy-side systems and electronic marketplaces.

Sell-side E-commerce

In order to determine how ocean carriers make use of the Internet, the Web sites of 66 of the largest ocean carriers have been scrutinized. Only two major carriers are lacking. China Shipping Container Lines does not seem to have a Web site and the Web site of MISC, part of the Grand Alliance, and a significant player in the market could not be viewed/entered. All the other container liner-shipping companies are present on the web. Figure 4.8 shows the results of this Web survey.

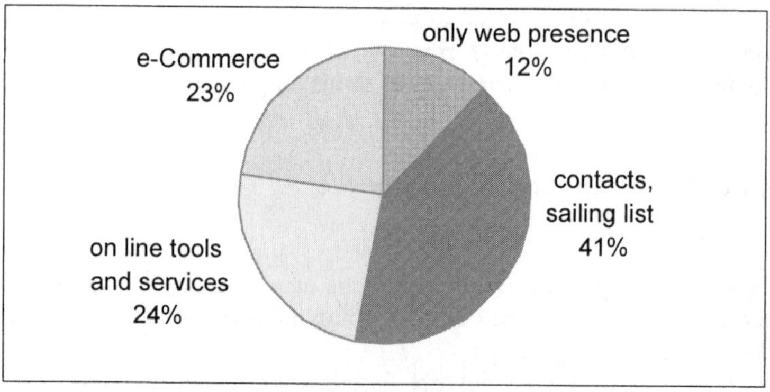

Figure 4.8 Level of sophistication of 66 carrier Web sites (mid-2000)

A minimal Web presence generally involves some company information such as an address and the ports of call. For 12 percent of the carriers, a simple presence on the Web is the most advanced stage they are in. Of the remaining carriers, a large number have included some additional shipping information such as sailing schedules and agents sometimes combined with e-mail addresses. This is how far 41 percent of the carriers have 'evolved'. The next stage consists of offering on-line tools and services such as tracking and tracing, Bill of Lading and account status. This is the level reached by another 24 percent of the carriers that were examined. The remaining 23 percent of the ocean carriers also provide on-line booking (via Bill of Lading preparation) for their customers, mostly accompanied with the possibility of printing out a legal copy of the Bill of Lading. ANZDL, APL, Cho Yang, Hanjin, HMM, OOCL, Maersk-Sealand and NYK are the leaders in e-commerce, with APL receiving the industry's only 'A' for Web-based business services (www.prnewswire.com). However, only small portions (under 2%) of all their transactions are done via the Internet. So, e-commerce has not been a tremendous success.

Buy-side and Marketplace E-commerce

Because the Internet also enables other companies to implement e-commerce, new parties are turning up. In the past two years, a number of Internet companies have entered the container shipping market. Some of them are container shipping specific, whereas others deal with inland and air transport as well. Moreover, it is not unthinkable that virtual non-vessel-owning common carriers (NVOCCs) will pop up on the Internet, or that some shipping agent's tasks will be taken over by dot-com companies. These third-party infomediaries may play an important role because shippers, when they require information on liner companies, equipment,

IT services, lanes, schedules etc., address them. Infomediaries can be categorized according to their nature:

- Buy-side infomediaries and auction marketplaces
  These sites have not been very popular with large ocean carriers, since the general belief is that they will lower prices. But, most of all, carriers feel that they 'don't have any space available in any trade lane which could be filled in this way without both destabilizing current customer relationships and worsening equipment imbalances'. Therefore, carriers are very hesitant to cooperate. Nevertheless, GoCargo quickly expanded into a company with more than 6,000 registered users including 40 carriers (R. Magown of GoCargo at the 'E-Business for Shipping' conference in Rotterdam, April 2000). However, it is difficult to predict how successful these buy-side infomediaries will be in the future. For example, GoCargo has ceased operations during the big shake out in the IT-sector, but other sites still exist. Also, auction marketplaces such as http://www.freightmarket.com/ and www.worldbid.com survived and are doing a lot better than container-shipping companies claimed they would.
- Non-auction marketplaces
  Besides buy-side infomediaries and on-line auctions, non-auction marketplaces have emerged. Buyers can search for quotations and, in some cases, track their cargo as well. Additionally, these sites offer extra information on shipping and provide on-line tools for price/contract negotiations. The most widely used marketplace-sites are www.etransport.com and www.celarix.com. But, the number of transactions on these two sites does not seem to imply much success in the future. However, both are heavily financed by venture capitalists; for example, E-transport is working with Cap Gemini. Other sites operating in this market include www.dartmaritime.com and www.pctb.com.

These organizations operate according to the 'net-broker concept' where the broker acts as facilitator and catalyst but does not take part in the (productive) activities of the virtual corporation.

In April 2000, a consortium of five major, European-based ocean carriers (Maersk-Sealand, P&O Nedlloyd, Hamburg Süd, MSC Mediterranean Shipping Co. and CMA CGM) launched an initiative to develop a B2B Internet portal for the container shipping industry. They agreed to join forces in order to create and financially support a transportation portal on the Internet. INTTRA (www.inttra.com) is the result of this initiative, and it was incorporated in January 2001. INTTRA is a neutral company, and none of the investors holds a majority ownership of the company shares. The initiative shows the distinct characteristics of the organizational variant of a virtual Web with a steering committee.

Later on Hapag-Lloyd and Kühne und Nagel became participants as well. Furthermore, INTTRA is open to non-participating ocean carriers, but these companies will be levied a fee. At the end of the year 2001, INTTRA was backed by 12 carriers that collectively serve more than 42 percent of the ocean-going transportation market. At that time, more than 1,400 registrants representing 400 companies in 50 countries had already managed more than 35,000 bookings and 60,000 containers via INTTRA. Because carriers sponsor INTTRA, there is no charge for shippers, freight forwarders, and consignees to access the service. In return, carriers realize savings when bookings, schedules, and event information are made available via the Web.

The advantage of INTTRA for shippers is that it provides a single integrated process for booking and tracking and tracing cargo across multiple carriers. The software enables customer interaction via multiple communication channels including the Internet, e-mail, phone and fax. In addition, on-line sailing schedules with comprehensive information (availability, departure/arrival dates, a complete list of ports, total transit time) are available to registered clients. Other options such as alerts, shipping instructions, reporting and contract management are under construction.

The openness of INNTRA is enhanced by a Web-based partner portal, which enables the system to work closely with various partners (exchanges, trading portals, ocean carriers etc.). Recently, a global provider of EDI and communication services for air cargo management (www.tradevision.net) connected INTTRA's ocean transportation network infrastructure directly to its air services portfolio.

4.4.2 Future Outlook

Since Information and Communication Technology is changing rapidly, it is difficult to predict the future use of the Internet in relation to the container shipping industry. However, it is obvious that ocean carriers will continue to maintain their own Web sites which can develop into more sophisticated stages of e-commerce offering on-line tools and services such as tracking and tracing, on-line booking and printing legal copies of the Bill of Lading. In addition, the initiative of 5 major ocean carriers to develop a joint portal will undoubtedly further strengthen the position of the liner shipping companies. The amount of information, partly confidential, is a unique asset. Since INTTRA became operational, many carriers, freight forwarders and shippers have joined.

Buy-side e-commerce could be successful, since price information is not revealed. Buyers simply make a bid for a specific transport and carriers can react without bidding against the competitors. Contrary to the airline industry, buy-side models in the container shipping sector have not boomed and it is doubtful whether they ever will. However, in specific niches, marketplaces and buy-side e-commerce may flourish, as GoCargo has shown.

Infomediary marketplaces have not shown impressive growth rates either, but they do have potential. As opposed to auctions that focus on the spot market, these infomediaries cover the whole container shipping market and could function as a starting portal or hub for shippers and carriers. A number of carriers already use infomediaries for tariff publishing and tracking and tracing services. As Internet awareness among shippers and carriers grows, companies such as E-Transport and

Celarix will continue to grow as well, possibly creating a strong portal/hub position. Meanwhile, they can serve as a marketing tool for ocean carriers.

In the years to come, most of the e-commerce volume will be generated on ocean carriers' websites and portals. On the one hand, this is based on the huge amounts of contract cargo increasingly being processed via the Internet. On the other hand, also occasional cargo will be booked this way (perhaps after checking information on an infomediary website first), simply because this is where ocean carriers too intend to pay most of their attention. Since carriers use different booking systems, software and formats, it will be very hard for auctions and infomediary marketplaces to provide these services. In this way, third-party marketplaces, auctions and buy-side e-commerce will be ousted to specific niche markets.

4.5 Conclusions

Due to the characteristics of the shipping network, maritime transportation generates huge information flows between the parties involved. Hence, Information and Communication Technology has always played an important role in the container shipping industry. Container shipping lines were among the first to embrace the medium of Electronic Data Interchange and most of them have implemented extensive (internal) Electronic Data Processing systems. However, e-commerce, i.e. the marketing, selling and buying of products and services via the Internet, has been far less applied. Although a survey among the major liner shipping companies, carried out in the first part of the year 2000, showed that most carriers own a website, this is still a long way from adopting fully integrated e-commerce systems. Over 75 percent of all ocean carriers only have a presence on the web (where some of them do offer more sophisticated services), whereas only the remaining quarter of the companies use the Internet for more advanced business transactions. Representatives of the container shipping lines that have adopted e-commerce (on-line booking and beyond) stated that their Web sites had resulted in cost reductions and efficiency improvements. Although on-line transactions amount to only 1-2 percent of the total transactions, this implies over 100,000 transactions per month in APL's case. If this number is multiplied with cost reductions per shipment, up to $135 according to Ryan (1999), huge cost savings are within reach. Bearing this in mind, it is amazing that the vast majority of container shipping lines had not yet reached this stage of e-commerce. Obviously ocean carriers will update and customize their websites in the years to come.

One of the reasons that e-commerce has not been implemented by the shipping industry, as happened in other industries (e.g. computing and electronics, utilities, paper and office products), might be connected with culture. The shipping industry is frequently perceived as a somewhat stubborn industry where executives firmly rely on their old beliefs and are not aware of the threats posed and opportunities created by the Internet. Another, more plausible reason might be

EDI. Because shipping lines were among the front runners in Information Technology, carriers are reluctant to move to the Internet. Value Added Networks and EDI are very costly to establish, and shifting from one technology to another could make all prior investments obsolete. Furthermore, EDI per se is not inferior to the Internet with respect to handling transactions and documentation. For large contract customers, the security and reliability of EDI might be preferable to the less expensive Internet.

However, the new Internet portal INTTRA can be considered a powerful attempt by the ocean carriers to catch up. The openness of INNTRA, being a virtual organization, is an important asset and various partners have already joined. The advantage of INTTRA for shippers is that it provides a single integrated process for booking and tracking and tracing cargo across multiple carriers.

Also, infomediaries and other providers of e-commerce have emerged in the container shipping industry. It is difficult to predict the success of third-party marketplaces, auctions and buy-side initiatives. From a traditional (carrier's) perspective they will not be very threatening, simply because they can only be successful if a lot of carrier companies are willing to join such an initiative. No large ocean carrier (or at least almost none) has shown any interest. Fear of price transparency, increased competition, unwillingness to see the market as a commodity market, and destabilization of customer relationships are the most important reasons. Forwarders are not willing to join either because they fear for their position of intermediary. Furthermore, it is almost impossible for auctions and buy-side Web sites to fully support on-line transactions, since there is no standardized booking system in the carrier industry.

The recent shakeout in the ICT-sector has left only a few infomediares on the market. However, the faith that venture capitalists and partners of some hubs/portals have, suggests a (potential) market for electronic marketplaces and auctions.

Reviewing the implementation process of e-commerce in the container shipping industry shows that the first attempts originated with parties not directly involved in the transportation process. Yet these initiatives have never assumed enormous proportions and, given the present situation in the ICT-sector, it is questionable whether they ever will. In relation to container shipping companies the adoption of e-commerce shows all the signs of the dialectics of progress, i.e. an initial lead in Information and Communication Technology turns into a developmental retardation. It took quite some time before the implementation of e-commerce started and even in the year 2000 the level of sophistication of ocean carriers' websites was still disappointing. However, this is fully offset by the carriers' joint portal INTTRA. In addition, the position of the liner shipping sector will be consolidated. Again cooperation between liner shipping companies pays off.

References

Aken J.E. van 1998: Strategievorming en Organisatiestructurering, Organisatiekunde vanuit Ontwerpperspectief, Kluwer Bedrijfswetenschappen, Deventer

Bakker S.H. 2000: E-Shipping, Exploration and Analysis of the Developments and Opportunities of E-Business in the Intercontinental Container Shipping Industry, MSc thesis, Delft University of Technology/Faculty of Technology, Policy and Management, Delft

Bakos J. 1991: Information Links and Electronic Marketplaces, Journal of Management Information Systems, Vol. 8, 2, M E Sharp Inc. Armonk, New York

Becker J.F.F. 2001: Virtual Organisation and Dynamism in Co-operation, TNO, Delft

Boyes J.R.C. (ed.) 1999: Containerisation International, 11, Informa Group plc, London

Centrum Transport Technologie 1996: Incomaas, Deelrapport Informatievoorziening, Rotterdam

Fingar P, Kumar H. and Sharma T. 2000: Enterprise E-Commerce, Meghan-Kiffer Press, Tampa

Franke U. and Hickmann, B. 1999: Is the Net-broker an Entrepeneur? What Role does the Net-broker Play in Virtual Webs and Virtual Corporations?, Journal of Organisational Virtualness, Vol. 1, 1, 120–139, VoNet (http://www.virtual-organization.net/)

GHR/CBS 1996: Port of Rotterdam Container Yearbook 1996, Gemeentelijk Havenbedrijf Rotterdam, Rotterdam

Goldman S.L., Nagel R. N. and Preiss K. 1995: Agile Competitors and Virtual Organization: Strategies for Enriching the Customer, Van Nostrand Reinhold, New York

Grenier R., Metes G. 1995: Going Virtual: Moving Your Organization into the 21st century, Prentice Hall, Upper Saddle River

Heaver T., Meersman H., Mogli F. and Voorde E. van de 2000: Do Mergers and Alliances Influence European Shipping and Port Competition?, Maritime Policy and Management, Vol. 27, 4, 363–373

Heaver T., Meersman H. and Van de Voorde E. 2001: Co-operation and Competition in International Container Transport: Strategies for Ports, Maritime Policy and Management, 2001, Vol. 28, 3, 293–303

Hengst-Bruggeling M. den 1999: Interorganizational Coordination in Container Transport: a Chain Management Design, Delft University of Technology, Delft

Katzy B.R. 1998: Design and Implementation of Virtual Organisations, Working Paper Series 98.002, Universität der Bundeswehr, Munich

Keen P.W.G. 1999: Competing in Chapter 2 of Internet Business, Eburon Publishers, Delft

Mertens P., Faisst W. and Ehrenberg D. 1998: Virtuelle Unternehmen und Informationsverarbeitung, Springer-Verlag, Berlin

Ryan R. 1999: Getting Satisfaction from EDI, in Containerisation International, issue 6, Boyes, J.R.C. (ed.), Informa Group plc, London
Stopford M. 1997: Maritime Economics, 2nd edition, Routledge, London

---

Endnote

[1] TEU = Twenty feet-Equivalent Unit

5 In Search of Urban Futures in the E-economy

Marina van Geenhuizen* and Peter Nijkamp**

*Delft University of Technology, Delft, The Netherlands
**Free University, Amsterdam, The Netherlands

5.1 Setting the Scene

The use of information and communication technology (ICT) in commercial transactions and work has grown tremendously since the mid-1990s. Most transactions using electronic commerce have been in business-to-business commerce, but business-to-consumer commerce is likely to play an expanding role in the coming years. An explanation of the popularity of the Internet and e-commerce can be found in network externality theory (e.g. Katz and Shapiro 1985; Capello 1994; Economides 1996).The positive network effects include increasing returns and first-mover advantages. In addition, various critical transaction and co-ordination advantages can be observed (e.g. Wigand 1997).

The changes in society embrace more than just faster and cheaper communication, with increasing intensity of transmission. With the growth of the Internet and e-commerce, a new architecture of society is under way. The changes – although not all clear, because of complex interaction and contradictory trends – point to a transformation of the organization of production (value chains), including the organization of firms, management structures and labour relations (Dussart 2000; Shapiro and Varian 1999; Westland and Clark 1999). There has been considerable speculation about the spatial impacts of this transformation, in particular the impacts on cities. Two extreme scenarios and one moderate scenario have been developed in this respect.

In the first, extreme, scenario it is emphasized that modern ICT makes it possible to overcome barriers of space and time. Firms and households become footloose, which means greater freedom in choosing their location. Homes as 'electronic cottages' are (re)designed to be access nodes for households as consumers, workers, etc. In this situation of the 'death of distance', place and location no longer matter in interaction and, as a consequence, large cities weaken their hold over economic and social life (Cairncross 1997). The basic hypothesis here is the ubiquitous access of firms and households to fast telecommunication networks.

The other extreme scenario says that large cities strengthen their position at the expense of smaller ones. This is based on the recognition of the role of large corporations in decision making in the global economic system and the fact that their decision making – requiring face-to-face contact – may hinder any trend for spread. Most probably there are only a few of such 'global urban command centres' in which highly qualified professional skills and tacit knowledge are produced and circulate, and access is available to the fastest global electronic networks (see, e.g. ,Gillespie and Williams 1988; Graham and Marvin 1996; Sassen 1996). ICT helps to extend the dominance of these cities by enabling action at a distance and remote control of activities in smaller towns and rural places. In addition, there are various generic arguments underpinning agglomeration under the influence of ICT. First, the use of ICT has enabled flexible specialization in manufacturing. This development has stimulated the spatial clustering of supplying firms due to the need to build and sustain trust and to transfer tacit knowledge. Secondly, many economic activities in cities are strongly interrelated with the knowledge that the workers there have (e.g. Duranton 1999). The local linkages of these workers are reinforced by the increasing demand for the fast establishment of face-to-face contacts in high-density knowledge transfer with customers and suppliers of services. A third argument in favour of agglomeration is based on the institutional and social-cultural dimension of cities, which act as a kind of 'glue' between firms and organizations, making them perform better. It is plausible that the social-cultural dimension of cities increasingly acts as a counterbalance to the impersonal and straightforward nature of electronic communication.

In the third scenario, the possibility of access for various selected medium-sized towns and smaller settlements to global decision making networks is recognized. In this vision, ICT has no deterministic impacts but works through the way in which it is 'constructed' in society. Socio-economic actions influence the application of the new technology and may change its impacts over time. The relation between the development of cities and the use of ICT is thus complex and non-linear, leading to a variety of results in different places and times (Graham and Marvin 1996; Malecki and Gorman 2001; Mitchell 1999).

The future of cities is all the more interesting because a general trend for spatial dispersal has recently been identified (e.g. Groth 2000; Lambooy 1998). A 'polycentric model' seems to be emerging, with smaller towns and rural areas exhibiting a stronger employment growth than large cities. The relative stagnation of large cities may be attributed so far to certain agglomeration diseconomies (Manshanden et al. 2000). The key issue in the context of the e-economy is whether the concomitant changes reinforce ongoing trends of spatial dispersal or weaken them. Against this background, the following questions will be addressed:

- Which opportunities and constraints in telecommunication infrastructure supply and in electronic communication can be identified that influence communication in distributed interaction?

- What impacts of the e-economy on the value chain may lead to changes of urban functions and transport? Will there be dispersal or concentration? What uncertainties are involved in terms of the direction and magnitude of these changes?

The basis for unravelling the impacts of the e-economy on urban functions is a scan of the recent literature and an empirical study of a sector subject to the strong influence of ICT and the e-economy, i.e. financial services (van Geenhuizen and Nijkamp 2001). The results of this analysis can be seen as 'educated guesses'. The chapter is structured as follows. First, in Section 5.2, there is an exploratory analysis of the spatial pattern of telecommunication infrastructure supply in the Netherlands in order to identify urban differences in access to such infrastructure and services. This is followed, in Section 5.3, by an analysis of the state of the art of opportunities and constraints in the electronic communication of different types of knowledge. Next, in Section 5.4, our attention turns to a number of emerging adaptations in the value chain. These adaptations are then linked with important actual and potential implications for urban functions and transport flow, including an assessment of the uncertainty involved in Section 5.5.

5.2 Limited Access to Global Grids

In this section, the position of Dutch cities in (inter)national telecommunication networks is explored in order to identify whether access to such networks is ubiquitous or limited to certain places. For this purpose a few selected fiber-optic grids are analysed, i.e. EuroRings (established by the partnership of companies KPN and Qwest), SURFNET 4/5 (subsidized by the Dutch government), and CityRings (established by company KPN). Table 5.1 shows the degree to which different classes of cities in the Netherlands – according to decreasing size – are connected with these grids. It appears that Amsterdam, as the largest city, has the best connections. Rotterdam is also in a favourable position. For example, the EuroRings network – seven grids connecting European agglomerations – connects both Amsterdam and Rotterdam with other European centres and, in the near future, with the largest cities in the United States, using the Qwest grid (KPN 2000). However, the remaining large cities: The Hague and Utrecht, are not connected. EuroRings represents international grids established in recent times in which market demand determines the location of the connections, i.e. the largest cities.

A different international network is SURFnet. It connects universities and research centres in the Netherlands with various parts of the world and offers the possibility for large enterprises to subscribe. All four large cities are connected, aside from eight mostly medium-sized university towns. To date, SURFnet4 is being replaced by SURFnet5, meaning a hundredfold capacity increase, as well as access to new generations of Internet. In the Netherlands the project – named Gigaport – is heavily subsidized by the government in order to push the

Netherlands internationally. By making use of Gigaport, knowledge institutes and enterprises can participate in advanced research based on new broadband technologies that are not available anywhere else, as in research of fast, three-dimensional simulation, international tele-consulting and data-mining. The third grid used here to identify the position of cities is the national CityRings (established by company KPN), i.e. rings around relatively large cities to increase transfer and connection capacity. All large cities and a considerable number of medium-sized towns in the Netherlands appear to be connected in this way, i.e. 85 percent of the 20 largest cities.

Table 5.1 Position of 20 largest urban agglomerations in the Netherlands in ICT networks (selected indicators) (1999/2000)

Urban agglomeration according to size	**EuroRings (KPN-Qwest)**	**SURFnet 4/5**	**CityRings (KPN)**
Largest (a)	100%	100%	100%
4 largest (b)	50%	100%	100%
10 largest	20%	80%	100%
20 largest	10%	55%	85%

a) Amsterdam
b) Amsterdam, Rotterdam, The Hague and Utrecht

If we focus in on the international position of Dutch cities based on various supply-side indicators, it appears that only Amsterdam plays an important role. In a ranking based on the price of services, choice of available connections and availability of the most advanced connections, Amsterdam is fifth in Europe, after London, Stockholm, Paris and Frankfurt (Graham 1999). According to an indicator concerning the use of Internet, however, Amsterdam is third in the world. Of the 50 most frequently used Internet routes, Amsterdam is end-station of seven routes in the world, against nine each for London and New York (Abramson 2000). As a node in Internet traffic, Amsterdam thus holds a top position in Europe.

Although the evidence is fragmentary, the conclusion may be drawn that there is no 'death of distance' in terms of access to the fastest global grids. Nodes and services are only created in those places where the potential market is largest based on the sheer size of cities or specialized research functions.

5.3 Limits of Electronic Communication

When considering opportunities and constraints of electronic communication, it is common practice to distinguish between codified knowledge and tacit knowledge. Contextual communication has been added as a separate class more recently

(Bolisani and Scarso 2000; Foray and Lundvall 1996). *Codified knowledge* is based on objective facts understandable for all actors that know the language (codes) and, consequently, it is transferable between different organizations (Table 5.2). In contrast, *tacit communication* deals with subjective knowledge, like personal experience, ideas and specific routines. Much of this communication is by chance, such as during work, at lunch or coffee breaks. A basic requirement is that the persons involved understand each other by using similar conventions and routines in understanding, usually merely by existing within one and the same organization. The third form, *contextual communication*, differs from the other two in terms of the aim of communication: namely transfer (exchange) of new meanings and new interpretative frameworks. There is a concomitant need to avoid predefined methods and outcomes, etc. in communication that basically serves a long-term process of interactive learning.

Table 5.2 Forms of communication and their characteristics

	Codified	**Tacit**	**Contextual**
Content	Objective knowledge in facts, figures, formulas	Subjective knowledge connected with ideas, perceptions and personal experience	Subjective knowledge about perspectives and interpretative frameworks
Environment	Within and between organizations	Mainly within organizations and between organizations that share culture	Between organizations with a different culture and frame of references
Way of transfer	Standard codes	Observation, interactive participation and practice	Continuous real-time interactive learning
Background of transfer	Intentionally	Intentionally, but often by chance	Intentionally, but without defined aim

The crucial question is to what extent these three types of communication can be carried out electronically (with sufficient cost savings), and particularly whether there is substitution of face-to-face communication by electronic communication or complementarity between the two types of communication. The answer so far is that there is an increasing substitution of codified knowledge and this seems to be accompanied by some compensation. Tacit communication and contextual communication suffer from major constraints using electronic devices, although some progress is being made regarding the former.

Table 5.3 Forms of communication, their success factors and constraints

	Codified	Tacit	Contextual
Success factors	Uniformity of codes	Trust, share of interpretative framework	Different frames of interpretation
Focus	Standardization, replicability, reliability, speed, efficiency, automatic processing	Flexibility, variability, management of ambiguity	Innovative content, context exploration, richness of form and content
Constraints in electronic transfer	*Few* In EDI a shortage of formalization of communication	*Substantial* • Intelligent agents: problems of validity, learning in unexpected situations, trust in delegating tasks • Multimedia using interactive tools: impossible to integrate all functions with universal applicability	*Structural* • Frustration of basic needs: continuous real-time interaction, broad range of information without pre-defined language, variety of means • No predefined aims and expected results (and costs)

With regard to codified knowledge, there are no constraints as long as the communication is sufficiently formalized (Table 5.3). Thus, an increase of the use of Electronic Data Interchange (EDI) is only hampered by a shortage of universal languages in business communication. For tacit communication, there are much more constraints. Trust and similarity in the use of an interpretative framework are critical success factors, and these can only be established and maintained in face-to-face communication. From experiences with global virtual teams in geographically-distributed decision making, one may draw the conclusion that important decision making can be done using electronic devices, but that a need still remains for face-to-face communication from time to time to refresh relationships and to re-establish trust and confidence (e.g. Maznevski and Chudoba 2000; Mitchell 1999). Two approaches to deal with the specificity of tacit knowledge transfer, i.e. the use of intelligent agents and multimedia using interactive tools, have led to important progress – in fact by codification – but still suffer from serious shortcomings (Table 5.3). For example, the use of intelligent agents suffers from a lack of validity in the representation of mental models by codified software and a lack of trust by the user. In addition, multimedia approaches only partly make it possible to satisfy match with the need for the richest information content and for a whole array of possible functions with universal applicability. With regard to contextual communication, one may say that substitution of face-to-face contact also seems impossible, in the longer term.

The major reason for this is that some important points of departure are not defined, i.e. the aim of communication and the means (channels). This type of communication is a long-term interactive process using a great variety of means and content.

In fact, in economic interaction there is a continuous evaluation, selection and combination of communication modes (Table 5.4). On the basis of two dimensions, i.e. the time dimension in terms of synchronous and asynchronous, and the spatial dimension in terms of local and distributed, four different modes can be identified. It seems that all modes have their appropriate roles, leading to choices dependent on the needs and willingness to pay the associated costs in particular contexts (Mitchell 1999).

Table 5.4 Advantages and disadvantages of communication modes

	Synchronous	**Asynchronous**
Local	Requires transport Requires coordination Richness of communication (intense, personal) **Very high costs**	Requires transport Eliminates coordination Reduces costs
Distributed	Eliminates transport Requires coordination Requires additional modes in complex communication Reduces costs	Eliminates transport Eliminates coordination Limited to particular communication **Very low costs**

Source: Adapted from Mitchell 1999 (p. 138).

In general, there has been a move towards the very low cost, distributed asynchronous quadrant in the past decades, but in most interaction there is a combination of modes, such as in the previously indicated decision making by global virtual teams. We may conclude that there are still substantial barriers in electronic communication of tacit knowledge, despite some progress in codification, and that barriers in electronic communication of contextual knowledge are insurmountable. Furthermore, the question of substitution is complicated due to the use of combinations of communication modes.

5.4 Value Chain Adjustment and Urban Functions

In this section, we focus on various adjustments in the value chain. We explore changes in interaction concerning goods and persons. Note that the financial services chain, used here to illustrate some developments, deals with immaterial services and that the situation is more complex for chains including physical goods (Kenney and Curry 2001). We will discuss the increased customer-driven production and growing customer services, desintermediation and reintermediation, the emergence of e-firms, innovative activity and high-level management, and the increased flexibility in work locations (Table 5.5).

The phenomenon of increased customer-driven production needs to be seen in the context of mass-individualization. E-commerce matches the latter development because there are many opportunities for selection and price-reduction of goods and services, for example by using 'co-buying'. Consumers may scan offerings worldwide and negotiate prices using intelligent agents. Accordingly, consumers may select unique packages of products and services. This development is true for financial services, books, music, travel services, job intermediation services, consumer electronics, toys, apparel and self-assembly personal computers (ABN-AMRO 1999). A major consequence is the closing-down of retail outlets and service offices at the lowest level of transactions. With regard to financial services, known for the use of a wide range of electronic devices, we have seen a closing down of local offices in the Netherlands by 21 percent in the past seven years. Note that such a contraction can also partly be ascribed to rationalization following on mergers and acquisition. Furthermore, a true customer-driven system in goods chains means that risk and inventory, such as of books and computer components, can be removed from the system. The concomitant changes in logistic organization and transport are not known to date but what seems clear is an increase in the number of trips and of the distances involved, as well as changing roles of distribution centres. A major point of attention is to improve customer delivery which may lead to new types of logistic networks and nodes at the end of the chain.

The second development – an increased supply of customer services – is also evident in a number of chains. Customer services are often based on personal face-to-face contact and physical presence of consumers. Pre-sales, for example, deals with customer specific advice and user information. For particular goods, it is necessary that consumers touch, feel and test the product, or merely enjoy the atmosphere (i.e. 'shopping experience'). It is plausible that such services can only be provided in nodes (cities) at a higher level than the local. We may illustrate this with financial services. Electronic banks incur 65 to 70 percent less costs than existing 'bricks and mortar' banks (Westland and Clark 1999), meaning that they supply services at cheaper rates and capture customers from 'bricks and mortar banks' In this competitive environment, 'bricks and mortar' banks feel the need to preserve the loyalty of their clients and supply custom-made services based on a thorough knowledge of their clients' profile. In past years, therefore, places have been established where banker and clients meet, often in offices at the regional level. The rise of regional bank offices with extended personal services can be seen as a kind of compensation for the loss of face-to-face contact at the local level (Table 5.6).

Desintermediation often occurs at the end of the value chain, but in some chains a whole range of intermediaries may disappear. This is the case where musicians or writers (poets) send their work directly to customers, thereby circumventing producers, wholesale and retail. In other chains, one may observe the disappearance of physical sections like markets and purchasing agents, and the insertion of new, virtual sections in the chain. Internet business-to-business auctions are meant to purchase globally at the cheapest rates. Some firms have established vertically-oriented sites, like General Motors and Ford for car components and Boeing for aircraft components. There seem to be two impacts here. First, the disappearance of physical segments of the value chain may cause cities to decrease in importance as meeting places and trade and transaction centres, and may concomitantly lead to a reduction of the traffic concerned. Secondly, by using electronic auctions, the cheapest producers (suppliers) are identified and contracted on a global scale. This may lead to an increase of the distances over which goods (components) need to be transported. The latter impact, of course, does not apply to financial services. In these services, however, one may observe the emergence of virtual intermediaries that support consumers in information gathering and processing, e.g. in the process of selecting products (services), and in connecting with selected suppliers. The rise of these 'infomediaries' is typical for the Internet because of the abundant and open information supply. Infomedairies may also – if neutral – verify the identity of buyers and sellers, and certify crucial information transfer (Saanen et al. 1999). It may be that the function of cities as meeting and information places and places to create trust in less complex matters will be partly taken over by virtual intermediaries. However, at the same time, it seems that new types of physical meeting and information centres will arise in close interaction with electronic ones, with the important role to make electronic meetings and information handling run smoothly. Face-to-face contact and electronic communication seem inextricably intertwined (Mitchell 1999).

The rise of e-firms can be observed in many value chains. These firms not only save substantial costs of retail outlets, inventory and personnel, but they may cover the entire world as their market area. They have limited needs for physical location, which allows them to be relatively footloose. However, so far only a few e-firms have succeeded in surviving. Substantial entry barriers arise for followers in situations where 'first movers' have already established market positions, trade marks and customer relations, i.e. considerable costs of marketing. It is not easy to assess the impact of these developments on the function of cities and transport and traffic. In any case, we may expect some blurring of the central place structure with more diffuse market relations and thinner flows of goods. In addition, what may happen is that established 'bricks and mortar' firms – by cooperating with e-firms and establishing Internet activity by themselves – strengthen their position. Firms located in certain large and medium-sized towns may be ahead in these developments because they enjoy agglomeration and information economies available there.

This assumption touches upon the future innovation and management function of cities. One may draw the conclusion that large cities and some medium-sized towns remain attractive as a location of such functions, given opportunities for face-to-face contacts, quick access to the richest electronic information in these places, and a reservoir of specialized knowledge workers serving complex data-processing and application (Glaeser 1998).

Table 5.5 Impacts on urban functions and transport

Value chain changes	Urban functions	Transport	Uncertainty (a) Direction	Magnitude
Customer-driven production	Decrease of role as a node in distribution	More trips and higher frequency in delivery of goods	+	+
	Decrease of retail function and other services at the lowest level	Decrease of customer trips on level of living quarters and small centres	-	+
Customer management and services	Increase of retail (other) services at higher level	Increase of customer trips at level of sub-centres and medium-sized towns	-	+
Desintermedia-tion (elimination of physical mar-kets, auctions, etc)	Decrease of trade and transaction centre function at different levels	Decrease of trips for trade and transactions at different levels	+	++
Re-intermediation (insertion of intermediaries)	Some loss of function of meeting and information places but new forms	Some decrease of trips to city centre or sub-centre for meetings and information	++	++
	Dispersal of production due to increased global outsourcing etc.	Increased distances	+/-	++
Emergence of e-firms active in global markets	Blurring of the central place system (urban hierarchy), except for upper level	More dispersed flow of customer trips, work trips and goods	-	++
Innovative activity and management	Remaining strong role of large cities, and some medium-sized ones	No change	+	+
Flexibility of work locations	Partial loss of work function by central city, and gain by edge and suburbs	Less commuting trips (home-based telework) and on average shorter trips (centre-based telework). Smaller focus on central city	-	+
	Maybe spread of living function as secondary effect	On average longer trips	+	++

a. - = small; + = considerable; ++ = large.

If we take the development of Amsterdam as a financial centre as an example, there seems to be a trend for concentration, although this is not evident from all indicators used (van Geenhuizen 2001) (Table 5.6). The number of jobs in Amsterdam has increased, but the share in the national total has remained the same. In addition, there is a small trend for the dispersal of headquarter activity in the Netherlands, given a decrease in the share of all headquarters in the country in Amsterdam. An in-depth analysis, however, shows that this decrease can be partly attributed to a withdrawal of foreign banks due to internal reorganization caused by national financial crises (van Geenhuizen and van der Knaap 1998). A clear trend for concentration in Amsterdam can be observed when taking the size of the banks into account. There has been an increase of more than 10 percent in Amsterdam's share in the national balance sheet total.

Table 5.6 Indicators for spatial trends in financial services in the Netherlands

	1991	**1999**	**Change**
Offices at the local and regional level (a)			
Retail offices	3971	3140	-20.9%
Offices for special customer services	93 (b)	247	+165.6%
Amsterdam			
Number of jobs (x1000)	44,6	53,7	+20.4%
Share of jobs in national total (%)	22.6	22.1	+/-
Share of headquarters (%)	61.0 (b)	53.8	-7.2
Share of balance sheet total(%)	52.3	63.8	+11.5

(a) Retail offices for ABN-AMRO, ING-Bank and RABO-Bank, higher level offices only for ABN-AMRO and ING.

(b) 1989.

Source: van Geenhuizen 2001.

In the e-economy, work relations are becoming more flexible, including the location of the working place. We may think of a variety of alternative work locations, including the home, a telecentre, a vehicle, a client's office, a hotel room, a virtual drop-in office, and so on. We will focus our discussion on the common forms of home-based and centre-based telecommuting. Already for some years, the use of electronic communication has enabled employees to perform working tasks at home that would otherwise be performed in the working place (named telework or home-based telecommuting). It seems, however, that there is only a partial substitution, for a number of reasons. For instance, the need for social contacts in the workplace, the need for a certain travel time to 'disconnect' from home and work, and the need for face-to-face working contacts (Mokhtarian and Bagley 2000; Salomon 2000). With regard to the latter argument, it seems reasonable to assume that, in chain segments that include management tasks and creative exchanges of ideas, only a partial substitution of physical meetings and travel can be expected. Some disadvantages of home-based telecommuting are avoided in a second type of telecommuting, i.e. the centre-based form. In this type of telecommuting, employees make use of conventionally-equipped offices at the edge of agglomerations or suburban centres ('flex-offices'), leased by the firm, where they connect themselves electronically to the firm and its clients and other relations. This telecommuting is still at the experimental stage, and success in terms of achieving economic self-sustainability remains questionable (Moktharian and Bagley 2000). With regard to transport, we may say that a partial substitution of the home-work trip (home-based telecommuting) and shorter home-work trips (centre-based telecommuting) together cause a substantial reduction of commuter traffic. There are also secondary impacts to be expected, which may be as important as the primary ones. The reduction of the number of commuting trips and average distances involved enhances the attraction of residential locations at a greater distance from the city centre, thus increasing suburbanization over larger areas. However, such secondary effects have not yet been observed systematically because these occur with a time-lag.

5.5 Urban Futures and Uncertainty

If one thing is clear, it is that the impacts of the e-economy on cities are highly differentiated, with regard to urban functions and the spatial scale level involved. This is mainly as a result of different patterns of substitution of physical interaction between persons by electronic communication and on different needs for combinations of communication modes. There is also variation in uncertainty surrounding the images of changes in urban functions, in terms of the direction and magnitude of the changes. Five changes seem relatively clear in terms of direction, based on the increased use of electronic communication and e-commerce, particularly their distance-shrinking effects but also the concomitant need for compensation. We present these changes here as hypotheses that need to be elaborated and tested in further research:

- Business-to-consumer e-commerce leads to the erosion of urban nodes at the lowest (local) level.
- Business-to-consumer e-commerce at lower levels leads to compensation at a higher level of urban nodes.
- Business-to-business e-commerce opens up global markets, leading to the dispersal of production activity, the reorganization of logistic chains, and the erosion of cities.
- The emergence of e-firms active on global consumer markets leads to the blurring of the central place system.
- Telecommuting leads to a partial dispersal of work to the edge of cities and to suburban places.

High uncertainty – in terms of direction and/or magnitude – exists with regard to the impacts of the elimination of physical markets (auctions) and the insertion of new (information) intermediaries in the city. Traditional meetings and relatively simple information supply may disappear from the cities. However, new physical meeting places – closely related with electronic networks and clubs – may arise but these are largely unknown to date (Mitchell 1999). In addition, important secondary effects of telecommuting in terms of related changes in the location of residential areas are highly uncertain.

Uncertainty stems from various sources, i.e. the complexity of urban reality, the lack of knowledge about new forms of ICT and about the direct and indirect impacts that unfold after different lengths of time. In addition, interest groups respond quickly to new opportunities partly in response to fashion, but they may also easily shift attention. The latter turbulence increases complexity, particularly in terms of policy making. Another point that needs to be kept in mind is that the available knowledge is fragmented and not achieved in a coordinated way. In this context we can identify the following knowledge gaps:

- The extent to which tacit knowledge can become codified and transmitted electronically.
- Factors that influence the pace of change: To what extent are physical (infra)structures or ongoing organizational developments working as impediments to change, or as catalysts or movers of change?
- The comprehensiveness of the changes: Are the impacts general or specific for particular value chains?
- The causal background of the changes: To what extent are adjustments in the value chain caused by ICT and the e-economy, or by (autonomous) organizational changes with similar impacts?
- The nature of indirect effects on the city, and the different time scales involved.

The existence of these important unanswered questions has two implications for ICT policy and for urban policy. First, there is a need for increased research and experimentation efforts, and a better coordination of these efforts. Secondly, the manifold uncertainty needs to be dealt with in policy making, and one important way for this is the design of flexible policies that can be adapted in time. An adaptation of once-chosen policy lines is needed if actual developments turn out to be different from those expected and new results from research and learning experiments become available.

References

ABN AMRO 1999: Web or Shop? E-commerce and Retail in the Nederlands (in Dutch) Amsterdam

Abramson, B.D. 2000: Internet Globalization Indicators, Telecommunications Policy, Vol. 24, 69–74

Bolisani, E. and Scarso E. 2000: Electronic Communication and Knowledge Transfer, Int. Journal of Technology Management, Vol. 20 (1/2), 116–133

Cairncross F. 1997: The Death of Distance. How the Communications Revolution Will Change Our Lives, Harvard Business School, Cambridge, MA

Capello R. 1994: Spatial Economic Analysis of Telecommunications Network Externalities, Avebury, Aldershot

Duranton G. 1999: Distance, Land and Proximity: Economic Analysis and the Evolution of Cities, Environment and Planning A, Vol. 31 (12), 2169–2188

Dussart C. 2000: Internet: The One-Plus-Eight 'Re-volutions', European Management Review, Vol. 18,4, 386–397

Economides N. 1996: Network Externalities, Complementaries and Innovations, European Journal of Political Economy, Vol. 12, 211–233

Foray D. and Lundvall B-A 1996: The Knowledge-based Economy: from the Economics of Knowledge to the Learning Economy, in OECD Employment and Growth in the Knowledge-Based Economy, OECD, Paris, 11–32

Geenhuizen, M. van and Knaap G.A. van der 1998: Financial Services and the Urban System in The Netherlands, TPM Studies in Urban and Regional Development, no. 2. Delft University of Technology, Faculty of Technology, Policy and Management, Delft

Geenhuizen, M. van and Nijkamp P. 2001: Electronic Banking and the City System in the Netherlands, in Leinbach T. and Brunn S.D. (eds.), The Wired Worlds of Electronic Commerce, John Wiley, London, 181–201

Geenhuizen M. van 2001: Amsterdam as a Changing Node in Financial Services, Geographica Helvetica, 3

Gigaport 2000: www.gigaport.nl

Gillespie A. and Williams H. 1988: Telecommunications and the Reconstruction of Regional Comparative Advantage, Environment and Planning 20, 1311–1321

Glaeser E.L. 1998: Are Cities Dying? Journal of Economic Perspective, Vol. 12, 2, 139–160

Graham S. 1999: Global Grids of Glass; On Global Cities, Telecommunications and Planetary Urban Networks, Urban Studies, Vol. 36, 5–6, 929–949

Graham S. and Marvin S. 1996: Telecommunications and the City. Electronic Spaces, Urban Places, Routledge, London

Groth N.B. 2000: Urban Systems between Policy and Geography, Regional Studies, Vol. 34, 6, 571–580

Katz M.L. and Shapiro J. 1985: Network Externalities, Competition, and Compatibility, American Economic Review, Vol. 75, 424–440

Kenney M. and Curry J. 2001: Beyond Transaction Costs: E-Commerce and the Power of the Internet Dataspace, in Leinbach T. and Brunn S.D. (eds.), The Wired Worlds of Electronic Commerce, John Wiley, London, 45–65

KPN 2000: www.kpn.nl, KPN – De onderneming – Het bedrijf – Data

Laan L. van der 2000: Ruimtelijke Implicaties van Arbeidsmarktontwikkelingen, 's-Gravenhage, RPD Onderzoeks publikatie

Lambooy J.G. 1998: Polynucleation and Economic Development, European Planning Studies, Vol. 6, 457–466

Malecki E.J and Gorman S.P. 2001: Maybe the Death of Distance, but not the End of Geography: the Internet as a Network, in Leinbach T. and Brunn S.D. (eds.), The Wired Worlds of Electronic Commerce, John Wiley, London, 87–105

Manshanden W., Muskens J., and van der Vegt C. 2000: The Amsterdam Economy as a Model, in, Boekema F. and Graaff W. de (eds), New Ideas in the Netherlands Spatial Research, Policy for Large Cities, Transport and Infrastructure (in Dutch), Van Gorcum, Assen

Maznevski M.L. and Chudoba K.M. 2000: Bridging Space over Time: Global Virtual Team Dynamics and Effectiveness, Organization Science, Vol. 11, 5, 473–492

Mitchell W.J. 1995: City of Bits. Space, Place, and the Infobahn, The MIT Press, Cambridge (MA)

Mitchell, W.J. 1999: E-topia "Urban Life, Jim – But Not as We Know it", The MIT Press, Cambridge, MA

Moktharian P.L. and Bagley M.N. 2000: Modelling Employees' Perceptions and Proportional Preferences of Work Locations: The Regular Workplace and Telecommunication Alternatives, Transportation Research Part A, Vol. 34, 223–242

Saanen Y.A., Verbraeck, A., and Sol, H. 1999: Snapshot of E-commerce's Opportunities and Threats, Electronic Markets, Vol. 9, 3, 181–189.

Salomon I. 2000: Can Telecommunications Help Solve Transportation Problems?, in Button K.J. and Hencher D. (eds.), Handbook of Transport Modelling, Elsevier, Oxford

Sassen S. 1996: Cities in a Time for Paradigm Shift?: Rethinking Modern Urbanism, Proceedings of the Seoul Metropolitan Fora 96, Seoul

Shapiro C. and Varian H.R. 1999: Information Rules. A Strategic Guide to the Network Economy, Harvard Business School Press, Cambridge, MA

Westland J.Ch. and Clark T.H.K. 1999: Global Electronic Commerce. Theory and Case Studies, The MIT Press, Cambridge, MA

Wigand R.1997: Electronic Commerce: Definition, Theory and Context, The Information Society, Vol. 13,1, 1–16

PART B: New Forms of Travel Behaviour

6 Inertia of Travel Behaviour: A Stated Preference Analysis of Commuting

Job van Exel* and Piet Rietveld**

*Erasmus University, Rotterdam, The Netherlands
**Free University, Amsterdam, The Netherlands

6.1 Introduction

This chapter reports on a behavioural experiment of commuting behaviour. We apply a stated preference method to assess commuters' mode preferences. The aim of this study is to identify whether variables other than the standard variables in mode choice models, including travel time and costs, have a significant effect on the preference ordering of alternatives.[1] More specifically, we analyze the effect of current behaviour on preference elicitation.

Usually stated preference studies are conducted independently of respondents' actual behaviour, i.e. their revealed preference, in that same area. "There are reasons to believe that [...] behaviours may vary across individuals, depending on the past behavior trajectories of the respective individuals" (Kitamura 2000, p. 117). According to Louviere and Street (2000. pp. 133-134), "travelers make several sequential decisions before their ultimate travel choices (even if the sequence was in the remote past)" and "causally prior decision processes can play major roles in market choices, and hence are of policy interest in their own right". Therefore, "failure to take sequential and conditional decisions into account may yield incorrect inferences and conclusions about processes due to sample-selection bias" and "it may be difficult to predict future behavior and/or outcomes if causally prior choices are not taken into account". We attempt to link current behaviour and elicited preferences and, as such, introduce additional information to the commuting mode choice model to arrive at sharper estimates of model coefficients.

The behavioural experiment we conduct consists of two phases. In Phase 1 participants in the experiment receive an extensive questionnaire concerning, among others, personal characteristics and their current and past commuting behaviour. The responses to this first phase are used to develop a second phase questionnaire. In Phase 2, the same respondents are presented with each other's commuting alternatives and asked to rank the alternatives for each commuting trip *as if* it were their own. The objective of this experiment is to analyze the role of inertia in commuting behaviour. Do respondents make a fully rational deliberation on the alternatives for each commuting trip? Do they make their choices independent of their current commuting behaviour?

Section 6.2 below describes the method of analysis and describes the questionnaires used in this behavioural analysis of commuting mode choice. Sections 6.3 and 6.4 present the main results from, respectively, the Phase 1 and Phase 2 questionnaires, by means of descriptive statistics. The findings from these sections are used to develop hypotheses about potential inertia in commuting behaviour, to be tested in Section 6.5 with a random utility model. Section 6.6 concludes this article with the main findings from this behavioural experiment

6.2 Method

The behavioural experiment aims at identifying inertia in commuting mode choice. Bovy and Stern (1990) refer to inertia in terms of 'certain thresholds that need to be crossed before changing routine behaviour' (p. 32); "factors [...] which encourage keeping the status quo and oppose behavioural change" (p. 109); and, 'adherence to alternatives that have already been used' (p. 110). Likewise, Windervanck and Tertoolen (1998) claim that commuters evaluate alternatives once, when they change home or work address, and then just stick to this chosen mode unless they have a very bad experience with the mode or other major changes occur. In this chapter we will adhere to this broad conceptualization of inertia.

In traditional mode choice models (see Figure 6.1), the coefficients of the standard mode choice variables are estimated and the model is applied to derive implications of changing parameter values on future behaviour.

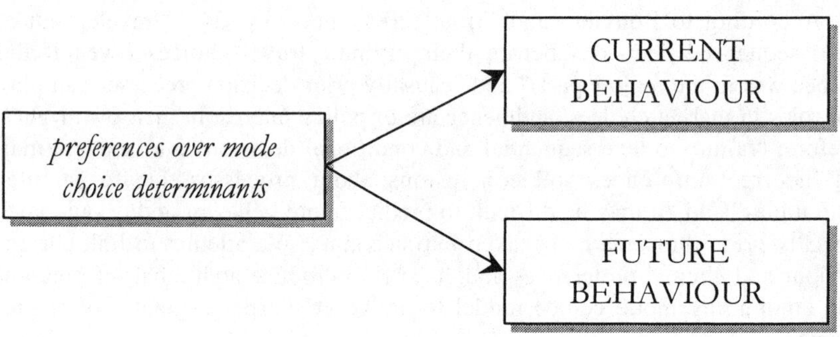

Figure 6.1 Standard behavioural model

The behavioural experiment discussed in this chapter is to investigate whether there are also other forces at work in determining travel behaviour, and whether a relation between current and future behaviour can be identified (see Figure 6.2). This relation could be either direct (❶), i.e. with current behaviour as a determinant of future behaviour, or indirect (❷), i.e. with current behaviour having an impact on the individual's preferences towards specific mode choice determinants and, through this, on future commuting behaviour. Both relations may lead to sluggish behavioural reactions to changes in choice determinants, and will thus be regarded in this behavioural analysis of the inertia of commuting behaviour.

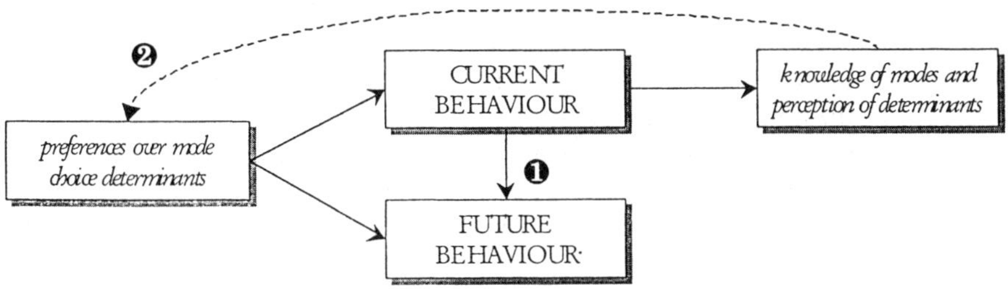

Figure 6.2 Alternative behavioural model

In this study, we apply a stated preference method to assess commuters' mode preferences. Stated preference 'typically predefines the characteristics of alternatives in the choice set and seeks behavioural responses from travellers in the form of either a preference ranking/rating or a choice selection' (Hensher, Barnard and Truong 1988, p. 45). A study comparing SP and RP methods provides 'evidence that individuals' stated preference among hypothetical travel scenarios are a reasonably accurate guide to true underlying preferences' (Wardman 1988, p.89). We conducted a survey of commuters and asked them to rank-order commuting alternatives that involve different combinations of prices and other characteristics on twenty different routes. 'A conceptually satisfying method for measuring consumer trade-offs among attributes is to have consumers rank-order several 'packages' that involve different combinations of prices and other characteristics' (Calfee and Winston 1998, p. 86).

The behavioural experiment consists of two phases. In Phase 1, respondents complete an extensive questionnaire concerning their travel behaviour, specifically their commuting behaviour. In Phase 2, respondents evaluate travelling alternatives on 20 commuting trips. The data collected with these questionnaires is applied to develop a random utility model of commuting behaviour and to analyze the potential inertia of this behaviour. Do respondents make a rational deliberation on the alternatives for each commuting trip and do they make their choices independent of their current commuting behaviour?

The Phase 1 questionnaire covers all standard travel behaviour determinants, i.e. variables referring to socio-economic background, transport system performance and individual perceptions thereof (Aarts, 1996; De Dios Ortúzar and Willumsen

1994; Stubbs, Tyson and Dalvi 1994; Rooijers 1992; Blaas, Vleugel, Louw and Rooijers 1992). The questionnaire consists of 44 questions, covering personal characteristics, current commuting behaviour, choice of job location, past commuting behaviour, attitudes towards alternative modes for the commuting trip, and willingness to participate in the Phase 2 questionnaires.[2] Most questions are either multiple-choice or scores on 5-point scales,[3] to facilitate the work of respondents and increase response. The questionnaire was distributed among 50 people up to the age of 32 years, all engaged in skilled labour in the same company and at the same office location. Consequently, only a limited amount of (younger) employees could be approached, to restrict the burden on regular activities in the company. The questionnaire was completed by 42 respondents (84% response). The respondents share one end of their commuting trip chain, and, at least in part, have the same set of travel alternatives to this shared end-node. This is important for our analysis because it helps us e the role of knowledge and experience with travel alternatives in commuting mode choice. On the one hand, the Phase 1 questionnaire gives us a clear and detailed picture of the complete set of travel alternatives people working at this particular office location have at their disposal and, on the other hand, the respondents will be reasonably informed about the alternatives they evaluate in the follow-up questionnaire.

Next, we selected 20 commuting trips from the Phase 1 responses for the phase two questionnaires, that is, 20 actual commuting trips from the Phase 1 sample-group. The objective was to present the respondents of Phase 1 with the travel alternatives on each others' commuting trips and, as such, perform a 'mirror-analysis' of their commuting choices. According to the distribution between urban and interurban trips in the sample group, 12 short-distance and 8 long-distance origin destinations were selected. Based on route-planners, public transport schedules and street plans of each trip, the most relevant travel alternatives were determined for each trip, in terms of mode, trip-chain, travel time and travel costs. For the 8 *long-distance* trips, car and public transport alternatives were considered; in the latter case mostly varying 1st and 2nd class train as the main mode and urban public transport and/or bicycle as the most relevant access and egress modes (Keijer and Rietveld 1999). For the 12 *local* trips, the modes car, urban public transport alternatives, cycling and walking to work were considered. Eventually, for 13 of the 20 commuting trips, we identified 4 travel alternatives, and for the other 7 – all short distance trips – we identified 3 relevant travel alternatives. For a full list of the provided alternative trip-chains per commuting trip, see Appendix 6.2.

Travel times for all origin-destination pairs (ODs) were determined with the assistance of route-planning software[4] and public transport travel information.[5] Travel costs were based on per kilometre costs for the different modes.[6] To facilitate their evaluation as potential commuting trips, the ODs were presented to the respondents graphically and by a description of the door-to-door trip-chain in terms of its modal components, single trip travel time and monthly costs. Thus, for each route the respondents received a map of the route and a template that they were asked to complete (see Appendix 6.4 for an example). After studying the map, the respondents were asked to indicate their experience with modes on the route in question and to rank the alternatives provided for the *as if* the commuting trip were their own. Respondents were asked to pick the routes randomly, to rule out any path dependency in eliciting preferences over alternatives.

After completing the full set of 20 templates, respondents were asked the following question: "What were, according to you, the main choice criteria you applied in ranking the alternatives for the 20 commuting trips?" The answers to this question were all recorded, and, based on this analysis, additional hypotheses were formulated about inertia in commuting behaviour.

The number of 20 commuting trips was chosen to limit the time effort for respondents. A pilot of the Phase 2 questionnaire showed that the time needed to complete the questionnaire was between 15 and 20 minutes. Together with the short concluding interview, the total time effort was thus about 20 minutes. Right after completing the Phase 1 questionnaire, almost 80 percent of the Phase 1 questionnaire respondents expressed willingness to participate in a follow-up round. When they were approached about six weeks later the willingness to cooperate had decreased significantly and a few respondents were no longer working with the same employer. In the end, 15 respondents were willing and able to complete the Phase 2 questionnaire, and 13 of them found their actual commuting trip among the selected set. We acknowledge that the final sample was small and that the results need to be interpreted with caution. However, we are interested in a specific subgroup of users and this small sub-population of 15 respondents is representative of the total sample taken from this subgroup in terms of age, gender, commuting distance, modes available, public transport season-ticket ownership, and working experience (see Appendix 1). We cannot make any firm statements concerning the representativeness of this population for the total of the subgroup (in the Dutch population), but will address this point in our discussion of results in Section 6.5 and in our conclusions.

6.3 Past and Current Commuting Behaviour

In this section, we discuss the results of the Phase 1 questionnaire, concerning individual characteristics, mode choice determinants and past and current commuting behaviour. The average age of the respondents is 27.4 years and 40 percent of them are female. One third lives in a one-person household, 55 percent in a two-person (all adult) household and 12 percent in a two-person household with one or more children. All respondents, but one, work and commute five days a week, with an average commuting distance of 19.3 kilometres and a commuting time of 31.1 minutes. This is slightly higher than the Dutch national average of 16.5 kilometres and 24 minutes (CBS 1997). This may be explained by the fact that skilled work and two-worker households are over-represented in the sample. In the sample group, 83.3 percent have the choice between two or more modes for their commuting trip, 47.6 percent have a car available, 76.2 percent a bicycle, and over 90 percent can make use of public transport for commuting. More detailed characteristics of the sample are included in Appendix 6.1.

6.3.1 Mode Choice

Table 6.1 below presents the average response frequencies for the use of alternative modes for commuting. It is notable that on average almost 30 percent of the answers fall into the middle two categories of 'often' and 'sometimes', implying that this group of respondents is not confined to a single commuting mode, but varies mode choice according to circumstances and their own or perhaps other household members' preferences.[7]

Table 6.1 Frequency of use of modes for commuting

Mode	(almost) always	often	sometimes	(almost) never
car (driver)	11.9%	4.8%	31.0%	52.4%
car (passenger)	n/a	n/a	4.8%	95.2%
public transport	28.6%	19.0%	33.3%	19.0%
bicycle	35.7%	21.4%	19.0%	23.8%
other	2.4%	n/a	7.1%	90.5%
average (overall)	15.7%	9.0%	19.0%	56.2%

Table 6.2 Daily mode choice for commuting and main mode

When mode choice is made	Average	Mode choice for commuting			Commuting distance	
		car	public transport	bicycle	≤ 8 kms.	> 8 kms.
I always travel to work in the same way	50.0%	100.0 %	73.3%	25.0%	37%	83%
In the morning I decide what is the best alternative for that day	50.0%	-	26.7%	75.0%	63%	17%

Moreover, with regard to the question of when actual commuting mode choice is made, exactly half of the respondents always claim to choose the same mode (see Table 6.2). It may well be that these respondents have a dominant best alternative for their commuting trip, or, as Kunert (1994) puts it "habits and routines which reduce the burden of constant decision making will establish some regular patterns of behaviour" (p. 273). The other half claims to choose the best alternative for the morning of the specific day. This notable flexibility in commuting mode choice is more frequent on shorter distances and among public transport and bicycle commuters, it increases with the number of modes available for the commuting trip and is inversely related to public transport season ticket ownership (see Figure 6.3). Inertia as a result of the pre-commitment from season ticket ownership was also demonstrated in the context of public transport strikes (Van Exel and Rietveld 2001).

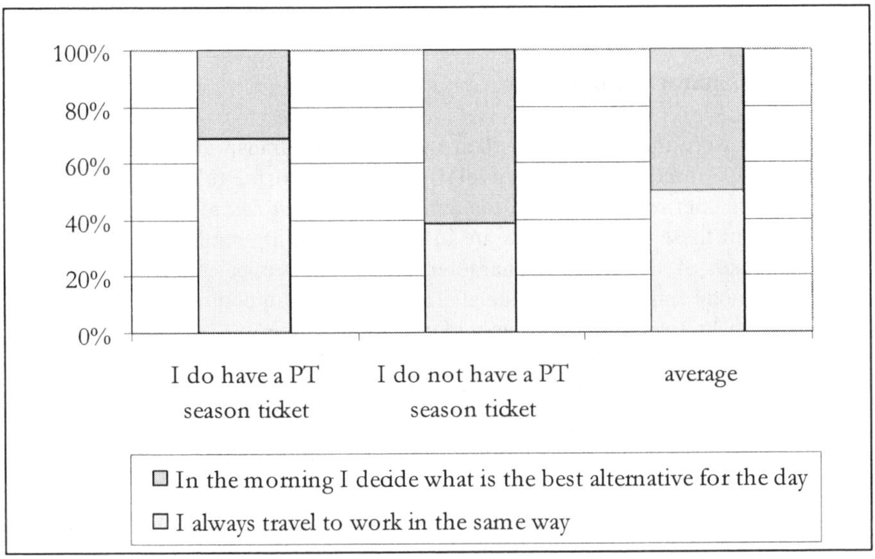

Figure 6.3 Mode choice for commuting and pre-commitment through PT season ticket

Table 6.3 shows the resulting modal split for the sample population. The popularity of the car is lower than one would expect on the basis of the Dutch national figures, both for short and long distances, implying that the respondents show slightly diverging behaviour compared with the average.

We also asked the respondents about other activities performed on working days and whether these trips were linked to commuting. About 40 percent of the respondents, mainly (65%) those with a shorter commuting distance, often make trips within their place of residence on working days for purposes other than commuting. One out of three states that these trips are never linked to commuting, and two out of three state that their mode choice for these trips is independent of commuting mode choice. Respondents who commute by car make fewer other trips during the week, but, in line with observations of Hensher and Reyes (2000), are much more likely to link them to commuting compared with those who commute by public transport or bicycle.

Table 6.3 Mode shares for commuting in sample (main mode)

mode	modal split (≤ 8 kms.)	modal split (> 8 kms.)	total
car	8.3%	29.2%	14.3%
public transport	25.0%	70.8%	38.1%
bicycle	66.7%	n/a	47.6%

6.3.2 Mode Characteristics

Respondents were interviewed for their opinion about transport modes on eleven characteristics (among others, travel time, travel costs, reliability, comfort, flexibility, weather resistance, and the ability to work *en route*). They were asked how important these characteristics are in their commuting mode choice, and how – on the basis of these same characteristics – they experience their current commuting mode and how they value alternative modes for commuting.

Travel time and reliability are considered most important in mode choice for commuting, whilst working *en route* and privacy are least important. The relatively high rank of weather is remarkable, and may be explained by the high modal share and weather sensitivity of the bicycle.[8] The bicycle is, however, praised for travel time and reliability, and also for flexibility, environmental friendliness and low travel costs. The car is valued negatively because it is not environmentally-friendly and travel time is wasted time, but scores high on flexibility, weather resistance, comfort, privacy and the ability to take along whatever you need for the day. Urban public transport is appreciated for its weather resistance, safety, environmental-friendliness and affordability, but comfort and privacy are real downsides in this mode. It appears that much of the substitution between bicycle and public transport on short distances may be explained by weather conditions and situational circumstances like having to take along a lot of things or dress neatly for a business appointment. Finally, the train also has its merits in terms of weather resistance, safety and environmental friendliness and, furthermore, it is said to be the best alternative for working *en route*. However, the train receives the lowest scores on travel time, reliability and flexibility, and does not really outclass the bicycle for comfort, the car for travel costs, or local public transport in privacy. Overall, the car is valued highest for commuting, followed by the bicycle, urban public transport and the train. On average, respondents value their actual commuting mode higher than the alternatives they have, suggesting most respondents use their most preferred mode for commuting.

6.3.3 Past Behaviour

Another potential influencing factor on commuting behaviour may be past travel behaviour, as travel choices and experiences from the past may have influence on attitudes and perceptions in the present (Kitamura 2000). Over 60 percent of the respondents were currently employed in their first job. Another feature of the sample is that 45 percent of the respondents had their last job or had studied in the same city in which they were currently employed, i.e. Rotterdam. On average the respondents made this previous commuting/school trip during a 3.5-year period. Then the average commuting distance was 14.5 kilometres, and the commuting time 33 minutes. This average distance was lower than the current commuting distance. 67 percent of the respondents stated that they often, or almost always, used the bicycle to travel to their previous occupation: 10% used the car and 40 percent public transport. This suggests that some respondents also switched commuting modes in the past. The much higher stakes for the bicycle and public transport account for the higher average commuting time, despite the lower average commuting distance. The high rating of public transport may be explained by the fact that half of the respondents could make free use of public transport, as 10 percent had public transport season-tickets reimbursed by the employer and 40 percent a students' public transport card.[9] About 40 percent of the respondents stated that their past travel behaviour did have an influence on current commuting behaviour, with diverging percentages between those previously using reimbursed public transport (52%) and those covering commuting expenses themselves (30%).

6.3.4 Perceptions

Individual perceptions of travel time and costs potentially also play a role in mode choice (Blaas, Vleugel, Louw and Rooijers 1992). Respondents were asked to estimate door-to-door travel time and costs for their current commuting trip by bicycle, car or public transport. These costs were converted into costs per kilometre by dividing them by the estimated commuting distance. As can be seen from Tables 6.4 and 6.5, respondents make high cost per kilometre and low travel speed estimates for almost all modes, and large differences can be observed between the different modes. On average, respondents give relatively favourable cost and speed estimates for their current commuting mode. Those commuting by bicycle give rather high-cost and low-speed estimates for car and public transport, while public transport commuters give relatively low-cost and high-speed estimates for all alternatives.

Table 6.4 Estimated travel costs per kilometre by respondents (in euros)

Mode	Average costs estimate	current commuting mode		
		bicycle	car	public transport
bicycle	0.13	0.19	0.08	0.02
car	0.53	0.74	0.44	0.27
public transport	0.39	0.53	0.43	0.19

Table 6.5 Estimated travel speed by respondents (in kilometres per hour)

Mode	Average speed estimate	current commuting mode		
		bicycle	car	public transport
bicycle	14.1	13.2	13.3	17.3
car	29.0	18.0	29.2	42.4
public transport	23.8	11.4	13.9	40.3

Many factors may influence these perception differences. Respondents may include factors that are specific to their situation (paid parking for bicycles at stations, parking fees or congestion for cars), or to regular use (discounts for season-ticket holders, variable car costs), but also may provide arbitrary estimates through lack of knowledge or experience (for example, based on hearing other people complain about costs of parking or congestion). These results cast some doubt on the usual assumption of mode choice models, i.e. travellers are well informed.

Finally, respondents were asked to indicate what measures they thought would influence their commuting mode choice. In decreasing order of importance, these are:

- Availability of good alternatives (83% of respondents);
- Employer reimbursements (64%);
- Parking fees (48%);
- Tolls or congestion pricing (38%);
- Public information campaigns (7%);
- Other (10%).

Respondents did not think very much of information campaigns as an instrument for changing their behaviour. These campaigns are, however, usually directed at changing attitudes and as such aim at an indirect effect on behaviour that is difficult for the respondents to estimate (and the same applies even for the developers of such campaigns). Monetary rewards and punishments are considered to be powerful, but especially the availability of good alternatives scores very high on behavioural change potential. This latter opinion is strongest among public transport users (88%), which may mean that they are most dissatisfied with their commuting mode, but are captive to it, or that current alternatives are unattractive as a result of, for example, congestion or parking problems and fees.

6.4 Stated Preference Experiment

In the second phase of the behavioural experiment the respondents were asked to evaluate 20 origin-destination combinations (ODs) and rank each of the provided door-to-door alternatives *as if* it were their own commuting trip.[10]

First, respondents are asked about their experience with modes on the route in question, to enable us to distinguish between well- and less-informed choices in the analysis. As can be seen in Table 6.6, respondents who commute over short distances (up to 8 kilometres) are better informed about the shorter, local ODs, whereas those who commute long distances are better informed about the long-distance, interlocal ODs. Moreover, respondents who commute over short distances are relatively better informed about longer distance ODs. This may be explained by their experience of business and leisure travel. Moreover, respondents are best informed about their own modes; car users are better informed about the long-distance ODs; and, bicycle users on average have more experience with cars than with public transport.

Table 6.6 Experience with modes on routes by current commuting distance[11]

Current commuting distance	OD distance		
	short	long	total
short	2,0	1,7	1,9
long	1,3	2,1	1,6
total	1,7	1,9	1,8

The respondents ranked the alternatives provided for each of the 20 ODs. Table 6.7 presents the percentage first ranking per alternative, split over short- and long-distance ODs. The car was ranked first on 11% of the short-distance ODs and on 40 percent of long-distance ODs. The bicycle, as the main mode of transport, is considered as first choice on 74 percent of the short-distance ODs.

Table 6.7 Frequency that specified alternatives rank first on short and long-distance ODs

OD distance	Alternative								
	car	train 2nd & urban PT	train 1st & urban PT	train 2nd & bicycle	bus	metro	tram	bicycle	walk
short	11%	n/a	n/a	n/a	n/a	15%	0%	74%	1%
long	40%	45%	3%	10%	2%	n/a	n/a	n/a	n/a

Tables 6.8 and 6.9 show the relative rankings of alternatives for long- and short-distance ODs. For example, on the long-distance Ods, the car is preferred by the respondents to the bus in 73 percent of pairwise comparisons and 2nd class train over 1st class train in 97 percent. On short distances the metro is preferred to the tram in 82 percent of the cases.

Table 6.8 Pairwise comparisons of alternatives on long distances (percentage of respondents who prefer alternative in first column)

Alternative	**Over alternative**			
	train 2nd & urban PT	**train 1st & urban PT**	**train 2nd & bicycle**	**bus**
car	36%	55%	49%	73%
train 2nd & urban PT	-	97%	20%	0%
train 1st & urban PT	3%	-	0%	60%
train 2nd & bicycle	80%	100%	-	53%

In Tables 6.10 and 6.11 the percentage of first rankings is related to the current mode choice for commuting. On the long distance ODs, current mode choice behaviour appears to be strongly correlated to the rankings. Car users have strong preferences for the car, and public transport users for public transport. The same cannot be said for the short-distance ODs. Whereas bicycle users appear pretty satisfied with their commuting mode, the intention to take the bicycle also appears very high with other mode users, especially among public transport users, whereas, the other way round, cyclists are quite undecided between car and public transport alternatives on both short- and long-distance ODs.

Table 6.9 Pairwise comparisons of alternatives on short distances (percentage of respondents who prefer the alternative in first column)

Alternative	Over alternative			
	metro	**tram**	**bicycle**	**walk**
car	27%	40%	13%	40%
metro	-	82%	19%	33%
tram	18%	-	17%	0%
bicycle	81%	83%	-	97%

Table 6.10 Percentage first ranking of alternatives on long distances versus current commuting mode

Current commuting mode	Alternative				
	car	**train 2nd & urban PT**	**train 1st & urban PT**	**train 2nd & bicycle**	**bus**
car	81%	19%	n/a	n/a	n/a
public transport	23%	55%	5%	13%	4%
bicycle	46%	42%	2%	10%	n/a

As mentioned earlier, the ODs were selected from the Phase 1 questionnaire and 13 of the 20 ODs concern commuting trips of respondents who completed the Phase 2 questionnaire. Table 6.12 and Table 6.13 compare the choices of respondents with, respectively, the choice of the 'owner' of the OD and their own current commuting behaviour.

Table 6.11 Percentage first ranking of alternatives on short distances versus current commuting mode

Current commuting mode	Alternative				
	car	**metro**	**tram**	**Bicycle**	**walk**
car	38%	21%	n/a	42%	n/a
public transport	2%	15%	n/a	81%	1%
bicycle	11%	13%	n/a	76%	n/a

Table 6.12 highlights the choices made by the respondents, including the 'owner' of the OD, as compared with the actual mode choice of the 'owner'. The choice frequencies on these 13 ODs can be said to provide a *mirror* to the behaviour of the 'owner' of the OD. The last column in Table 6.12 indicates the percentage consensus in choice between respondents and the 'owner' of the OD.

Table 6.12 Current mode choice of OD 'owners' versus mode choice frequency of the respondents

OD	Actual mode choice 'owner' of OD	Mode-choice frequency				consensus
		car	public transport	bicycle	walk	
1	public transport	67%	33%	n/a	n/a	33%
2	public transport	33%	67%	n/a	n/a	67%
3	car	53%	47%	n/a	n/a	53%
4	public transport	73%	27%	n/a	n/a	27%
6	public transport	13%	87%	n/a	n/a	87%
7	public transport	7%	93%	n/a	n/a	93%
9	bicycle	7%	27%	67%	n/a	67%
10	bicycle	n/a	n/a	93%	7%	93%
12	bicycle	n/a	7%	93%	n/a	93%
13	bicycle	n/a	20%	80%	n/a	80%
17	public transport	27%	13%	60%	n/a	13%
19	bicycle	33%	n/a	67%	n/a	67%
20	public transport	n/a	33%	67%	n/a	33%
Average						62%

As can be seen from the table, on average there is 62 percent consensus in commuting mode choice between the respondents. There is more than 50 percent consensus of choice on 9 of the 13 ODs (numbers 2, 3, 6, 7, 9, 10, 12, 13 and 19) and a high level of dissension on some other ODs (1, 4, 17 and 20). Table 6.13 shows the relation between the current commuting behaviour of respondents, as described by them in the Phase 1 questionnaire, and their stated preferences on the 20 ODs from the Phase 2 questionnaire. When analyzing how respondents have chosen, again there is great diversity in choice behaviour, but a high degree of consensus (70%) between respondents' current commuting distance and mode choice and their stated preference or intended behaviour on the 20 ODs.

Table 6.13 Current commuting distance and mode versus mode choice frequencies per respondent

Resp	Current commuting distance	Current commuting mode choice	Mode choice frequency						consensus on "own" distance *(column 2)*
			short distance ODs				long distance ODs		
			car	public transport	bicycle	walk	car	public transport	
1	long	public transport	n/a	8%	92%	n/a	n/a	100%	100%
2	short	bicycle	n/a	n/a	100%	n/a	38%	63%	100%
3	long	public transport	n/a	n/a	100%	n/a	25%	75%	75%
4	long	bar	33%	8%	58%	n/a	63%	38%	63%
5	short	bicycle	33%	8%	58%	n/a	88%	13%	58%
6	long	public transport	8%	n/a	92%	n/a	38%	63%	63%
7	short	public transport	n/a	33%	67%	n/a	n/a	100%	33%
8	short	bicycle	n/a	8%	92%	n/a	38%	63%	92%
9	short	bicycle	n/a	n/a	100%	n/a	75%	25%	100%
10	long	public transport	n/a	n/a	100%	n/a	13%	88%	88%
11	short	public transport	n/a	n/a	100%	n/a	38%	63%	n/a
12	short	bicycle	33%	17%	50%	n/a	38%	63%	50%
13	short	public transport	8%	67%	17%	8%	50%	50%	67%
14	long	bar	42%	33%	25%	n/a	100%	n/a	100%
15	short	bicycle	n/a	42%	58%	n/a	n/a	100%	58%
Average									70%

Table 6.14 Statements by respondents on mode choice determinants and frequency among respondents of similar statements (% of respondents)

Mode choice determinants	**Statements by respondents (examples)**	**Freq.**
travel costs	*"First class train is not worth the price difference"*	53%
	"In urban traffic public transport costs are not an issue"	
deliberation over choice variables	*"I am more sensitive to costs than to travel time"*	47%
	"Mode choice depends primarily on where I have to be, for what purpose, and at what time of day"	
	"Speed is most import and then costs. Except for local travel, then always public transport or bicycle because it doesn't make much difference"	
	"When I travel for business I always take the car. I can claim my expenses, and costs therefore are of no concern"	
comfort	*"Train is much more comfortable than bus, you can work during the trip in the train"*	40%
	"The convenience of cycling is important to me"	
reliability	*"I only take the car when it is really faster. Because of all the congestion, especially during peak hours, I double the travel time provided by route-planners"*	40%
	"Time of day is important because of density of public transport supply"	
	"I prefer not take the car to large cities, too many parking problems"	
transfers in public transport	*"The number of transfers is an important criterion for me"*	40%
	"The number of transfers is a measure of uncertainty"	
travel time	*"With public transport I feel I'm always waiting. Only metro is acceptable, quite frequent and mostly on time"*	20%
	"I only take the car when travel time differences become too big"	
safety	*"Later at night, I prefer to avoid stations and take the car"*	7%

After completing the Phase 2 questionnaires, the respondents were asked the following question: 'What are, according to you, the main choice criteria you applied in ranking the alternatives on the 20 commuting trips?' The responses to these questions have been recorded and analyzed. Table 6.14 includes the responses, which are related to the criteria of mode choice that are captured through the standard model (see Figure 6.1). Many of the responses do not, however, relate directly to these traditional criteria and are more like *preconditions* or *thresholds*. Preconditions are related to the context of choice. Thresholds concern minimum aspiration or maximum acceptable levels that, if not attained or exceeded, lead to the elimination of the matching alternative from the choice process. Preconditions and thresholds suggest a sequential or hierarchical choice process, rather than a simultaneous and compensatory one, as is traditionally assumed. In a sequential choice process, the choice set is either first submitted to some criteria, only after which are the remaining alternatives compared and evaluated in the traditional manner, or, alternatives do not enter the evaluation unless they fulfil some aspiration or acceptable level on the basis of key criteria (Tversky 1972a, 1972b; Bovy and Stern 1990; Louviere and Street 2000). One of the first and most prominent models of choice addressing this kind of decision process is the model of choice by elimination, or elimination by aspects model, introduced by Tversky (1972a, 1972b). The stated preconditions mostly relate to the suitability of modes,[12] whereas thresholds are almost all travel time-, distance- or costs-related.[13] All in all, the responses to the question on choice criteria imply that the sample group may not be representative of the general public in their preferences and travel behaviour. Kunert (1994) has studied variability in mobility patterns across life-cycle groups and found that the segmentation of samples into life-cycle groups is 'behaviourally relevant'. Thus, though the estimated model coefficients in Section 6.5 may not be representative of the general public, they are relevant for the subgroup in our study. Therefore, for the purpose of our analysis this is not disturbing.

In conclusion, the descriptive statistics of the Phase 1 and Phase 2 questionnaires, discussed in Sections 6.3 and 6.4 of this chapter, provide a broad picture of the factors that influence the commuting behaviour in this sample group. The following ideas are tested with a random utility model of commuting mode choice developed in the next section:

Current commuting behaviour has a direct influence on elicited preferences (❶ in Figure 6.2);

Current commuting behaviour has an indirect influence on elicited preferences through knowledge of modes and, consequently, perception of their characteristics (❷ in Figure 6.2).

6.5 The Model

Random utility choice theory assumes that, as a result of uncertainty, the value (or utility) of each alternative of choice undergoes random fluctuations, and that the decision maker at any point in time chooses the alternative with the highest value (Tversky 1972a). Thus, when deciding on how to make a specific trip, travellers will choose the alternative that maximizes their utility according to their preferences and the utility-providing characteristics of the available alternatives for that trip. In a random utility model, the utility of an alternative is decomposed into explainable and random components, because it is impossible to observe the true utility of any alternative. Therefore individual preferences are viewed as stochastic and choice is modelled as a probability (Louviere and Street 2000). Supposing that the only utility providing characteristics of modes are total costs (C) and travel time (T), then the utility function has the following general form:

$$U = f(C,T), \tag{6.1}$$

and, the linear function representing the utility of alternative *j* for individual *i* is:

$$U_{ij} = \alpha_i = \beta \cdot C_j + \gamma \cdot T_j + \varepsilon_{ij}, \tag{6.2}$$

where the constant α_i is a measure of the utility for individual *i* of the activity that is the goal of the trip and the coefficients β for costs and γ for time are expected to be negative, indicating that the utility of travelling to this activity is inversely related to the travel time and costs of the travelling alternative of choice.[14] An increase in time or costs of an alternative thus decreases the likelihood that this alternative will be chosen, all else being equal (Calfee and Winston 1998). The ratio of the time and cost coefficients γ/β represents the trade-off between time and money, the value of time. Finally, the variable ε_{ij} is a random disturbance term that captures the 'unexplainable' effects of unobserved tastes of the individual or attributes of the travelling alternative of choice.[15]

Individual *i* now chooses alternative *j* over alternative *k* if and only if U_{ij} (the utility that individual *i* derives from choosing alternative *j*) exceeds U_{ik} (the utility that individual *i* derives from choosing alternative *k*), that is, when:

$$\begin{aligned} &U_{ij} \succ U_{ik} \Leftrightarrow U_{ij} - U_{ik} \succ 0 \\ &\Leftrightarrow U_{ij} - U_{ik} = (\alpha_i - \alpha_i) + \beta \cdot (C_j - C_k + \gamma \cdot (T_j - T_k) + (\varepsilon_{ij} - \varepsilon_{ik}) \succ 0. \end{aligned} \tag{6.3}$$

The constant α_i which for individual *i* is identical for all modes on the same trip, drops out of the equation. When ε_{ij} and ε_{ik} are drawn from independent and identical Weibull distributions, the parameters β for costs and γ for time can be estimated by using the logit model (Cramer 1991).

In the second phase questionnaire, respondents were asked to evaluate 20 origin-destination combinations, of which 13 have 4 and 7 have 3 alternative trip-chains, ultimately allowing for 1485 pairwise comparisons of alternatives to use in the model estimation.[16] The estimation results for the formulation of mode choice according to Equation (6.3) are presented in Table 6.15 below.

The coefficients of the estimation have the expected negative signs and are both significant. The corresponding value-of-time estimation is €7.29, a value that is in line with the recommendations by the Dutch Ministry of Transport for evaluation studies in the transport sector (Ministry of Transport 1998).[17]

Table 6.15 Logit model estimation of mode choice for commuting: base model

Variable	**Coefficients**[18]
Total costs	-0.0042
	(0.0004)
Total travel time	-0.0450
	(0.0060)
Number of observations	1485
Percentage correct	58.8%
-2 log likelihood	1950.1
Cox & Shell r^2	0.070
Nagelkerke r^2	0.094

This estimation is based on total costs and travel time for all the alternatives.[19] Sometimes it is claimed that individuals are poorly informed about the costs of alternatives and that they compare the car with alternative modes not based on total costs but, in the case and in favour of the car, based only on variable costs (Blaas, Vleugel, Louw and Rooijers 1992; Rietveld and Verhoef 1998; Windervanck and Tertoolen 1998; see also Section 6.3.3 of this chapter). In the description of alternatives, the total, fixed and variable costs have been provided to the respondents. Therefore, lack of knowledge of the total costs should not apply here It may, however, still be questioned how the information provided is actually applied. The next estimation tests the use of different notions of costs among car owners and non-owners, i.e. assuming that respondents who have a car use the variable costs information (C^V), whereas those who do not, use total costs (C^T), (see (6.4) and Table 6.16).

$$U_{ij} = \delta \cdot \beta \cdot C_j^V + (1-\delta) \cdot \beta \cdot C_j^T + \gamma \cdot T_j + \varepsilon_{ij}, \tag{6.4}$$

This model performs significantly better than the former one, providing support for the hypothesis that past travel choice behaviour, and, in this particular case, car ownership, functions as a precondition to how modes are evaluated. Thus, in our model, the costs of owning a car are included in C^T for non-car owners and disregarded for car-owners.

Table 6.16 Logit model estimation of mode choice for commuting; model with variable car costs for car owners, total car costs for non-owners, and total costs for all other modes

Variable	Coefficients
Variable costs for car owners and total costs otherwise; total costs for other modes	-0.0048 (0.0005)
Total travel time	-0.0111 (0.0042)
Number of observations	1485
Percentage correct	63.5%
-2 log likelihood	1933.9
Cox & Shell r^2	0.081
Nagelkerke r^2	0.107

One can also debate whether individuals always value time in the same way. There is ample and long-standing evidence which suggests that travellers value differently the time spent in different transport modes, as well as total travel and transfer times (Beesley 1965; Quarmby 1967; Starkie 1971). In our study, the commuting alternatives are presented as door-to-door trip-chains, including the time spent on each modal component of the chain. In the estimation presented below, the aggregate travel time per trip in (4) is replaced by the modal components, i.e. car, train (2nd class), train (1st class), urban public transport, bus bicycle and walking segments (see Equation (6.5) and first column (i) of Table 6.17):

$$U_{ij} = \delta \cdot \beta \cdot C_j^V + (1-\delta) \cdot \beta \cdot C_j^T + \gamma \cdot T_j^C + \lambda \cdot T_j^{T2} + \mu \cdot T_j^{T1} + \quad (6.5)$$

$$\nu \cdot T_j^{UPT} + \tau \cdot TB_j^B + \varphi \cdot T_j^{BK} + \omega \cdot T_j^W + \varepsilon_{ij},$$

where T^C = travel time segment by car; T^{T2} = travel time segment by train (2nd class); T^{T1} = travel time segment by train (1st class); T^{UPT} = travel time segment by urban public transport; T^B = travel time segment by bus; T^{BK} = travel time segment by bicycle; T^W = travel time segment by walking.

Following this formulation of the choice model, the valuation of travel time across modes when ranked in decreasing order of attractiveness is: bicycle, urban public transport, walking, train (2nd class), car, train (1st class) and bus. Although, from a general public (or international) point of view this sequence may seem a bit strange, it is consistent with the statements by the respondents and the descriptive statistics presented earlier. An interim conclusion we can draw from this is that the members of the sample group belong to a sub-culture with divergent preferences with respect to their travel behaviour. This is not so strange when recalling that respondents are all from more or less the same age group and work in the same high-skilled profession at the same firm. A cautious typology of this sub-culture would include pro-environmental, pro-physical exercise and (moderate) anti-luxury attitudes emphasising the functional aspects of travel modes for commuting, i.e. travel time, flexibility, reliability and the possibility to take along various things.

Next, transfer time is added to the model, with T^{TRF} as the transfer time segment of the total travel time (see (6.6) and (ii) in Table 6.17). Transfer time is the period between the arrival and departure of two consecutive public transport links in the trip-chain.[20] Though the coefficient for transfer time is not significant, it has the expected sign and the coefficients of the other variables remain practically unchanged. Transfer time is retained in the model:

$$U_{ij} = \delta \cdot \beta \cdot C_j^V + (1-\delta) \cdot \beta \cdot C_j^T + \gamma \cdot T_j^C + \lambda \cdot T_j^{T2} + \mu \cdot T_j^{T1} + \\ + \nu \cdot T_j^{UPT} + \tau \cdot T_j^B + \varphi \cdot T_j^{BK} + \omega \cdot T_j^W + \varphi \cdot T_j^{TRF} + \varepsilon_{ij}, \tag{6.6}$$

where: T^{TRF} = travel time segment of transfer time.

Starting from the first model estimation (see Table 6.16), the differentiations of travel costs and travel time significantly increased the explanatory power of the model, raising the average r^2 from 0.082 to 0.218 in (ii) and the percentage of correct predictions from 59 to over 72. The model as formulated in (6) is taken to be the standard model from which the formulated hypotheses will be tested and will henceforth be referred to as 'SM (6)'.

Because the standard model actually already incorporates the *precondition* of car ownership, we first introduce current ownership of a public transport season ticket to the model (see (7). The results of the estimation of this model ((iii) in Table 6.17) show that the current ownership of a public transport season ticket has a significant influence in explaining mode choice.

$$U_{ij} = SM(6) + \xi \cdot EXP_{ik} + \varepsilon_{ij}. \tag{6.7}$$

where: $PTST_{ik}$ = 1 if respondent has a public transport season ticket, otherwise 0.

Next, experience with modes is added to the model (see Equation (6.8)). As can be seen from the templates of the Phase 2 questionnaire (see Appendix 6.4), respondents were asked to state how well they are acquainted with car, public transport and bicycle on the OD in question. Experience is, therefore, defined here as the frequency with which the specific trip has been made with the same mode within the last three months. Estimation of this model including mode experience (see (iv) in Table 6.17) and of a model combining models (6.7) and (6.8), i.e. including both public transport season-ticket ownership and mode experience (see (v) in Table 6.17), both make a significant contribution to the explanation of commuting travel behaviour of the respondents. Taking this into account, the preconditions of car and public transport season ticket ownership, as well as experience with a specific mode on the OD in question, can be said to have an influence on commuting mode choice, i.e.:

$$U_{ij} = SM(6) + \xi \cdot EXP_{ik} + \varepsilon_{ij}. \tag{6.8}$$

where: EXP = experience on OD with mode.

Next, we introduce a variable that represents *current travel behaviour.* The variable $MYMODE_{ij}$ is a binary variable that gets a value of 1 whenever the considered alternative *i* is both respondent *j*'s actual mode for commuting in the same commuting distance class and respondent *j* always travels to work in the same way, otherwise 0 (see (6.9). In the Phase 1 questionnaire, respondents have been asked how they make their commuting mode choice. As can be seen from Table 6.2, every day, half of the respondents choose the best commuting option for that specific day, whereas the other half claim to always commute in the same way. For the latter group, one may expect that there is a stronger relation between past and current behaviour, above all because they may have less knowledge of the characteristics of other modes. Hence:

$$U_{ij} = SM(6) + \xi \cdot MYMODE_{ik} + \varepsilon_{ij}. \tag{6.9}$$

where: MYMODE = 1 if mode *i* is also the actual commuting mode and respondent *j* always commutes to work in the same way, otherwise 0.

As can be seen from (vi) in Table 6.17, the impact of this current travel behaviour variable is significant and has the expected sign. The combination of the models (6.7), (6.8) and (6.9) has also been estimated (see (vii) in Table 6.17). The estimated values of time for the different modes in (vii) are (per hour) €42.2 for car, €35.6 for train (2nd class), €38.4 for train (1st class), €18.3 for urban public transport, €35.0 for bus €10.8 for bicycle and €13.5 for walking. The corresponding choice ordering of the alternatives is: bicycle, walking, urban public transport, bus, train (2nd class), train (1st class) and car. The time values and corresponding choice sequence support the suggestions from the interviews with the respondents that the members of the sample group belong to a sub-culture with divergent preferences with respect to their travel behaviour.

The estimated values of time are significantly higher than Dutch averages and deserve some additional attention. First of all, a commuter's value of time tends to rise with income and trip duration. As the respondents are all engaged in skilled labour and travel above average commuting distances, value of time estimates may be expected to exceed national averages. According to a large number of studies conducted in the past three decades, the value that commuters are willing to pay to save travel time is generally between 20 and 100 per cent of their gross wage rate, primarily depending on the type and circumstances of the study. Moreover, mode choice models, like the one discussed here, tend to yield higher values than route choice models because the estimates also reflect the relative comfort and convenience of alternative modes (Calfee and Winston 1998).

Table 6.17 Logit model estimation of mode choice for commuting

Variable	Coefficients						
	(i)	(ii)	(iii)	(iv)	(v)	(vi)	(vii)
Variable costs for car owners and total costs	-0.0040	-0.0040	-0.0036	-0.0035	-0.0031	-0.0041	-0.0032
else, total costs for other modes	(0.0006)	(0.0006)	(0.0006)	(0.0006)	(0.0006)	(0.0006)	(0.0006)
Travel time by car	-0.1412	-0.1421	-0.1867	-0.1522	-0.1990	-0.1428	-0.1986
	(0.0153)	(0.0164)	(0.0178)	(0.0167)	(0.0183)	(0.0164)	(0.0183)
Travel time by train (2nd class)	-0.1355	-0.1359	-0.1589	-0.1428	-0.1677	-0.1361	-0.1673
	(0.0149)	(0.0151)	(0.0155)	(0.0153)	(0.0158)	(0.0151)	(0.0158)
Travel time by train (1st class)	-0.1492	-0.1495	-0.1712	-0.1578	-0.1813	-0.1491	-0.1806
	(0.0151)	(0.0153)	(0.0157)	(0.0155)	(0.0160)	(0.0153)	(0.0160)
Travel time by urban public transport	-0.0471	-0.0474	-0.0805	-0.0531	-0.0878	-0.0456	-0.0861
	(0.0159)	(0.0161)	(0.0170)	(0.0163)	(0.0173)	(0.0161)	(0.0173)
Travel time by bus	-0.1497	-0.1496	-0.1577	-0.1562	-0.1654	-0.1484	-0.1644
	(0.0165)	(0.0165)	(0.0160)	(0.0168)	(0.0163)	(0.0165)	(0.0163)
Travel time by bicycle	-0.0385	-0.0391	-0.0512	-0.0395	-0.0521	-0.0374	-0.0509
	(0.0090)	(0.0097)	(0.0100)	(0.0098)	(0.0101)	(0.0098)	(0.0101)
Travel time by walking	-0.0639	-0.0641	-0.0644	-0.0642	-0.0645	-0.0624	-0.0634
	(0.0098)	(0.0099)	(0.0102)	(0.0100)	(0.0103)	(0.0100)	(0.0103)
Total transfer time	-	-0.0028	-0.0628	-0.0016	-0.0636	-0.0058	-0.0645
		(0.0184)	(0.0200)	(0.0187)	(0.0202)	(0.0185)	(0.0203)
Public transport season ticket	-	-	1.0101	-	1.0270	-	1.0132
			(0.1380)		(0.1388)		(0.1393)
Experience with mode on origin/destination OD	-	-	-	0.3074	0.3260	-	0.3233
				(0.0672)	(0.0693)		(0.0693)
MYMODE*	-	-	-	-	-	0.3782	0.2286
						(0.1841)	(0.1916)
Number of observations	1485	1485	1485	1485	1485	1485	1485
Percentage correct	72.7%	72.4%	71.4%	71.2%	71.1%	72.1%	71.3%
-2 log likelihood	1741.1	1741.1	1684.0	1719.0	1660.6	1736.9	1659.2
Cox & Shell r^2	0.193	0.193	0.223	0.204	0.235	0.195	0.236
Nagelkerke r^2	0.257	0.257	0.297	0.273	0.314	0.260	0.314

*The variable gets a value of 1 whenever the considered alternative *i* is both respondent *j* 's actual mode for commuting in the same commuting distance class [short vs. long] and respondent *j* always travels to work in the same way, otherwise 0

Though the survey questionnaires did not include questions about income, the occupation and years of experience of all respondents is known and the average hourly income is estimated at between €20 and €30. Consequently, the values of time estimated in our study are in the top half of Calfee and Winston's spectrum. Moreover, taking into account that our study introduces information into the model about current behaviour, the relative higher knowledge and experience with one of the alternatives may increase its relative desirability and thereby the stated values of time for other modes. Yet another interpretation for the high time values may be that the value of time estimates is higher because our study sample mainly consists of young professionals in their first job. As they may not yet be certain about either their residential or working locations, some of them may have rather inefficient commuting patterns. Moreover, Kunert (1994) finds that this life-cycle group has a high level of mobility as a result of the number of activities they are engaged in. This also puts more pressure on time value. In time, as they settle, they will adapt their pattern of activity and mobility to one that is justified by more general values of time. Finally, we do not rule out that the modest sample, in combination with the modelling technique, could lead to problems. High correlation between travel time and cost variables may give rise to higher values of time. Besides, respondents may concentrate on the first choice and then lose concentration when they have to rank the non-preferred alternatives. It may, therefore, be questioned whether a model based on the ranking information is as valuable as one using only the most-preferred alternative information. Moreover, we increased the amount of information by extrapolating binary choices from ranking information. As a result, we cannot assert that all observations are independent.[21] We intend to investigate these modelling problems in future research.

6.6 Conclusions

This chapter presents a stated preference analysis of commuting behaviour. Contrary to the independence axiom pertaining to rational behaviour and – often implicitly – adopted in stated preference studies, we found evidence of a relationship between current commuting behaviour and elicited preferences on the 20 commuting trips evaluated. This relation is shown to be direct as well as indirect, through altering preferences or the choice set. We found that ownership of a car or a public transport season ticket, as well as the use of modes affect travel choices, and that experience with a transport mode increases the probability that this mode may be chosen for commuting. Following the definition by Stigler and Becker (1977) – use grows with exposure – we may state that travel modes are addictive (or habituating).

This study focuses on commuting behaviour. Because of its repetitive nature, commuting is an important source for the formation of preferences, habits and (pre-)commitments in travel behaviour. The respondents were asked to complete each of the commuting templates *as if* it were their commuting journey. It is, however, relatively easy to imagine that the choices made are applicable to a much wider range of behaviour than only commuting behaviour, as the experience acquired through commuting most probably will spill over to travelling for other motives.

Some observations concerning the outcomes of this behavioural analysis are applicable. First, it has become clear throughout the study that the sample population is peculiar in some aspects. The strong preference for bicycle and public transport over cars and of 2nd class over 1st class train, as well as the statements concerning the relative importance of mode characteristics, reveal an attitude towards travelling that is different from that of the general public. This may be considered an obvious result of the sample population selection. A cautious typology of this sub-culture would include pro-environmental, pro-physical exercise and (moderate) anti-luxury attitudes, emphasising the functional aspects of travel modes for commuting, i.e. travel time, flexibility, reliability and take-along aspects. However, this primarily has an impact on the choice order of alternatives and not necessarily on the hypothesized relationships. Moreover, all respondents are young professionals, from whom, to some extent, capricious behaviour may be expected. This is visible from the relatively large percentage of respondents who shift between modes in such a relatively regular behaviour as commuting. Therefore, the inertia of behaviour revealed in this behavioural study may be much more omnipresent in reality, as commuting trips on average have been made for much longer periods and thus, inertia has had much better opportunity to become entrenched.[22]

Important implications of these findings for future research include that stated preference studies which do not pay attention to the demonstrated effect of current behaviour on elicited preferences may yield misleading results. Individuals respond from their current situation, i.e. interpret questions in the context of their own established inert/habitual patterns.

This finding also has strong policy implications, since it emphasises once more that price measures alone, may not be sufficient to generate envisaged behavioural change. It is also necessary to open up behavioural alternatives, as significant parts of the population may picture themselves captive to a narrow choice-set, built up through years of experience.

Acknowledgements

Job van Exel wishes to acknowledge the employees and management of NEI in Rotterdam (NL) for their cooperation in this study. We wish to thank two anonymous referees and Michel Beuthe for their helpful comments and suggestions.

References

Aarts H. 1996: Habit and Decision Making – the Case of Travel Mode Choice, PhD thesis, University of Nijmegen, Nijmegen

Beesley M.E. 1965: The Value of Time Spent in Travelling: Some New Evidence, Economica, Vol. 32, 126, 174–185

Ben-Akiva. M. and Lerman S.R. 1985: Discrete Choice Analysis: Theory and Application to Travel Demand, The MIT Press, Cambridge, MA

Blaas E.W., Vleugel J.M., Louw E. and Rooijers T. 1992: Car Ownership, Car Use and Driving Behaviour; Determinants of Energy Consumption of Car Traffic (in Dutch), Delft University Press, Delft

Bovy P.H.L. and Stern E. 1990: Route Choice – Wayfinding in Transport Networks, Kluwer Academic Publishers

Calfee J. and Winston C. 1998: The Value of Automobile Travel Time – Implications for Congestion Policy, Journal of Public Economics, Vol. 69, 83–102

Cramer J.S. 1991: The Logit Model – An Introduction for Economists, Hodder & Stoughton

De Dios Ortúzar J. and Willumsen L.G. 1994: Modelling Transport, John Wiley & Sons

Golob T.F., Kim S. and Ren. W. 1996: How Households Use Different Types of Vehicles – a Structural Driver Allocation and Usage Model, Transportation Research A, Vol. 30, 2, 103–118

Hensher D.A., Barnard P.O. and Truong T.P. 1988: The Role of Stated Preference Methods in Studies of Travel Choice, Journal of Transport Economics and Policy, Vol. 22, 1, 45–58

Hensher D.A. and Reyes A.J. 2000: Trip Chaining as a Barrier to the Propensity to Use Public Transport, *Transportation*, 27, 341–361

Keijer M.J.N. and Rietveld P. 1999. How Do People Get to the Railway Station? The Dutch Experience, Transportation Planning and Technology, Vol. 23, 215–235

Kennedy, P. 1998: A Guide to Econometrics, 4th edition, The MIT Press, Cambridge, MA

Khattak A.J. and Palma A. de 1997: The Impact of Weather Conditions on the Propensity to Change Travel Decisions – a Survey of Brussels Commuters, Transportation Research A, Vol. 31, 3, 181–203

Kitamura, R. 2000: Longitudinal Methods, in Hensher D.A. and Button K.J. (eds.) Handbook of Transport Modelling, Pergamon

Kunert U. 1994. Weekly Mobility of Life Cycle Groups, Transportation, Vol. 21, 271-288

Louviere, J. and Street D. 2000: Stated Preference Methods, in Hensher D.A. and Button K.J. (eds.) Handbook of Transport Modelling, Pergamon

Ministry of Transport. 1998: Recommendations Regarding Valuations of Time of People (in Dutch), Adviesdienst Verkeer en Vervoer, Ministry of Transport, Rotterdam

Polak J. and Heertje A. (eds.) 1993: European Transport Economics, ECMT, Blackwell

Quarmby D.A. 1967: Choice of Travel Mode for Journey to Work, Journal of Transport Economics and Policy, Vol. 1, 3, 273–314

Rietveld P. and Verhoef E.T. 1998: Social Feasibility of Policies to Reduce Externalities in Transport, in Button K.J. and Verhoef E.T. (eds.) Road Pricing, Traffic Congestion and the Environment – Issues of Efficiency and Social Feasibility, Edgar Elgar

Rooijers A.J. 1992: Beliefs and Motives with Respect to Car Use for Commuting (in Dutch), Verkeerskundig Studiecentrum, University of Groningen, Groningen

Starkie D.N.M. 1971: Modal Split and the Value of Time: A Note on 'Idle-Time', Journal of Transport Economics and Policy, Vol. 5, 2, 216–20

Stigler G.J. and Becker G.S. 1977: De Gustibus non est Disputandum, American Economic Review, Vol. 67, 2, 76–90

Stubbs P.C., Tyson W.J. and Dalvi M.Q. 1984: Transport Economics, George Allen & Unwin.

Tversky A. 1972a: Choice by Elimination, Journal of Mathematical Psychology, Vol. 9, 341–367

Tversky A. 1972b: Elimination by Aspects: A Theory of Choice, Psychological Review, Vol. 79, 4, 281–299

Van Exel N.J.A. 1995: Modal Split – Explanation and Environmental Consequences of Mode Choice in the Netherlands (in Dutch), Doctoral Thesis, Faculty of Economics, Erasmus University, Rotterdam

Van Exel N.J.A. and Rietveld P. 2001: Public Transport Strikes and Traveller Behaviour, Transport Policy, Vol. 4, 237–246

Wardman M. 1988: A Comparison of Revealed Preference and Stated Preference Models of Travel Behaviour, Journal of Transport Economics and Policy, Vol. 22, 1, 71–91

Windervanck C. and Tertoolen G. 1998: Can Car Use be Steered? (in Dutch). Adviesdienst Verkeer en Vervoer, Ministry of Transport, Rotterdam

Wooldridge J.M. 1999: Introductory Econometrics: a Modern Approach, South-Western College Publishing

Appendix 6.1 Sample population characteristics

	Phase one questionnaire	Phase two questionnaire
Respondents	42	15
response	84.0%	45.5%
Gender (% female)	40.5%	40.0%
Age (average)	27.4	27.6
≤ 27	42.9%	40.0%
> 27	57.1%	60.0%
Household type		
single adult	33.3%	20.0%
two adults	54.8%	66.7%
other	11.9%	13.3%
Working days per week	4.9	4.9
full-time	95.2%	93.3%
Commuting distance (kilometres)	19.3	20.3
≤ 8 kilometres	71.4%	66.7%
> 8 kilometres	28.6%	33.3%
Commuting time (minutes)	31.1	36.5
≤ 30 minutes	71.4%	60.0%
30 to 60 minutes	19.0%	33.3%
> 60 minutes	9.5%	6.7%
Number of transport modes available for commuting		
one	16.7%	20.0%
two	47.6%	40.0%
three	26.2%	33.3%
four	9.5%	6.7%
Type of transport modes available for commuting		
car	47.6%	60.0%
public transport	90.5%	86.7%
bicycle	76.2%	66.7%
other (walking, rollerblades/rollerskates)	14.3%	13.3%

Appendix 6.2 Trip chains[1]

OD[2]	Alternative 1	Alternative 2 (ODs 1-8)/ Alternative 6 (ODs 9-20)	Alternative 3 (ODs 1-8)/ Alternative 7 (ODs 9-20)	Alternative 4 (ODs 1-8)/ Alternative 8 (ODs 9-20)	Alternative 5 (ODs 1-8)/ Alternative 9 (ODs 9-20)
1	car	bus-train(2nd)-metro-walk	bus-train(1st)-metro-walk	bus-train(2nd)-bicycle	
2	car	walk-metro-train(2nd)-metro-walk	walk-metro-train(1st)-metro-walk	walk-metro-train(2nd)-bicycle	
3	car	bus-train(2nd)-metro-walk	bus-train(1st)-metro-walk	bus-train(2nd)-bicycle	
4	car	walk-bus-train(2nd)-metro-walk		walk-bus-train(2nd)-bicycle	walk-bus-bus-walk
5	car	bus-train(2nd)-train(2nd)-metro-walk	bus-train(1st)-train(1st)-metro-walk	bus-train(2nd)-train(2nd)-bicycle	
6	car	train(2nd)-metro-walk	train(1st)-metro-walk	train(2nd)-bicycle	
7	car	bicycle-train(2nd)-metro-walk	bicycle-train(1st)-metro-walk	bicycle-train(2nd)-bicycle	
8	car	bus-train(2nd)-metro-walk	bus-train(1st)-metro-walk	bus-train(2nd)-bicycle	
9	car	walk-metro-walk		bicycle	
10	car			bicycle	walk
11	car	walk-metro-walk		bicycle	walk
12	car	metro-walk	tram-walk	bicycle	
13	car	walk-metro-walk	tram-walk	bicycle	
14	car	tram-metro-walk	tram-walk	bicycle	
15	car	walk-metro-walk		bicycle	
16	car	walk-tram-metro-walk	walk-tram-walk	bicycle	
17	car	tram-metro-walk		bicycle	
18	car	bus-metro-walk		bicycle	
19	car	walk-bus-metro-walk		bicycle	
20	car	metro-walk		bicycle	

[1] It should be recognized that all trip-chains involve some amount of walking, for example walking to or from your car or bicycle. However, in this analysis walking has only been included where the amount is considered significant, that is, more than 3 minutes.

[2] Origin-destination pairs (ODs) numbers 1 to 8 are long-distance trips, numbers 9 to 12 are local trips

Appendix 6.3 Characteristics of origin-destination combinations[3]

Origin-Destination			Alternative 1		Alternative 2		Alternative 3		Alternative 4		Alternative 5		Alternative 6		Alternative 7		Alternative 8		Alternative 9	
Nr	Origin of commuting trip	Distance[4]	Single trip travel time (minutes)	Trip costs per month (NLG)	Single trip travel time (minutes)	Trip costs per month (NLG)	Single trip travel time (minutes)	Trip costs per month (NLG)	Single trip travel time (minutes)	Trip costs per month (NLG)	Single trip travel time (minutes)	Trip costs per month (NLG)	Single trip travel time (minutes)	Trip costs per month (NLG)	Single trip travel time (minutes)	Trip costs per month (NLG)	Single trip travel time (minutes)	Trip costs per month (NLG)	Single trip travel time (minutes)	Trip costs per month (NLG)
1	Voorhout	53.2 (28) 3.8f	53	vast 426 var 319 tot 745	74	370	74	511	70	363										
2	Amsterdam	85.7 (80) 3.8f	66	vast 686 var 514 tot 1200	104	500	104	729	100	493										
3	Valkenburg	48.8 (38) 3.8f	53	vast 390 var 293 tot 683	73	370	73	511	69	363										
4	Dordrecht	24.6 (17) 3.8f	35	vast 197 var 147 tot 344	63	233			59	226	65	200								
5	Amsterdam	86.7 (80) 3.8f	71	vast 694 var 520 tot 1214	106	506	106	718	115	499										
6	Leiden	45.6 (38) 3.8f	48	vast 365 var 273 tot 638	53	314	53	455	49	307										

[3] Total travel times include walking to and from transport modes at the start and end of the trip and transfer times between modes.

[4] Distance by car, with the distance on the highway between brackets. Distances marked with an "f" concern distances by bicycle, where different from distance by car.

Origin-Destination			Alternative 1		Alternative 2		Alternative 3		Alternative 4		Alternative 5		Alternative 6		Alternative 7		Alternative 8		Alternative 9	
Nr	Origin of commuting trip	Distance[4]	Single trip travel time (minutes)	Trip costs per month (NLG)	Single trip travel time (minutes)	Trip costs per month (NLG)	Single trip travel time (minutes)	Trip costs per month (NLG)	Single trip travel time (minutes)	Trip costs per month (NLG)	Single trip travel time (minutes)	Trip costs per month (NLG)	Single trip travel time (minutes)	Trip costs per month (NLG)	Single trip travel time (minutes)	Trip costs per month (NLG)	Single trip travel time (minutes)	Trip costs per month (NLG)	Single trip travel time (minutes)	Trip costs per month (NLG)
7	Utrecht	53.9 (48) 3.8f	47	vast 431 var 324 tot 755	50	376	50	526	59	369										
8	Delft	22.9 (17) 3.8f	29	vast 183 var 138 tot 321	42	207	42	283	38	200										
9	Rotterdam	6.5 4.5f	19	vast 78 var 52 tot 130									19	60			21	30		
10	Rotterdam	1.8	7	vast 22 var 14 tot 36													8	30	22	0
11	Rotterdam	1.9	4	vast 23 var 15 tot 38									18	60			9	30	24	0
12	Rotterdam	2.1	9	vast 25 var 17 tot 42									11	60	19	60	9	30		
13	Rotterdam	3.0	10	vast 36 var 24 tot 60									16	60	20	60	14	30		
14	Rotterdam	7.2	17	vast 86 var 58 tot 144									32	60	40	60	33	30		

Origin-Destination			Alternative 1		Alternative 2		Alterna-tive 3		Alterna-tive 4		Alterna-tive 5		Alterna-tive 6		Alterna-tive 7		Alterna-tive 8		Alterna-tive 9	
Nr	Origin of commuting trip	Distance[4]	Single trip travel time (minutes)	Trip costs per month (NLG)	Single trip travel time (minutes)	Trip costs per month (NLG)	Single trip travel time (minutes)	Trip costs per month (NLG)	Single trip travel time (minutes)	Trip costs per month (NLG)	Single trip travel time (minutes)	Trip costs per month (NLG)	Single trip travel time (minutes)	Trip costs per month (NLG)	Single trip travel time (minutes)	Trip costs per month (NLG)	Single trip travel time (minutes)	Trip costs per month (NLG)	Single trip travel time (minutes)	Trip costs per month (NLG)
15	Rotterdam	5.2	13	vast 62 var 42 tot 104									21	60			24	30		
16	Rotterdam	10 (8) 6.5f	15	vast 80 var 60 tot 140									31	60	35	60	30	30		
17	Rotterdam	10.5 (8) 7.0f	15	vast 84 var 63 tot 147									28	60			32	30		
18	Rotterdam	5.2	16	vast 62 var 42 tot 104									27	100			24	30		
19	Rotterdam	6.0	14	vast 72 var 48 tot 120									28	100			28	30		
20	Rotterdam	6.1	15	vast 73 var 49 tot 122									19	60			28	30		

Appendix 6.4 Template Phase Two questionnaire (example short distance OD)

Origin-Destination: 13
From: Rotterdam/IJsclubstraat
To: Rotterdam/Brainpark

EXPERIENCE				
Have you made this trip or a large part of it in the last three months, with as main mode of transport:	0 times	1 to 2 times	3 to 5 times	more than 5 times
a. car	❑	❑	❑	❑
b. public transport	❑	❑	❑	❑
c. bicycle	❑	❑	❑	❑
d. other, ...	❑	❑	❑	❑

ALTERNATIVES		Single trip travel time (minutes)	Travel costs per month (NLG)	Ranking (1 = best alternative)
Car	Distance 3.0 kilometres, of which 0 on highways. If you do not have a car, you need to purchase one.	10	fixed 36 variable 24 total 60	
Metro	walk to metro station 7 min. metro to Kralingse Zoom 2 min. walk to Brainpark 7 min.	16	60	
Tram	walk to tram stop 3 min. tram to Erasmus tram stop 5 min. walk to Brainpark 12 min.	20	60	
Bicycle	To Brainpark by bicycle (3.0 km.)	14	30	

Endnotes

[1] We acknowledge that many factors have an influence on the travel choices people make and that it is important to take as many of these as possible into account when modelling travel behaviour. In this chapter we, however, confine ourselves to the question of whether people's daily travel behaviour has an effect on preference elicitations in hypothetical travel situations. The remainder of our modelling is therefore pretty straightforward.
[2] The questionnaire has not been included for reasons of length, but can be obtained (in Dutch) from the contacting author.
[3] E.g. totally agree – totally disagree; very important – very unimportant; always – never (see, e.g., Rooijers 1992).
[4] ALH routeplanner, available from www.alh.nl.
[5] Time schedules from different public transport companies (NS rail, RET Rotterdam) as well as the Internet (www.ov9292.nl) and telephone (0900-9292) travel information services.
[6] For the calculation of travel costs, the following assumptions are used:

- car: for urban trips fixed costs €0.14 per kilometre; variable costs €0.09 per kilometre; total costs €0.23 per kilometre (various sources); for inter-urban trips fixed costs €0.09 per kilometre; variable costs €0.07 per kilometre; total costs €0.16 per kilometre (various sources);
- train: for trips up to 40 kilometres €0.14 per kilometre for a 1st class ticket; €0.09 per kilometre for a 2nd class ticket; for trips of more than 40 kilometres €0.10 per kilometre for a 1st class ticket; and €0.07 per kilometre for a 2nd class ticket (source NS, Dutch railways);
- public transport: season tickets for public transport are €27.2 per month for urban trips and €45.4 per month for suburban trips (source GVB, urban public transport company of the city of Amsterdam; RET, urban public transport company of the city of Rotterdam). In combination with a train season ticket, the tariffs are €16.8 and €25.4 per month (source NS, Dutch Railways)
- bicycle: purchase €363.0 (lifetime 8 years), maintenance 10 percent, paid parking at train stations €9.1 per month (source NS, Dutch railways).

[7] For a comprehensive analysis of household multi-vehicle usage behaviour, see, e.g., Golob, Kim and Ren (1996).
[8] Khattak and de Palma (1997) discuss commuters' response to adverse weather.
[9] In the Netherlands, students receive a public transport travel card. They can choose between a week or a weekend card, with free use of all public transport in the part of the week of their choice and a 40 percent (interurban) to 50 percent (urban) discount on public transport in the remainder of the week.
[10] The full set of trip-chains per origin-destination combination is included in Appendix 6.2. A description and characterization of the 20 commuting origin-destination combinations is included in Appendix 6.3. An example of the templates is included in Appendix 6.4.

---

[11] No experience in the last three months with a specific mode on the route received a score of 1; one or two times in the last three months a score of 2; three to five times in the last three months a score of 3; and, more than five times in the last three months a score of 4 (see Appendix 6.4).

[12] Some examples: "I only look at travel time", "I find comfort, by which I mean the car, very important", "I would never use the bicycle before or after transport", "In the city I always take the bicycle, never the car", "Cycling is not suitable for days when you need to wear a suit or to carry a lot of stuff", "Car is the last option for me, I'd never buy one", "I am allergic to buses, too much waiting", "I may be prejudiced, but no tram or bus for me", "I never take the bus unless it is the only option" and "In the train, I'll always choose 2nd class".

[13] Some examples: "I only use the car when travel time differences become too big", "In case of many transfers I prefer the car", "For local trips price is irrelevant", "Environmental aspects are the most important choice factor for me", "I only travel 1st class on train journeys longer than 30 minutes", "I'll use the bicycle for trips up to 30 minutes, otherwise the car", "When cycling takes more than 15 minutes I choose public transport", "Physical exercise is a positive asset of cycling, but not for longer than half an hour" and "Time of day is important because of the frequency of public transport supply".

[14] Mode specific considerations can be included in two ways: via the constant (i.e. α_{ij}), and via the intercept. We have tried to use the two approaches simultaneously, but that led to unstable results. Finally, have chosen the second approach because this seems to make more sense when complex transport chains are considered (as we do here). Note that the specification in (6.5) weighs the importance of the various transport modes in a chain according to the time spent in each of the modes.

[15] For an extensive discussion of the background and technical aspects surrounding random utility choice models the reader is referred to Ben-Akiva and Lerman (1985), Cramer (1991) or De Dios Ortúzar and Willumsen (1994) and, more specifically in relation to stated preference research, to Henscher, Barnard and Truong (1988) or Louviere and Street (2000).

[16] Three alternative modes for a single trip; say j, k and l, result in three pairwise comparisons of modes for that trip; namely j versus k; j versus l; and k versus l. Likewise, four alternatives results in six pairwise comparisons. Thus, altogether the respondents make $15 \cdot (13 \cdot 6 + 7 \cdot 3) = 1485$ pairwise comparisons of modes.

[17] Those who want to recalculate the values of time have to take into account that in the templates of the Phase 2 questionnaire, the costs are expressed in Dutch guilders per month, whereas travel time is expressed in single-trip minutes. Hence, to arrive at euros per hour time value, the quotient of time and costs coefficients γ/β has to be multiplied by 1.5 (single-trip to round-trip (multiply by 2), from day to month of average 20 working days (multiply by 20) and from minutes to hours (divide by 60)) and converted to euros (1 Dutch guilder (NLG) = 0.4538 euro).

[18] Standard errors in parentheses.

[19] We chose to present monthly travel costs in combination with single-trip travel times, as our feeling is that respondents are most familiar with these conceptualizations of costs and time in relation to their commuting travel.
[20] Transfers between public transport and bicycle or walking, before or after using the transport mode, are not included in this transfer time, but in the access or egress time, which also includes the time actually walking or on the bicycle.
[21] We owe these last comments to an anonymous referee.
[22] For an analysis of inertia of public transport commuters following a public transport strike, see Van Exel and Rietveld (2001).

7 The Same Procedure as Last Weekend: Routines and Leisure Mobility

Martin Lanzendorf

Urban Research Centre, Utrecht, The Netherlands

7.1 Introduction

Recently, the question of how to achieve a sustainable transport system has attracted wide interest (OECD 1997; SCAST 1999). More kilometres are travelled by motorized vehicles every year, with consequently more fuel consumption and greenhouse gas emissions worldwide. As transportation statistics from many countries reveal, many of these kilometres are travelled for leisure activities. Nevertheless, as a review of the literature shows, little research has been undertaken on leisure travel, either on a European or a national level (ECMT 1998; Braunolte et al. 1999; Lanzendorf 2001).

This chapter reports analyses of the decision making of individuals on leisure travel and describes their use of routines. Following the definition of Holzapfel et al. (1996: 50), we take leisure to mean that part of the day which individuals use for their own needs beyond their daily obligations and duties (job, housework, necessary physical regeneration), and which depends on the social conditions. Thus, while no choices are available for obligatory activities, individuals freely choose their leisure activities from a broad range of alternatives. Consequently, we expect the decision making for leisure travel to be more complex and involve fewer routines than for shopping or commuting.

Routines are an essential tool for individuals to simplify their daily decision making (see, e.g., Cullen and Godson 1975; Cullen 1978; Giddens 1984; Esser 1991). If a choice situation is well known to someone and a specific behaviour has been found to fulfil the situation-specific needs, then this same behaviour will be reapplied in the next comparable situation, thereby reducing the information seeking or time costs of making a new decision. This behaviour, which immediately follows the perception of a situation and precedes a choice between alternatives, is referred to as a *routine*. Although some social scientists stress the importance of routines in an individual's behaviour, they are rarely included in explanations of travel behaviour. Some of these explanations involve qualitative methods, or refer to the general knowledge of human behaviour to shed light on the importance of routines (see, e.g., Beckmann 1988; Turrentine 1994; Knie 1997; Krämer-Badoni and Wilke 1997). Others show the importance of routines for mode choice decisions by modifying the theory of planned behaviour (Bamberg 1996a); some social psychologists undertake experiments (Verplanken et al. 1998, Aarts and Dijsterhuis 2000). Unfortunately, the research reported to date is restricted to mode choice, fails to include activity or destination choices, and does not consider leisure travel.

For the purpose of this study, some rational choice models recently developed and found useful in the social sciences have been used (Esser 1991, 1993; Friedrichs et al. 1994; Lindenberg 1989). The main one (Esser 1991) integrates routines in a two-step rational choice model; it is discussed in a later section. Furthermore, three sub-decisions of the decision for leisure travel have been distinguished and analysed in detail: the activity; place (destination); and, transport mode choices.

The chapter addresses three key questions: (i) What is the relationship between an individual's travel patterns (that is, mode choice and travel distance) on a workday and for leisure at the weekend? Does this relationship give any indication of the importance of routines?; (ii) Which elements of the decision making (routines, for example) are important in leisure travel including the choice of activity, place, and mode?; and (iii) Do external factors (residential neighbourhood, car ownership, mobility style) have any influence on the decision making process?

This chapter is organized as follows. Section 7.2 describes the theoretical background of the study, including the rational choice model used and its application to the decision making process in leisure travel. Thereafter, Section 7.3 describes the methodology employed and the neighbourhoods in which the research was carried out. Section 7.4 compares workday and weekend leisure travel patterns and Section 7.5 discusses the importance of routines and other factors with an impact on the travel decisions. Finally, the impact of residential neighbourhood, car ownership and mobility style on the decision making process is considered in Section 7.6. The chapter ends with some conclusions.

7.2 Theoretical Background: A Rational Choice Model

The theoretical model employed in the analysis has its roots in the recent advances made in rational choice theory (see, e.g., Becker 1976; Simon 1982; Lindenberg 1989; Coleman 1990; Esser 1991; and, for a comprehensive overview, see Kirchgässner 1991; Kunz 1997); it has been applied in different forms to transportation (see Bamberg 1996b; Braun and Franzen 1995; Franzen 1997) and environmental research (see Diekmann and Preisendörfer 1992, 1993; Diekmann 1996; Lüdemann 1993, 1997). Henceforth, we refer to the individual actor, as suggested by the subjective expected utility theory (SEU theory), as a Restricted, Resourceful, Expecting, Evaluating and Maximizing Man (RREEMM):

(R)estricted in the capability of perceiving information and alternatives; (R)esourceful in finding new solutions; (E)xpecting outcomes of alternative actions on the basis of incomplete information and subjective preferences; (E)valuating the alternatives by their expected utility; (M)aximizing this utility by the choice made. The SEU theory assumes an actor's rationality to be bounded, so that only a limited number of alternatives are perceived to reduce costs for gathering and processing new information. The SEU theory stresses the subjectivity of the evaluation of alternatives and thus the selection of the alternative with the highest subjective utility. Moreover, the actor stops searching for a better choice if an alternative satisfies the actor's needs (principle of satisfycing).

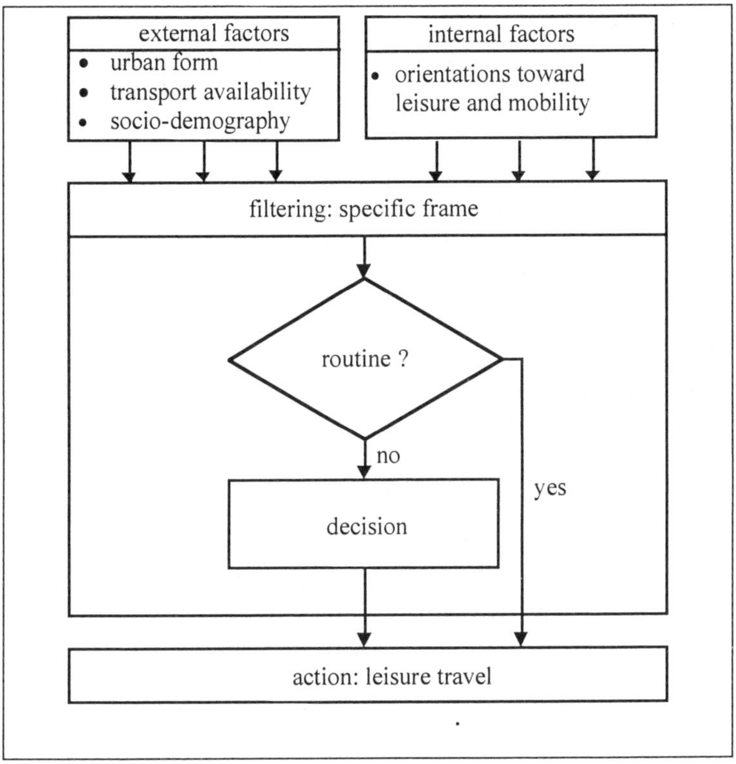

Figure 7.1 General decision model explaining leisure travel

Some rational choice theorists (e.g. Esser 1991; Friedrichs et al. 1994) argue that individuals reduce the complexity of their decision making as suggested by the SEU theory by using routines and frames. Routines are bundles or sequences of actions which an actor chooses in a given situation after receiving some specific hints. Thus, routines are used without alternatives being evaluated. To include routines, standard SEU theory is modified into a two-step decision process (see Figure 7.1): first, the actor decides whether a routine corresponds with the situation-specific aims. If this is the case, the routine is chosen and no further decision making has to be undertaken. This type of choice is frequent for most everyday activities. Second, if no routine is available, an alternative is chosen by the standard SEU evaluation scheme mentioned above. Frames are an individual's superior aims that exist prior to the decision making and have an impact on it. They are cognitive structures that reduce the number of aims in a choice situation and so reduce its complexity. For example, the aim of spending some time with other family members can be a frame for the choice of a leisure activity on a Sunday afternoon. Both routines and frames reduce the effort necessary for decision making: in a choice situation, routines simplify the structure of means (that is, the number of perceived alternatives); frames simplify the structure of aims (that is, the number of evaluation criteria).

For the application of this general decision making model to the travel choices described in the remaining part of the chapter, we make three assumptions. Firstly, an individual's travel choices form an integral part of his or her everyday activities, so that the model is compatible with the human activity approach, although focusing on decision making. Secondly, we consider external factors such as urban form, availability of transport modes, and socio-economic characteristics as external constraints for the decision making. Finally, we include an individual's orientations towards leisure and mobility as internal factors for the choice process (see Lüdtke 1995).

Recent applications of the rational choice theory to travel behaviour can be divided into two groups: those analyzing the impact of environmental concern on travel behaviour (Diekmann and Preisendörfer 1992; Diekmann 1996; Franzen 1997); and those explaining mode choice by the general social-psychological theory of Fishbein/Ajzen, which suggests that attitudes, subjective norms, the relative importance of attitudinal and normative considerations, as well as the frequency of previous mode use, explain the actual behaviour (Bamberg and Schmidt 1994; Bamberg 1995, 1996a; Prognos & Institut für Angewandte Verkehrs- und Tourismusforschung 1997). Unfortunately, these researchers are focused on the mode choice and have neglected the activity and destination choices. Furthermore, the mode decisions explained are only for commuting or educational travel and are unsuitable for leisure travel, which is characterized by a broader freedom of choice (that is, of activities, destinations, and modes).

To overcome the restriction to mode choices alone, we observed individual travel decisions in the framework of three sub-decisions: the activity, the destination; and, the mode choice (Figure 7.2). These seem to be the most important for travel patterns and impacts, although other choices are related to travelling, such as route choice, accompanying people, luggage (like sports equipment, waterproof clothing, food and drink), or whether there are extra activities while travelling (like talking, reading, listening to the radio or a CD). The route choice – frequently included in transportation models – is of minor importance in accounting for leisure travel in comparison with the three choices mentioned above. Moreover, external factors (urban form, availability of transport modes, socio-economic status) and internal ones (orientations, preferences) are included in the choice model.

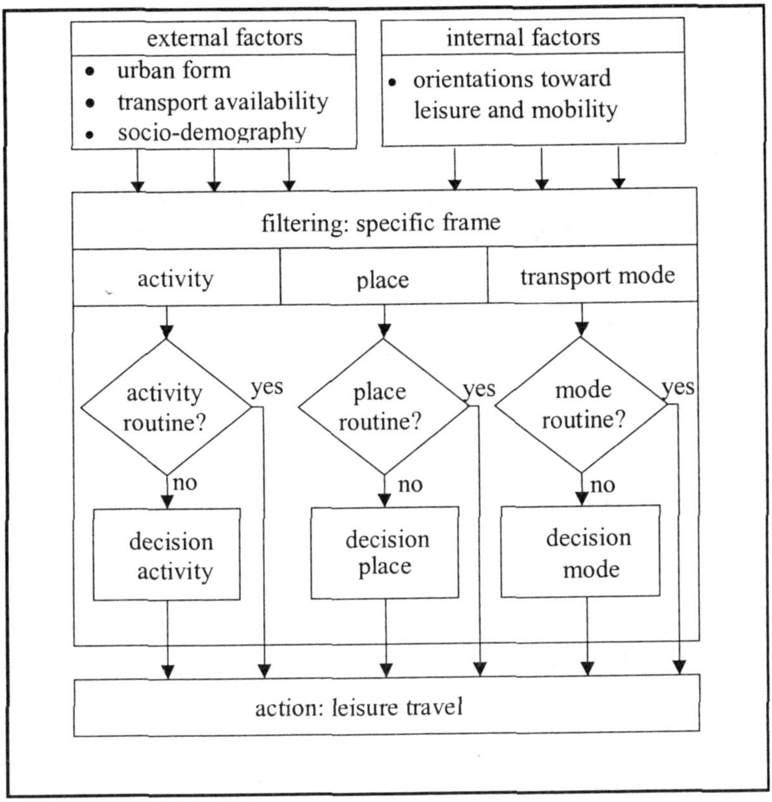

Figure 7.2 Specific decision model explaining leisure activity

7.3 Data and Methodology

The decision making of individuals was a major part of a case study of several aspects of leisure mobility in Cologne (for details, see Lanzendorf 2001). A survey was conducted from April to July 1997 in four residential neighbourhoods of Cologne with a representative sample of the adult inhabitants (18 years and older) with German citizenship. The municipality of Cologne provided the sample addresses. Eventually, 949 people participated in the face-to-face interviews; the response rate of 46 percent was satisfactorily high.

The respondents completed a questionnaire and kept a trip diary; additionally, one-third of them were interviewed about their decision making for leisure travel.

First, the questionnaire included questions dealing with personal and household characteristics, urban form elements (garden ownership, housing type, and so forth), the availability of transport modes (car ownership, for example), leisure, and mobility orientations. Second, the participants noted all trips made during the previous weekend and also on the last regular workday (we excluded Fridays and holidays). For each trip, the place and time of departure and arrival, the modes of transport, and the purpose were noted for each trip. The diaries had enough space to note four modes and two purposes for each trip. The third, qualitative part of the interview was structured around the decision making for the leisure travel undertaken on the previous Sunday. This semi-structured part asked about the activity, place, and mode choice. It was tape-recorded and later coded according to an evaluation scheme. On average, the last part of the interview took between five and ten minutes.

The neighbourhoods surveyed (Figure 7.3) differ with respect to the distance from the inner city, the housing stock, or the availability of leisure facilities and parks. *Belgisches Viertel* is located closer to the city centre than the other areas; it forms a part of the inner city. Apartment houses dating from the beginning of the 20th century ('Gründerzeit') are frequently encountered in this urban area. It is attractive, both as a living environment and for spending free time, as there are many recreational facilities (bars, cafés, restaurants, theatres, cinemas, shopping streets, and so forth). Inner-city parks are within walking distance and provide some green areas.

Zollstock, formerly a working-class neighbourhood, is also within easy reach of the inner city. Housing associations own the majority of the dwellings; these were mainly built between 1920 and 1960. Some leisure facilities (cafés, restaurants, shops), parks and green areas are within walking distance.

Longerich is further away from the inner city than either Belgisches Viertel or Zollstock. Its main part was built in the 1950s. A mixture of single-family terraced houses and apartment blocks characterizes the housing stock. Only a few leisure facilities such as cafés or restaurants are located in Longerich; some pleasant, green areas are within walking or cycling distance.

Rath, the neighbourhood located the furthest away from the inner city, is close to a large, attractive recreational area, the Koenigsforst, which borders the east of Rath. Rath consists of two parts: first, a former village with one and two-family houses; second, the 'Goettersiedlung', built in the 1920s, with detached single-family houses surrounded by large gardens.

Although all four neighbourhoods have access to the public transport system by at least one tram-line, the quality of this transport opportunity differs. For instance, the inner city location of Belgisches Viertel has access to several tram and subway lines, while Rath on the outskirts of the city has only one tramline connecting it with the city centre. The household and population characteristics in the four neighbourhoods differ as well. The population density decreases with increasing distance from the inner city, and the car density increases. In Belgisches Viertel more than two-thirds of all household types are singles and three-quarters of the population are between 18 and 59 years old. Rath and

Longerich have fewer single households (38%), but more children and elderly people. In Zollstock, age and household size are both average.

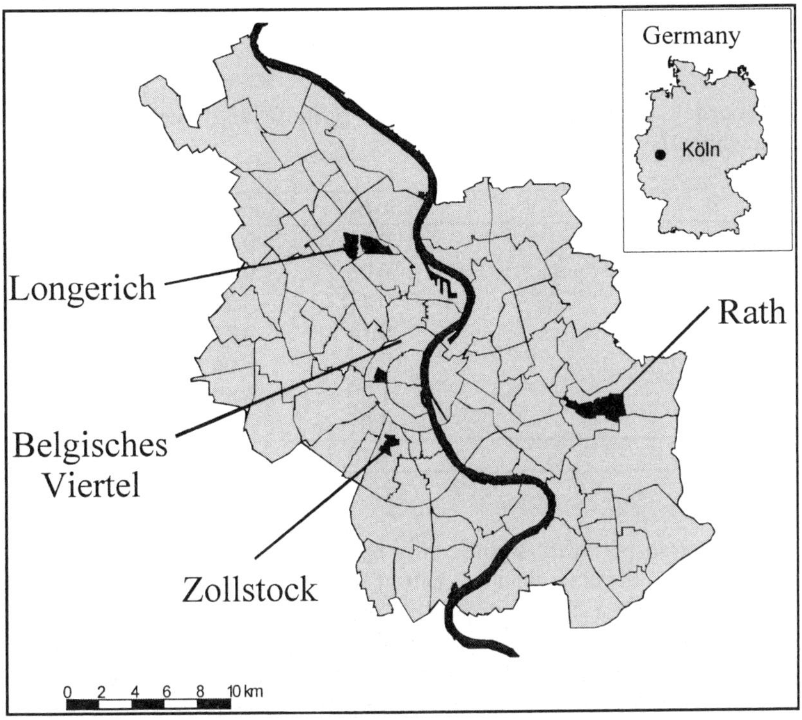

Figure 7.3 Location of the Cologne neighbourhoods surveyed, 1997

Another methodological remark needs to be made before the travel diaries are considered in more detail. The travel patterns were analyzed on the basis of the journeys made; these comprised all trips, from leaving until returning home. This method has the advantage of shedding light on the different purposes combined in trip-chains, although more work is required to prepare the data set for further analysis.

7.4 Comparing Workday Travel with Weekend Leisure Travel

At weekends, leisure is relatively more important than other travel purposes. In our case study, we found that on a workday 28 percent, on a Saturday 48 percent, and on a Sunday 75 percent of all journeys are for one of the five leisure purposes (Table 7.1: see Lanzendorf 2001, for details): (i) leisure facilities (sport, recreational shopping, pubs and bars, restaurant, theatre, cinema, fun-parks, etc.); (ii) social contacts (visits to friends and relatives); (iii) leisure facilities and social contacts combined; (iv) recreation (travelling to some recreational area, usually parks, forests, allotments, or second homes); (v) just driving/walking around (travel by car or bicycle for fun).

Table 7.1 Journey frequencies per person by purpose and weekday, selected Cologne neighbourhoods, 1997

N=949	*workday*	*Saturday*	*Sunday*
leisure facilities	0.13	0.23	0.21
leisure facilities & social contacts	0.06	0.14	0.15
social contacts	0.11	0.21	0.20
recreation	0.14	0.21	0.25
just driving/walking	0.05	0.06	0.08
obligations	1.20	0.89	0.24
overnight stay	0.01	0.03	0.06
Total	1.70	1.76	1.20

The modes of transport vary significantly by the leisure activity (Figure 7.4). While the car is used for more than one half of the journeys for social contacts, its share is much lower for leisure facilities, recreation or just walking/driving around. Instead, for recreation and just walking/driving around, non-motorized modes prevail. Public transport is less important for leisure travel. Its highest share is for leisure facilities (15%), which are frequently located in central city areas with good accessibility by public transport.

In the rest of this section, we examine the relationship between travel behaviour on workdays and for leisure at weekends. People who travel by car frequently, or for long distances on workdays might be expected to do the same at weekends: on the other hand, they might prefer to stay at home for recreational purposes. For the analysis of this relationship, we first defined mode types from the workday trip diaries for every person interviewed: 6 percent of them did not travel at all ('non-mobile'), 25 percent only cycled or walked ('non-motorized'), 44 percent used a car, but not public transport ('car'), 16 percent used public transport, but not a car ('public transport'), 7 percent used both a car and public transport ('car and public transport'), and 2 percent used other modes ('other'). Second, we defined a mode type for the weekend leisure travel in the same manner by using the information from the diaries (Table 7.2).

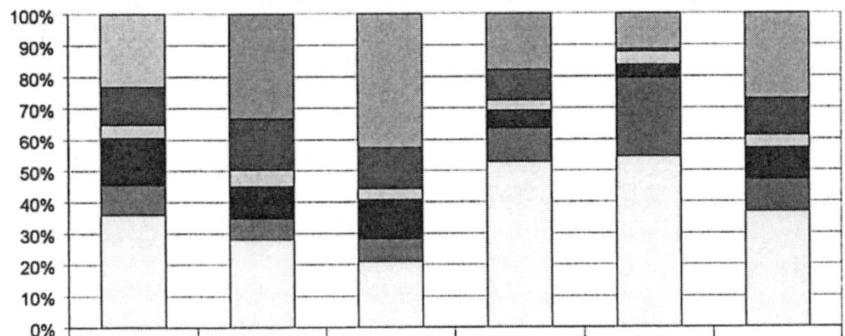

	leisure facilities	leisure facilities & social contacts	social contacts	recreation	just driving/walking	leisure (total)
car, driver	23%	33%	43%	18%	11%	27%
car, passenger	12%	16%	13%	10%	1%	12%
others	4%	5%	3%	3%	4%	4%
public transport	15%	10%	12%	6%	4%	10%
bicycle	10%	7%	7%	11%	25%	10%
walk	36%	28%	21%	53%	54%	37%

Figure 7.4 Transport modes by journey purpose for leisure activities on weekends

The workday and the weekend leisure travel are compared in two steps: first, the workday mode type with respect to the mode type, the frequency of car use, and the car distances for leisure travel on weekends; second, the car distances on workdays with those for leisure on weekends. First, starting with the comparison of workday and weekend leisure mode types, we find that the share of non-mobile persons almost triples on weekends whereas the public transport and the car become less important (Table 7.2). However, 40 percent of the respondents are of the same mode type on both workdays and weekends (Figure 7.5).

Table 7.2 Mode types on workdays and for leisure at weekends

N=949	*workday (all purposes)*	*weekend leisure*
non mobile	6%	17%
non-motorized	25%	24%
Car	44%	40%
public transport	16%	11%
public transport & car	7%	8%
Other	2%	1%
Total	100%	100%

Taking a closer look at the mode types, workday car users stay more frequently with the same mode for weekend leisure than workday public transport users.

From the car users, 59 percent keep the car for weekend leisure, whereas from the public transport users only one quarter keeps to the same mode. Instead, another quarter of them uses the car for weekend travel (Figure 7.5).

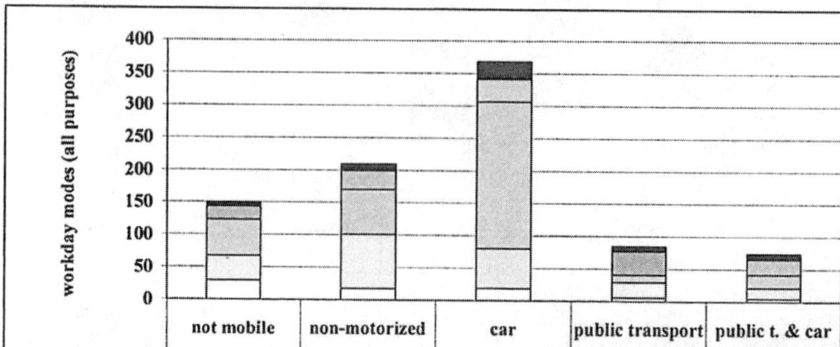

	not mobile	non-motorized	car	public transport	public t. & car
■ public t. & car	7	10	27	8	9
▨ public transport	20	30	35	36	23
□ car	56	69	226	12	21
□ non-motorized	38	84	61	23	16
□ not mobile	29	17	19	6	5

weekend modes (leisure only)

Figure 7.5 Respondents by workday and weekend leisure mode types (N=887)

We explain the differences between workday car and public transport users on weekends by the distinct supply of modes for leisure travel. The public transport services are limited during non-peak hours, especially at weekends or at night-time, and for peripheral regions, which are frequent destinations for recreational travel. Conversely, car users enjoy less congestion and better parking opportunities in cities at weekends than on workdays, possibly only with the exception of Saturday morning's shopping hours. Therefore, the accessibility of many destinations for leisure activities is better by car than by public transport. Additional requirements, such as the transport of luggage, or travelling in a small group with other family members, can make the car substantially more convenient than public transport for leisure travel. We conclude that there is no reason for car users to change their mode habits on weekends, whereas public transport users face some disadvantages for leisure travel on weekends, which can be overcome by using a car.

As expected from the previous paragraph, car users travel more frequently and longer distances by car for weekend leisure than others (Table 7.3). For social contacts, however, workday public transport users travel above average car distances at weekends, which is surprising given the low car distances of this group for other leisure purposes. We suppose that visits to friends and relatives are difficult to maintain by public transport at weekends, because the homes of people are dispersed over the whole urban area and do not follow a hierarchical place system, as, for example, the shopping locations or some leisure facilities do. Therefore, the destinations of social contacts are difficult to reach within a reasonable travel time by public transport, which is mainly organized to serve the hierarchical system of places. We conclude that the maintenance of social contacts is an important reason for individuals to use a car.

Table 7.3 Frequency and car distances for leisure travel at weekends by modes of transport on a regular workday

		weekend (leisure purposes only)											
		frequency of journeys				*car distances (km)*							
		all modes		*car only*		*all leisure purposes*		*leisure facilities*		*social contacts*		*recreation just driving / walking*	
workday (all purposes)		F=5.68*** η=0.16		F=28.95*** η=0.34		F=4.87*** η=0.15		F=2.48** η=0.11		F=3.22** η=0.12		F=4.03*** η=0.14	
	N	m	s	m	s	m	s	m	s	m	s	m	s
non mobile	53	0.94	1.17	0.28	0.57	12	44	4	21	7	39	1	3
non-motorized	224	1.71	1.25	0.38	0.66	20	67	3	15	4	19	4	22
car	387	1.86	1.42	0.99	1.00	45	98	6	21	17	56	12	43
public transport	144	1.79	1.26	0.37	0.60	24	68	1	10	11	47	1	8
public t. & car	62	1.82	1.22	0.89	1.03	42	83	8	26	8	24	13	55
(other)	(17)	(1.71)	(1.10)	(0.53)	(0.72)	(38)	(67)	(9)	(35)	(3)	(9)	(15)	(50)
total	887	1.75	1.33	0.68	0.89	33	83	5	19	11	44	8	35

m denotes the average value and s the standard derivation. All mean values in the table are significantly different from each other (analysis of variance with *** p<0.01, ** p<0.05). The journey purpose 'leisure facilities & social contacts' is omitted because the analysis of variance showed no significant differences. The workday mode type 'other' was excluded from the analysis of variance because of the low number of cases (17).

Finally, when comparing the car distances covered on workdays with those for leisure at weekends, surprisingly, no relationship between these two variables appears. Here, remains a question, which could be solved only by more extensive surveys.

7.5 Decision Making and Routines for Leisure Travel

7.5.1 Four Types of Decision Making for Leisure Travel

For the analysis of individual decision making for leisure travel, we first consider how the three sub-decisions on leisure travel (activity, place, mode choice) intervene and whether the choices are made at the same time, in sequence, or some

other logical order. Following the evaluation of the qualitative interviews we distinguished four types of decision making:
(i) the first type, A-P-M, is defined by three sequential sub-decisions: first, activity; second, place; third, mode decision. At least in theory, at each stage of the decision an alternative action exists. This type is found in 28 percent of all the decisions analysed. The following situation exemplifies this. Somebody decides to watch a certain film in the evening (activity); then chooses a cinema (place); and finally, decides how best to reach this place (mode of transport).
(ii) The second type of decision making AP-M, summarizes the decisions that are made in two steps: initially, the activity and place choices; next, the mode choice. At 53 percent, this is the most frequent type of all. The decision to meet up with a neighbourhood group may be of this kind when the group meets at a fixed place (in a neighbourhood centre, for example). Hence, the decision to participate in such a meeting already includes the decision of the meeting place. A similar example is a visit to friends or relatives. Having decided to visit somebody, there is usually no choice about the place of the visit. There is, however, still the choice of the transport mode. But not every visit to friends or relatives has to be of that kind. For example, in the case where you decide to visit some friends on a Saturday afternoon and phone round to ask which of your friends is at home and would be glad for you to pay a visit, the decision making has three stages: first the activity; next, the place; and finally, the mode of transport. This situation belongs again to the first case: A-P-M.
(iii) The third type, AM-P, starts with a simultaneous decision on activity and mode of transport. In a second step, a decision on the place is made. An example of this type can be the decision to go hiking, cycling, or for a drive (just walking, riding, driving). Of all the decision making, 11 percent is of this type.
(iv) Finally, the fourth type of decision making, AMP, concerns the situation where the activity, place, and mode are chosen simultaneously. 8 percent of the cases analyzed are of this type. Only the 'stay at home' activity in our sample fitted this case. No further place or mode decisions are involved. We concluded that the activity decision always took place first and that, in some cases, the activity choice already directly involved the mode and destination choices.

To avoid misunderstandings, it should be mentioned that, for the distinction between consecutive and simultaneous sub-decisions, only the activity definition of the respondent was used but no additional information about available alternatives on modes or places. Thus, as explained above, the visit of a friend can be of the decision type AP-M or A-P-M depending on whether or not the activity 'visiting' already includes a particular friend or not for the respondent. Furthermore, in the perspective of this paper, the mode decision can be independent of the combined activity/place decisions even if only one mode exists by which a place can be reached. In the latter case, the mode choice can be a separate sub-decision, albeit a trivial one.

7.5.2 Types of Decision Making and Activities

The types of decision making described above interact with the leisure activities (Table 7.4): type A-P-M is frequently used for going out to bars and restaurants (84%), meeting friends outside the house (47%), and going for a walk (47%); type AP-M is the most frequent for visiting friends (94%) or relatives (89%), street fairs (64%), weekend and holiday trips (63%), allotments (100%), or sports events (100%); type AM-P is most frequently used for going for a walk in a recreational area (58%) and for going for a drive or cycle ride with no particular destination (58%); type AMP only applies to staying at home (100%).

Table 7.4 Leisure activities by type of decision making

	N	A-P-M	AP-M	AM-P	APM	*total*
bar, restaurant, pub, etc.	N=55	84%	15%	2%		100%
meet friends (not at home)	N=58	47%	45%	9%		100%
walk around, strolling	N=17	47%	29%	24%		100%
second home, allotment	N=12		100%			100%
visit sports event	N=10		100%			100%
visit friends (at home)	N=52	6%	94%			100%
cemetery	N=13		92%	8%		100%
visit relatives	N=83	7%	89%	4%		100%
neighbourhood groups, social activities	N=34	29%	71%			100%
club, fair, or street party	N=33	36%	64%			100%
weekend trip, (short) holidays	N=27	33%	63%	4%		100%
cinema, theatre, theatre, concert, etc.	N=15	47%	53%			100%
sports (active)	N=24	33%	50%	17%		100%
go for a walk (in recreational area)	N=53	32%	9%	58%		100%
drive around without a destination	N=12	33%	8%	58%		100%
stay at home	N=36				100%	100%
other	N=22	50%	45%	5%		100%
total	N=430	28%	52%	11%	8%	100%

The leisure activities are grouped according to the most important decision type. Leisure activities with less than 10 situations are grouped in the category 'other'. If there were 2 activities in one situation of decision making, we included the situation in each of the activities.

7.5.3 Importance of Routines for the Decision Making of Individuals

When coding the interviews, we defined as routines all those statements showing that the respondents made their decisions without thinking about them. We also found some cases in which people argued cogently why they had made a particular choice, but said at the same time that they made the same choice on every similar occasion: they always performed a particular activity, went to a certain place, or travelled by a particular mode. In this chapter, we refer to this behaviour as a routine as well.

Hence, the survey shows that 31 percent of the activity choices, 35 percent of the place choices, and 78 percent of the mode choices are routines (Table 7.5). We conclude, first, that routines are of great importance for travel-related decisions, and, second, that mode choices most often are routines, while 'real' decisions are only made in exceptional cases, as suggested by the standard SEU theory. We expect these observations also to be valid for most other travel purposes, in particular for repeated activities like shopping or commuting.

Table 7.5 Use of routines with sub-decisions on activity, place and mode of transport

	activity	*place*	*mode of transport*
	N=430	N=169	N=346
routine	31%	35%	78%
no routine	67%	41%	16%
missing values	2%	24%	7%
total	100%	100%	100%

7.5.4 Routines in the Context of Types of Decision Making

Next, we analyzed how routines at the sublevels of decision making (that is, the sub-decisions for activity, place, and mode) fit with the four types of decision making introduced above. We distinguished four major cases (Table 7.6): (i) all the sub-decisions (activity, place, and mode) are routines (21% of all cases); (ii) only the mode choice is a routine and the other choices are not (38%); (iii) no routines are involved at all (24%); (iv) all other cases (17%).

Table 7.6 Type of decision making by the use of routines

N=430	*overall routine*	*only mode routine*	*no routine*	*other*	*total*
A-P-M	11%	29%	22%	38%	100%
AP-M	24%	57%	16%	4%	100%
AM-P	35%	-	31%	33%	100%
APM	22%	-	78%	-	100%
total	21%	38%	24%	17%	100%

The importance of routines varies according to the type of decision making (Table 7.6). The largest share of complete routines is for type AM-P (35%) and the lowest for type AP-M (11%). Type AM-P lies between the other two cases (24%), but it has the highest number of decisions with mode-only routines (57%). Decisions without any routines are highest for the type APM (78%), which consists only of the activity 'staying at home', and should be considered as a special case.

7.5.5 Routines and Journey Purposes

The journey purpose and the importance of routines are interrelated (Table 7.7). As previously shown, decisions for social contacts are mainly associated with the decision type AP-M which, again, frequently involves mode routines but not activity or place routines. Hence, it is not surprising that social contacts are associated with a high share of mode-only routines. This relationship can readily be explained since, for visits to friends or relatives, the way to get to them is usually well known, and a best mode of transport has already been found. In total, 76 percent of all social contact journeys include at least a mode routine.

Table 7.7 Journey purposes by the use of routines

		overall routine	*only mode routine*	*no routine*	*other*	*total*
leisure facilities	N=85	15%	40%	22%	22%	100%
social contact & leisure facilities	N=71	24%	37%	23%	17%	100%
social contact	N=109	15%	61%	17%	6%	100%
recreation nature/ driving/walking	N=102	32%	22%	22%	25%	100%
overnight stay	N=27	15%	52%	4%	30%	100%
stay at home	N=36	22%		78%		100%
total	N=430	21%	38%	24%	17%	100%

To summarize, although routines are expected to be more a part of obligations on workdays and not of leisure time, the importance of routines for leisure travel at weekends has been shown. Routines are important for activities, destinations and, most frequently, for mode choices. Even when people journey to places never previously visited, they do not necessarily decide between different modes of transport in search of the best suitable. Instead, they prefer the mode of transport they always use for that type of journey. Consequently, mode routines are important even if decisions for new destinations are made. Turning the argument around, we may even presume that mode routines have an impact on the decisions for activities and destinations to visit. A destination may not be chosen if the mode routine does not allow it to be reached easily. For instance, the non-availability of parking areas or of a public transport supply for certain destinations will then affect the decision to choose that destination.

Before comparing the sub-decisions on activity, place, and mode of transport in detail, we will first consider some of their differences. In theory, the number of available options for leisure activities is extremely large. Not only is there a broad range of activities (cafés, dancing, sports, cinema, theatre, strolling around, meeting people, etc.), but additionally for most activities there is also a choice between available options (dancing, for example, offers different styles, varying between traditional and modern, while distinct locations can be identified for various types). Furthermore, with the differentiation of society and increasing individualization, we observe the same pattern for leisure activities (see Schulze 1995; Beck and Beck-Gernsheim 1994; Gross 1994). That individuals can, and must, choose between multiple options is a characteristic of the post-modern society. To reduce the over-complexity of available options, individuals use filter processes before making decisions. These filters consist first of ideas of what is important in leisure time (like quietness, recreation, social contacts, action, fun). Second, through previous experiences, people know how to reach what they want most. Hence, they probably most frequently undertake leisure activities which are already familiar to them.

The available opportunities for places are comparatively fewer than for the activity choices. Individuals acquire a specific knowledge and perception of opportunities in their potential activity spaces through their previous activities in time and space (Dijst 1999; Carlstein et al. 1978). By using this knowledge, the space for available opportunities for place choices is reduced substantially. Only a very few activities, such as going for a drive for fun, involve a search for new places. We expect the number of new places perceived to be limited.

The case for the choice of transport modes differs from the other two cases. No potentially broad range of opportunities seems to exist. Instead, the number of available modes is limited if we do not go into too much detail (by considering different types of car, for example) or do not consider all theoretically possible special cases (sailing, rowing, skiing, roller-skating, and so forth). If we restrict our decision choice to the categories of car (including passenger and driver), public transport, plane, walking, and cycling, we obtain a small set of choices. Even when we consider the most frequent combinations of different transport modes on one journey in our choice set, we do not get many more choices. Hence, we have a limited, well-known available choice set. Although individuals do not perceive each of these categories as equally relevant to their choices, at least they know that they exist.

Taking into account these differences between the choice sets for activity, place, and mode, we can explain some of the differences between the related decisions that we observe (Table 7.8). For the activity choice, alternatives and their evaluations are stated relatively seldom, whereas frames and factors impacting on the decision making are frequently mentioned. We conclude that this situation is related to the necessity to reduce the complexity of the decision situation by using filter processes. Either the activity choice is a routine (31 percent of all cases), or it is influenced by frames and other factors, such as discussions with family members. A decision process, such as that assumed to lie at the core of the rational choice theory (i.e. in the SEU theory), can only rarely be observed.

Table 7.8 Characteristics of the sub-decisions on activity, place and mode of transport

	activity	*place*	*mode of transport*
	N=430	N=169	N=346
frames	41%	12%	12%
routine	31%	35%	78%
alternatives stated	16%	14%	25%
evaluation criteria stated	13%	46%	50%
factors for the decision making stated	70%	36%	9%

Independent place decisions (that is, decisions which do not form part of the activity decision) are a routine in more than one-third of all cases. Here, we consider the preceding activity choice as a frame for the place decision. Only exceptionally do the respondents state alternative places (14%), but they frequently mention evaluations of alternatives (46%) and other factors impacting the decision (36%). Therefore, respondents give reasons more frequently for place than for activity decision.

Finally, the mode decision is, for the most part, a routine (78%) if it has not formed part of an activity choice. Only in a few cases are factors stated for the decision made (9%), which reinforces the importance of routines for the mode choice. The preceding activity decision always provides a frame for the mode decision, but almost no other frames are stated. Compared with the other decisions, alternatives (25%) and evaluations (50%) are most frequently mentioned. The limited choice set seems to make it easier for individuals to give reasons for their choices.

7.6 The Impact of Neighbourhood, Car Ownership and Mobility Style on the Decision Making Process

In this section, the relationships of external and internal factors impacting on the process of decision making for leisure travel are discussed. We considered three factors known to have a significant impact on leisure travel patterns (see Lanzendorf 2001) and which we expected to affect the decision process: (i) the residential neighbourhood as an indicator of spatial opportunities; (ii) car ownership as an indicator of the availability of transport modes; (iii) mobility styles constructed by an individual's orientations toward leisure and mobility (see the note below Table 7.9 for further reference) as an internal factor.

Table 7.9 Decision making for leisure travel by mobility styles

		FAMOS	*MULTIS*	*CARCULTS*	*STROLLERS*	*QUICKFITS*	*DOMOS*	*SELFCULTS*	*total*
		N=98	N=39	N=52	N=52	N=93	N=24	N=72	N=430
type of decision making (χ^2=38.1 p<0.01)	*A-O-V*	24%	21%	33%	19%	29%	4%	47%	28%
	AO-V	54%	64%	54%	58%	47%	63%	42%	52%
	AV-O	14%	10%	6%	17%	13%	8%	6%	11%
	AOV	7%	5%	8%	6%	11%	25%	6%	8%
type of routine (χ^2=27.8 p<0.1)	*overall routine*	19%	18%	13%	31%	19%	50%	17%	21%
	only mode routine	33%	51%	46%	33%	41%	21%	38%	38%
	no routine	26%	15%	27%	19%	24%	21%	32%	24%
	other	22%	15%	13%	17%	16%	8%	14%	17%
activity (χ^2=17.6 p<0.01)	*routine*	36%	28%	21%	44%	28%	54%	21%	31%
place (χ^2=0.6 n.s.)	*routine*	37%	33%	35%	32%	38%	33%	32%	35%
mode of transport (χ^2=10.8 p<0.1)	*routine*	71%	88%	73%	85%	80%	94%	69%	77%

Remark: The mobility styles are derived from orientations toward leisure and mobility and can be characterized as follows: FAMOS with family-orientation and non-motorized travel; MULTIS with multiple activities and modes; CARCULTS with high culture and car, STROLLERS with family values and strolling; QUICKFITS with sport (fit) and speed (quick); DOMOS with domestically-modest and no special modes and, finally; SELFCULTS with movies, theatre, friends and speed. A more detailed account of the mobility styles can be found in Lanzendorf 2001.

From these three factors only the mobility styles show a significant impact on the decision making process (Table 7.9); neither residential neighbourhood nor car ownership affects it significantly. We find that the type of the decision process, which is itself closely related to the activity performed (see previous section), changes with the mobility style. On the one hand, the DOMOS, persons with a domestically-modest style, have a very low share of the decision type A-P-M (4%), which is frequently related to activities in leisure facilities as stated above. On the other hand, the SELFCULTS and the CARCULTS, two more self-expressive mobility styles, have an above-average share of this type A-P-M (47% and 33%, respectively). In contrast, the share of routine activity decisions is above the average for the DOMOS and low for the individualistic SELFCULTS and CARCULTS.

The results of the previous analysis show a close relationship between orientations toward leisure and mobility and decision making. In particular, the importance of routines and the structure of the decision process differ between the mobility styles.

7.7 Conclusions

For the purpose of this study, we analyzed a combination of quantitative and qualitative data to study in depth the decision making for leisure travel. Esser (1991) and Friedrichs et al. (1994) have suggested some modifications to rational choice theory that allow the inclusion of routines and frames as filters in the decision process, thereby making the theory more realistic. We expected to find more cases of clear decisions between distinct alternatives for leisure travel than for obligatory travel.

First, we compared workday travel with leisure travel at weekends and found a strong relationship between the travel patterns. Workday car users drive more frequently and cover longer car distances than others for leisure at weekends. This is, however, not true for workday public transport users who change their mode choice for leisure at weekends. In particular, public transport users cover long distances by car at weekends for social contacts.

Second, by analyzing the decision process for leisure travel we found that most choices were not made on the basis of the evaluation of alternatives as suggested by the SEU theory (subjective expected utility theory). Instead, most decision making includes the filter processes of routines and frames. Routines reduce the structure of means in a decision situation, because a bundle of actions is activated after some specific hints are received. Although at first glance, it may seem counterintuitive, many of the activities, places, and modes chosen for leisure on one weekend are the same as they were in the previous weekend. The mode decisions in particular are mainly routines. Even when people want to reach places they have never seen before, they do not necessarily try to find out the best choice of transport mode. In many cases, they already know their mode before they decide where to go. Consequently, mode routines can have an impact on the decision making for activities and destinations: an activity or a destination is not chosen if the mode routine does not allow them to be accessed. For instance, a destination without a parking area is difficult for habitual car drivers to reach, and locations without public transport accessibility are difficult to reach for those with public transport routines.

Frames are a second concept structuring individual's decision making; they reduce the number of aims in a specific situation. Just as for routines for the mode choices, frames are mainly important for the activity choices by pre-structuring the decision making. Although we found other factors of importance in the process of decision making at weekends, we assert that the most important part of the decision making is done not through the evaluation of two or more alternatives, but by frames and routines. Moreover, since routines and frames can be expected to be more important for other travel purposes, we suggest that they are important for all travel decisions.

In the final section of the chapter, we considered the impact of mobility styles, residential neighbourhood, and car availability on the decision making. While the latter have no impact, we found a close relationship between mobility styles and the decision making, in particular the importance of routines for a given type of decision making. Thus, the mobility style – as an internal factor with an impact on the decision making – reflects the different types of decision making better than the two external factors (residential neighbourhood, car ownership) included in the analysis.

Acknowledgements

The author would like to thank Jillian Anable, Michel Beuthe and the two anonymous reviewers for their useful comments.

References

Aarts H. and Dijsterhuis A. 2000: The Automatic Activation of Goal-directed Behaviour: The Case of Travel Habit, Journal of Environmental Psychology, Vol. 20, 75–82

Bamberg S. 1995: Wie Bekommt Man den/die Autonutzer/-nutzerin in den Bus? Probleme und Ergebnisse einer Anwendung der Theorie geplanten Verhaltens im Kontext praktischer Verkehrsplanungsfragestellungen, Zeitschrift für Sozial-psychologie, Vol. 6, 243–262

Bamberg S. 1996a: Habitualisierte Pkw-Nutzung: Integration des Konstrukts 'Habit' in die Theorie des geplanten Verhaltens, Zeitschrift für Sozialpsychologie, Vol. 27, 295–310

Bamberg S. 1996b: Zeit und Geld: Empirische Verhaltenserklärung mittels Restriktionen am Beispiel der Verkehrsmittelwahl, ZUMA-Nachrichten, Vol. 38, 7–32

Bamberg S. and Schmidt P. 1994: Auto oder Fahrrad? Empirischer Test einer Handlungstheorie zur Erklärung der Verkehrsmittelwahl, Kölner Zeitschrift für Soziologie und Sozialpsychologie, Vol. 46, 80–102

Beck U. and Beck-Gernsheim E. (eds.) 1994: Riskante Freiheiten. Individualisierung in modernen Gesellschaften, Suhrkamp, Frankfurt

Becker G.S. 1976: The Economic Approach to Human Behavior, University of Chicago Press, Chicago

Beckmann K.J. 1988: Vom Umgang mit dem Alltäglichen. Aufgaben und Probleme der Infrastrukturplanung, Institut für Städtebau und Landesplanung der Universität Karlsruhe, Karlsruhe

Brannolte U., Axhausen K., Dienel H.-L. and Rade A. (eds.) 1999: Freizeitverkehr. Innovative Analysen und Lösungsansätze in Einem Multidisziplinären Handlungsfeld. Dokumentation eines Interdisziplinären Workshops des BMBF am 10. und 11. Dezember 1998 im Hotel Bristol in Bonn, TU Berlin Universitätsbibliothek, Berlin

Braun N. and Franzen A. 1995: Umweltverhalten und Rationalität, Kölner Zeitschrift für Soziologie und Sozialpsychologie, Vol. 47, 231–247

Carlstein T., Parkes D.N. and Thrift N.J. (eds.) 1978: Timing Space and Spacing Time (3 volumes), Arnold, London

Coleman J.S. 1990: Foundations of Social Theory, Belknap Press of Harvard University Press, Cambridge, MA and London

Cullen I.G. and Godson V. 1975: Urban Networks: The Structure of Activity Patterns, Progress in Planning, Vol. 4, 1–96

Cullen I.G. 1978: The Treatment of Time in the Explanation of Spatial Behaviour in Cities, in Carlstein T., Parkes D.N. and Thrift N.J. (eds) Human Activity and Time Geography. Timing Space and Spacing Time Vol. 2, Arnold, London, 27–38

Diekmann A. 1996: Homo Ökonomicus. Anwendungen und Probleme der Theorie rationalen Handelns im Umweltbereich, in Diekmann A. and Jaeger C.C. (eds.), Umweltsoziologie, Sonderheft 36/1996 der Kölner Zeitschrift für Soziologie und Sozialpsychologie, Westdeutscher Verlag, Opladen, 89–118

Diekmann A. and Preisendörfer P. 1992: Persönliches Umweltverhalten: Diskrepanzen zwischen Anspruch und Wirklichkeit, Kölner Zeitschrift für Soziologie und Sozialpsychologie, Vol. 44, 226–251

Diekmann A. and Preisendörfer P. 1993: Zur Anwendung der Theorie rationalen Handelns in der Umweltforschung. Eine Antwort auf die Kritik von Christian Lüdemann, Kölner Zeitschrift für Soziologie und Sozialpsychologie, Vol. 45, 125–134

Dijst M. 1999: Two-earner Families and Their Action Spaces: A Case Study of Two Dutch Communities, GeoJournal, Vol. 48, 195–206

ECMT (European Conference of Ministers of Transport) 1998: Transport and Leisure. Conclusions of Round Table 111, Paris, 15-16 October 1998 (http://www.oecd.org/cem/conclus/rt111e.pdf)

Esser H. 1991: Alltagshandeln und Verstehen. Zum Verhältnis von erklärender und verstehender Soziologie am Beispiel von Alfred Schütz und 'Rational Choice', Mohr, Tübingen

Esser H. 1993: Soziologie. Allgemeine Grundlagen, Campus, Frankfurt, New York

Franzen A. 1997: Umweltbewußtsein und Verkehrsverhalten. Empirische Analysen zur Verkehrsmittelwahl und der Akzeptanz umweltpolitischer Massnahmen, Rüegger, Chur, Zürich

Friedrichs J., Stolle M. and Sapouridis K. 1994: Situation und Entscheidung. Empirisches Praktikum SS 1993 und WS 93/94, Forschungsinstitut für Soziologie der Universität Köln, Köln. (unpublished research report)

Giddens A. 1984: The constitution of society. Outline of the Theory of Structuration, Polity Press, Cambridge

Gross P. 1994: Die Multioptionsgesellschaft, Suhrkamp, Frankfurt a.M

Holzapfel H., Münsterjohann U., Lehmann F., Reesas G. and Wohltmann H. 1996: Freizeitmobilität. Freizeit und Freizeitverkehr – Eine Bestandsaufnahme, in Forschungsverbund, Ökologisch verträgliche Mobilität in Stadtregionen (ed.), Rahmenbedingungen von Mobilität in Stadtregionen. Teilprojekt 5: Arbeits- und Freizeitmobilität. Wuppertal Institut, Wuppertal, 39–100

Kirchgässner G. 1991: Homo Oeconomicus. Das ökonomische Modell individuellen Verhaltens und seine Anwendung in den Wirtschafts- und Sozialwissenschaften. Mohr, Tübingen:

Knie A. 1997: Eigenzeit und Eigenraum: Zur Dialektik von Mobilität und Verkehr, Soziale Welt, Vol. 47, 39–54

Krämer-Badoni T. and Wilke G. 1997: Städtische Automobilität zwischen Autobesitz und Autolosigkeit, Mitteilungen 1/97 aus dem Forschungsverbund Ökologisch verträgliche Mobilität, Wuppertal Institut, Wuppertal

Kunz V. 1997: Theorie rationalen Handelns. Konzepte und Anwendungsprobleme, Leske + Budrich, Opladen

Lanzendorf M. 2001: Freizeitmobilität. Unterwegs in Sachen sozial-ökologischer Mobilitätsforschung. Materialien zur Freizeit- und Fremdenverkehrsgeographie 56, Geographische Gesellschaft, Trier

Lindenberg S. 1989. Social Production Functions, Deficits, and Social Revolutions: Prerevolutionary France and Russia, Rationality and Society, Vol. 1, 51-77

Lüdemann C. 1993: Diskrepanzen zwischen theoretischem Anspruch und forschungspraktischer Wirklichkeit. Eine Kritik der Untersuchung über Persönliches Umweltverhalten: Diskrepanzen zwischen Anspruch und Wirklichkeit, Kölner Zeitschrift für Soziologie und Sozialpsychologie Vol. 45, 1, 116–124

Lüdemann C. 1997: Rationalität und Umweltverhalten. Die Beispiele Recycling und Verkehrsmittelwahl, DUV, Wiesbaden

Lüdtke H. 1995: Zeitverwendung und Lebensstile. Empirische Analysen zu Freizeitverhalten, expressiver Ungleichheit und Lebensqualität in Westdeutschland, Marburger Beiträge zur Sozialwissenschaftlichen Forschung 5, Institut für Soziologie, Marburg

OECD (Organisation for Economic Cooperation and Development) 1997: Towards Sustainable Transportation. The Vancouver Conference, OECD, Paris

Prognos AG and Institut für Verkehrs- und Tourismusforschung 1997: Modellversuch Mobiles Schopfheim – zur Veränderung von mobilitätsbezogenen Einstellungen und des Verkehrsverhaltens. Ergebnisse der Begleitforschung im Auftrag des Ministeriums für Umwelt und Verkehr Baden-Württemberg und der Stadt Schopfheim. Research report published by Prognos, Basel, Heilbronn

SCAST 1999: Conference on Social Change and Sustainable Transport. Organised by the European Science Foundation and the U.S. National Science Foundation, February 1999 in Berkeley, USA

Schulze G. 1995: Die Erlebnisgesellschaft. Kultursoziologie der Gegenwart, Campus, Frankfurt, New York

Simon H. A. 1982: Models of Bounded Rationality, MIT Press, Cambridge, MA

Turrentine T. 1994: Lifestyles and Life Politics: Towards a Green Car Market, Ph.D. Dissertation UCD-ITS-RR-94-30, Institute of Transportation Studies, University of California, Davis

Verplanken B., Aarts H., van Knippenberg H. and Moonen A. 1998: Habit versus Planned Behaviour: a Field Experiment, British Journal of Social Psychology, Vol. 37, 111–12

8 Parking Policy Measures and Their Effects on Shopping and Work Trips

Jari Kurri

Helsinki Institute of Technology, Helsinki, Finland

8.1 Introduction

The supply and pricing of parking places and facilities in city centres have an important effect on travel demand in general and modal split and trip distribution in particular. While parking cost and some other parking-related variables are standard elements in urban travel demand models, there are important gaps in our knowledge of the ways in which different parking policy measures affect travel demand. Discrete choice models provide powerful and theoretically attractive tools for impact analysis. Nevertheless, there are problems in modelling that are partly related to supply-side and partly to demand-side information about parking. The present study is an attempt to fill some of those gaps.

During the last 15 years, so-named 'stated preference techniques' have been used frequently in transportation analysis. Nevertheless, there are not very many stated preference surveys that deal with factors affecting choice of parking location. In a recent paper, Hensher and King (2001) investigate, in the context of a present-day trip to the city centre of Sydney, how parking cost and parking location affect mode and parking choices. Respondents were asked to consider six alternatives, including driving to the city centre, with three different parking locations, as well as park and ride, public transport, and forgoing the trip. An earlier paper by Axhausen and Polak (1991) describes stated preference exercises that were carried out in Karlsruhe and Birmingham and that dealt with choice of parking location (including illegal parking!). Besides parking cost and walking time, the parking alternatives were characterized by two additional variables, the time spent searching and queuing for the parking space and the time spent travelling from home to the starting point of the search process.

The present chapter concentrates on the methods used in, and results of, a stated preference survey that was carried out in two case study areas in the Helsinki metropolitan area in Finland (Kurri and Laakso 2001). In the first area, in the city centre of Helsinki, there are about 36,500 parking places, of which 45 percent are on the streets, 11 percent in parking facilities (underground and multi-storey car parks etc.), and 44 percent on private sites. On-street parking is restricted to one, two, or four hours, while private sites serve mainly long-term employee or residential parking. The second case study area consists of the district centre in Tapiola, located 10 km west of the city centre of Helsinki. Tapiola new town is the commercial and recreational centre of Espoo, the second largest city in Finland. The central part of Tapiola is quite small, about 0.5 km^2 in area, and there are free parking places within walking distance. On-street parking is mostly prohibited, but there are about 2,800 parking places in parking facilities.

The general goal of transport policy in Helsinki concerns the public transport share of all motorized trips to or from the city centre (the peninsula of Helsinki). In 2000, the modal share of public transport was 63.0 percent (goal 64%) for the whole day, and 69.5 percent (goal 70%) for trips to the city centre during the morning peak hours, from 6 to 9 a.m. In the last 20 years, these shares initially dropped from 67 to 60 percent and from 75 to 69 percent, but during the last 5 or 10 years they have risen slightly. Besides a relatively good public transport system and a restrictive parking policy, there are also some external factors that have affected this development. The share of public transport decreased continuously during the 1970s and 1980s as a result of growing car ownership. During the first part of the 1990s, however, average car ownership decreased as a result of deep economic depression and high unemployment in Finland. In recent years, the population of the Helsinki metropolitan area has grown rapidly, mostly in consequence of internal migration from other parts of Finland. Many of the new citizens of Helsinki are young city dwellers who live alone and do not have a car. Factors that may explain the high share of public transport also include recent increases in the price of petrol, as well as the introduction of some new low-priced public transport tickets.

One of the objectives of the parking policy in Helsinki is to discourage long-term parking in the city centre, and consequently to favour public transport as a viable choice for car owners who work in the city centre. In Finland, cities have a monopoly on town planning, and therefore the city can regulate the supply of public parking spaces, as well as the supply of parking spaces for new building projects. Nevertheless, the cities have no direct control over existing off-street and private parking. Almost half of the parking spaces in the city centre of Helsinki are located on private sites, and hence most parking policy measures have only a limited effect on the use of parking spaces.

8.2 Data and Methods

8.2.1 Stated Preference Survey

The term 'stated preference (SP) techniques' refers to a number of different approaches, all of which use people's statements of how they would respond to different hypothetical situations presented to them ('people state their preferences') (Pearmain et al. 1991). Unlike traditional revealed preference (RP) data, SP data do not describe actual behaviour. Therefore, the consistency of SP data with actual market behaviour is always questionable. On the other hand, RP data have many deficiencies that SP data do not. For instance, measurement of explanatory variables for RP models often involves many simplifying assumptions, such as spatial and temporal aggregation, whereas with SP data the values perceived by the respondent equal the values presented to the respondent and used in modelling. Moreover, by choosing an appropriate experimental design, the researcher can, to some extent, avoid some of the typical drawbacks of RP data: (multi)collinearity (two or more variables are linearly correlated, so that it is not possible to isolate their effects), and lack of variance (in reality, travel times, for example, may not vary sufficiently).

In order to minimize various biases inherent in an SP exercise, it is important that the hypothetical situations presented to the respondents are as realistic as possible and that the attribute values relate to the respondents' actual experience. In order to achieve this, the stated preference exercises were customized to actual trips, which required a two-stage survey. During the first stage, in May 2000, car parkers were recruited by putting a questionnaire between the windscreen and the windscreen wiper of cars parked on selected streets and off-street (excluding residential parking). This means that all the respondents are current car users. The response rate in the first stage was 35 percent (of the 3,139 questionnaires distributed, 1,087 were returned by post). Altogether, 862 persons (79%) expressed a willingness to take part in the second part of the survey. However, not all of them were accepted for the second stage. For instance, if their trip was included in a complex trip chain, it was screened out. During the second stage, between late October and early November 2000, questionnaires including customized stated preference questions were sent by post to 549 persons. In all, 382 (70%) questionnaires were returned by post. Parking cost, walking distance, and other variables included in the stated preference exercise were tailored to the trip the respondent reported during the first stage. Furthermore, the names and types of origin and destination, as well as the departure time of the trip, were printed out on the first page of the questionnaire.

The SP experiment was arranged in two parts. The first part dealt with general factors that have an influence on the choice of parking location. It comprised eight choices between two hypothetical parking options characterized by four variables or attributes: walking distance from the parking location to the destination of the trip, the cost of parking, the type of parking, and the time needed to find a free parking place (Figure 8.1). The type of parking could be a parking facility (underground or multi-storey car park etc.), on-street parking, a public parking area (surface car park), and a private parking place (on a private site).

The second part included eight choices between the use of a car and public transport (Figure 8.2). The car alternative was characterized by four attributes: travel time; fuel and other variable costs; walking distance from the parking location to the destination of the trip; and parking cost. The public transport alternative was characterized by fare and a list of succeeding stages of the trip:

- walking time in minutes to the stop or station;
- type of vehicle (bus, tram, metro, (local) train);
- headway of the vehicle in minutes;
- in-vehicle time in minutes;
- walking time (in minutes) to the destination;
  or (if there was a transfer)
- walking time (in minutes) to the next stop or station (and then type of vehicle etc.).

Choice task 2			
Alternative A		Alternative B	
Walking distance	100 m	Walking distance	100 m
Parking cost	24 FIM	Parking cost	24 FIM
Type of parking	parking facility	Type of parking	on-street
Search time	6 min	Search time	2 min
I choose	☐ (alternative A)		☐ (alternative B)

Figure 8.1 Example of stated preference task: choice of parking location

Choice task 8		Public transport (succeeding stages and cost)	
Car		walking distance 150 m	
Travel time	20 min	by bus (headway 20 min) 14 min	
Fuel cost	14 FIM	walking distance 400 m	
Parking cost	30 FIM		
Walking distance	100 m		
		Fare	8 FIM
I choose	☐ (car)		☐ (public transport)

Figure 8.2 Example of stated preference task: choice of travel mode

The values of the attributes or variables were varied according to an experimental design that was partly based on fractional factorial design (for more information about stated preference designs, see, e.g., Louviere et al. 2000). Since each attribute had two levels and the levels of four variables were varied, there were, in total, $(2^4)^2 = 16^2 = 256$ different combinations in the full factorial design. Nevertheless, some of these combinations were mirror images in which alternative

A and alternative B were reversed. A fractional factorial design that reduced the number of combinations from 16 to 8 was used to select the levels (or values) of the attributes for the left-hand-side alternative (alternative A or car). The levels for the right-hand-side alternative were chosen separately, using the same fractional factorial design and some randomization in order to avoid correlation between variables.

Table 8.1 shows an example of how the levels for alternative A were varied in the first part of the stated preference exercise. (The actual experimental design was randomized, however.) Depending on the type of the attribute in question, either '0' or '1' was selected as the base level corresponding with the base value of that attribute. For instance, parking cost was increased and not decreased, and therefore '0' was selected as the base level. The base values equalled, or were at least close to, the present values. In most cases, the other value was calculated as a percentage change with respect to the base value. Absolute values were shown to the respondents, however. Because all the respondents were current car users, the car alternative was usually made worse, and public transport better, than it is now. Small changes in time components were not allowed.

Table 8.1 An example of the experimental design for the part of the exercise concerning the choice of parking location

	1	2	3	4	5	6	7	8
Walking distance	0	0	1	0	0	1	1	1
Parking cost	0	1	0	1	1	0	1	0
Type of parking	0	0	1	1	0	0	0	0
Search time	0	1	1	1	0	1	1	0

Although factorial designs are routinely used in the design of experiments in many fields of science and the arts, they are not very well suited to logit analysis. Unlike classical regression analysis, in the case of maximum likelihood estimation and logit models the so-called 'design matrix' is a function of the unknown parameters, and therefore simulation and other techniques may be needed in order to find the best design. Nevertheless, it should also be remembered that statistical optimality (uncorrelated variables, etc.) is not the only criterion, since the values of the attributes are presented to humans and therefore there are some constraints, such as plausibility and feasibility.

Public transport trips often include several stages with different vehicles. In the present SP survey, the maximum number of transfers was one, and therefore the maximum number of succeeding stages of the trip was seven (walk, wait, ride, walk, wait, ride, walk, excluding the concealed waiting time that may occur at the origin or destination of the trip). The number of transfers and succeeding stages and vehicles were chosen separately for each trip and each respondent. However, they were not varied across the eight choice tasks presented to each respondent.

Although these simplifications were necessary for practical reasons, they have some undesirable effects on modelling. In addition, walking times to or from metro or railway stations were not varied, since the present survey does not deal with the expansion of the metro network. For similar reasons, car travel time was not varied.

8.2.2 Modelling

Multinomial logit models were used to analyse the responses (for more information on discrete choice models and standard logit mode choice models with linear-in-parameters utility functions, see, e.g., Ben-Akiva and Lerman 1985 or Ortúzar and Willumsen 1994). Qualitative or discrete choice models are suitable for cases where the dependent variable is not continuous and quantitative, but can have only a finite number of different qualitative values. Furthermore, random utility models, such as a logit model, make use of an abstract construct called utility, which represents the relative attractiveness of each alternative. The greater the utility of the alternative, the higher the probability that the alternative will be chosen. The utility is assumed to consist of two additive parts, a deterministic part and a random part:

$$U_{in} = V_{in} + \varepsilon_{in}, \tag{8.1}$$

and the choice probability is, according to the principle of utility maximization:

$$P_n(i) = \Pr\left(U_{in} \geq U_{jn}, \forall j \neq i, j = 1,\ldots,m\right)$$

where

$P_n(i)$ = probability of choosing alternative i for observation n

U_{in} = utility of alternative i for observation n (8.2)

V_{in} = deterministic part of the utility

ε_{in} = stochastic part of the utility

m = the number of alternatives (here two).

If the random disturbance terms ε_{in} are assumed to be independently and identically distributed, and to follow an extreme value distribution named after Gumbel, it can be shown that the choice probability is:

$$P_n(i) = e^{V_{in}} \Big/ \sum_{j=1}^{m} e^{V_{jn}}. \tag{8.3}$$

Note that the stochastic terms should be independent with respect to different observations as well as to different alternatives. This assumption was not entirely fulfilled, since we used stated preference data that included several choices from each respondent.

We used utility functions that are linear with respect to the parameters to be estimated:

$$V_{in} = \beta_{i,0} + \beta_{i,1}x_{in,1} + \beta_{i,2}x_{in,2} + \cdots + \beta_{i,k}x_{in,k} \quad , \tag{8.4}$$

where $x_{in,l}$ ($l = 0,\ldots,k$ and $x_{in,0} = 1$) are variables and $\beta_{i,l}$ are unknown coefficients or parameters to be estimated by the method of maximum likelihood (the subscript *i* denotes the alternative, and *n* denotes the observation). The utility measure is without unit; that is to say, V_{in} is a pure number. Therefore, if the variable $x_{in,l}$ is expressed in minutes (e.g. travel time), then the unit of the corresponding coefficient $\beta_{i,l}$ must be 1/min. A variable associated with the parameter $\beta_{i,l}$ is called generic if the same parameter is used for all alternatives; in other words $\beta_{1,l} = \beta_{2,l} = \cdots = \beta_{m,l}$. However, most of the variables in the present study are alternative-specific. In mode-choice analysis, one of the alternative-specific constants, $\beta_{r,0}$, serves as a reference alternative since it is not identifiable and must be fixed to zero. Therefore, there are at most $m-1+mk$ parameters to be estimated, if all variables are alternative-specific. On the other hand, the models for choice of parking location do not include any alternative-specific constants, because parking alternatives A and B were indistinguishable from one another except for their attributes.

In order to study the effects of different parking policy measures on modal split, we applied the models to the actual trips on which the stated preference exercise was based. In other words, the deterministic utility components, V_{in}, were calculated according to Equation 8.4 and substituted in Equation 8.3 to get the choice probabilities of each alternative and each observation. The base values of the explanatory variables, $x_{in,l}$, that were used were as close as possible to the actual values reported by the respondents. However, the values of the search time variable were estimated. The sums over observations of the choice probabilities give the shares of the alternatives. The modal split resulting from each action (e.g. an increase of 30 percent in parking cost) was then compared with the reference case, where there was no change in any variable. Besides modal effects, we were also interested in the trade-offs between two factors, such as travel time and cost, which gives an estimate of the value of travel time. If $\hat{\beta}_{\text{car,time}}$ and $\hat{\beta}_{\text{car,cost}}$ are the parameter estimates for travel time and travel cost, the value of time is:

$$\text{VoT} = \hat{\beta}_{\text{car,time}} \big/ \hat{\beta}_{\text{car,cost}} \,.$$

In the present study, we have only two parking and two mode alternatives. As to binomial choices, we can operate with the differences in utilities. For example, the choice probability is:

$$P_n(1) = e^{V_{1n}} \Big/ \sum_{j=1}^{2} e^{V_{jn}} = e^{V_{1n}} \Big/ \left(e^{V_{1n}} + e^{V_{2n}}\right) = 1 \Big/ \left[1 + e^{-(V_{1n} - V_{2n})}\right]. \tag{8.5}$$

The right-hand side of the equation has the form of a logistic curve. In addition to logit (logistic probability unit) models, some probit (normal probability unit) models were estimated as well. Probit models are based on the assumption that the stochastic terms are normally distributed. However, we will not present any comparisons between probit and logit models, since the results were quite similar.

Parking and public transport costs were presented either as monthly costs or for a single trip. We could have simply divided monthly parking costs by 22 (the average number of working days per month) and monthly ticket price by 44 (2x22 commuter trips per month). However, another solution is to use a method, proposed by Ben-Akiva and Morikawa (1990), that is aimed at combining information from multiple data sources (e.g. revealed and stated preference data). The idea is to scale the utility functions of different data sources in such a way that the variances of random terms, or levels of randomness, are equal. This is achieved by multiplying the utility functions of all data sources (except one) by the scaling factors that are to be estimated. Unfortunately, the introduction of scaling factors makes the utility functions non-linear in their parameters. However, Bradley and Daly (1991) have developed an estimation technique that makes it possible to estimate all parameters simultaneously. This joint estimation method can be visualized by means of an artificial tree structure (Figure 8.3). The utilities of the artificial alternatives on the right-hand side of the tree have the same variances as the unscaled left-hand-side alternatives. In the present study, we have two kinds of data (either monthly or single-trip costs were presented), and therefore only one scaling factor or unknown parameter, θ, is needed.

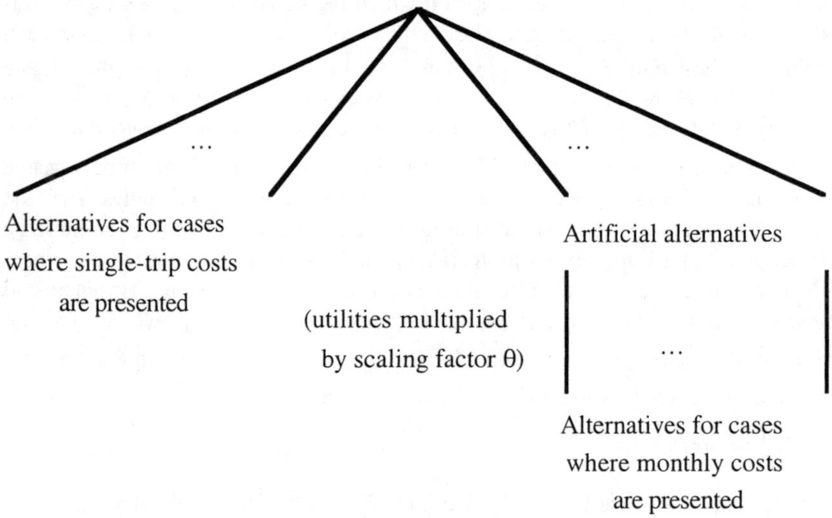

Figure 8.3 Structural model with an artificial tree

According to a key assumption behind the joint estimation method, the ratio of any two estimates of parameters that are common to both main branches (data sources) should have the same order of magnitude if calculated from separate models for each data source. In other words, the absolute values of estimates may vary, but the trade-offs, such as the value of time, should not be too unequal. This condition is not fully met since the car drivers who were shown monthly costs were not selected randomly. Monthly costs were, in most cases, related to the trip to work, and there is reason to believe that commuters have different preferences to other car drivers. On the other hand, logit modelling is always more or less based on the implicit assumption that all the respondents are homogenous with respect to their preferences, which is of course not true.

It was quickly observed that a large number of respondents found the questions irrelevant to their trip, which caused some modelling difficulties, such as high variances in the estimates of the parameters or even indications that the parameters might be wrong (wrong signs of parameters). Many respondents explained why they had made their choice. There were some exogenous (to the experimental design) explanations that occurred again and again. The situational constraints affecting respondent decisions which were not properly taken into account in the SP survey included trip chaining, need for the car during the day, heavy shopping bags or heavy objects to be carried, bad weather, and deviations in the route because of the transport of children to or from a day nursery. Therefore, the observation was not used in the mode choice modelling if the person clearly indicated that they had based their choice on some of the exogenous factors listed above and not on the level-of-service variables (the choices of 22 out of 374 respondents were rejected).

Some of the modelling difficulties were probably caused by the general heterogeneity of the observations, in particular by the great variance in parking costs. Therefore, some kind of segmentation was needed. Unfortunately, the number of observations does not enable more than two segments. Separate models were estimated for home-based work trips and other trips.

8.3 Results

8.3.1 Models

Choice of Parking Location

The respondents were asked, in the first-stage questionnaire, who paid the parking fee (the respondents themselves or the employer, etc.). It is clear that responses to changes in parking cost are quite different if the employer pays. However, special attention must be paid to cases where the respondent reported that the parking cost was zero (RP cost = 0), but did not indicate who would pay if parking was not free. There are at least four different circumstances that can be separated (based on the type and location of the parking place):

1. free parking for everyone;
2. free parking for customers purchasing goods worth more than a certain amount of money (e.g. 200 FIM);

3. employee parking where the respondent does not know or remember how much he or she pays for parking, or parking is non-transparently subsidized by the employer;
4. free parking on a private site (e.g. business meeting), excluding employee parking.

Since there were only a few observations for customer (2 above) and visitor parking (4 above), these two groups were combined.

Table 8.2 Model for choice of parking location

Variable	Coefficient	t-value
Walking distance (m)		
part between 0 and 500 m	–0.003624	–9.3
part over 500 m	–0.002307	–4.5
Parking cost		
per parking (FIM)		
self-paid, parking not free (RP cost > 0)	–0.05640	–5.0
employer pays, parking not free (RP cost > 0)	–0.02451	–3.8
free parking (RP cost = 0), would be self-paid	–0.02135	–1.0
free parking (RP cost = 0), paid by employer	–0.06231	–7.7
free parking (RP cost = 0), not clear who would pay		
parking is free for everyone	–0.2058	–4.2
customer or visitor parking	–0.1061	–5.2
employee parking	–0.1105	–11.0
other	–0.1511	–6.7
per month (FIM/month)		
paid by the respondent	–0.01352	–3.2
paid by the employer	–0.001361	–4.5
Type of parking (dummies)		
facility (reference)	0	—
on-street	–0.6303	–2.0
parking area	0.2410	0.8
private (not identifiable)	0	—
same as before (inertia)		
facility	0.4463	1.4
on-street	0.9294	2.8
parking area	–0.1626	–0.5
private	–0.04475	–0.4
Time needed to find a free parking place (min)	–0.2756	–11.0
Structural coefficient (θ)	1.601	2.9 θ=1
3035 observations	$\rho^2 = 0.1421$	

1 euro is 5.94573 Finnish marks (FIM)

Table 8.2 shows the coefficients of the best parking location model found. The parking costs that were presented to the respondents were either for a single trip or monthly costs. For self-paid parking, the ratio of the coefficients is quite low (–0.05640/(0.01352×1.601) = 2.6). This suggests that there are some non-linear effects concerning how respondents react to parking costs, since the average actual monthly cost was about 6.4 times higher than the average single-trip cost. For parking costs paid by the employer, the same figure is 17.4, whereas the ratio of the parameters is 11.2. In most cases, monthly parking costs are related to employee parking.

The coefficients for parking cost are less negative if the employer pays, as it should be. However, this is not the case for the third and the fourth cost coefficients, which were used if the actual cost was zero ('free parking (RP cost = 0)'). About one-third of the parkers who reported zero costs indicated that their parking cost was or would be paid by someone else (their employer, etc.). It is possible that many of them suspected that they themselves would have to pay for parking in a hypothetical situation where parking is no longer free. Unfortunately, the first-stage questionnaire did not make a difference between who paid and who would pay, and therefore the label of the fourth cost coefficient is 'free parking (RP cost = 0), paid by employer', and not 'free parking (RP cost = 0), would be paid by employer'. It is suggested that instead of these two cost coefficients, the cost coefficients for cases where parking was not free are used.

There are several trade-offs that can be derived from the model that are perhaps not very reasonable. For instance, (–0.05640)/(–0.003624) = 16 m/FIM shows that one Finnish mark (about 0.17 euro) equals about 16 metres in walking distance. In other words, car drivers are willing to pay about 1 euro to save 100 m of walking distance. However, the effect of walking distance on the choice of parking location is assumed to be piecewise linear in such a way that it is a concave function with respect to distance.

The time needed to find a free parking place varied between 0 and 9 minutes. The value of time that can be derived from the coefficients is unrealistically high, 49 euros/hour, or 293 FIM/hour ((–0.2756)/(–0.05640)). Therefore, the results should not be extrapolated for times exceeding 10 minutes.

The effect of the type of parking is modelled with several dummy variables. The parking facility was chosen to be the reference whose dummy was constrained to zero. The dummy for on-street parking is negative, as might be expected (car drivers prefer facilities to on-street parking, if all else is equal). It should be noted that, in the city centre of Helsinki, the alternative to a parking facility is usually on-street parking and, in Tapiola, a parking area. In addition, there were three 'inertia variables', which, by definition, are 1 if the alternative has the same type of parking as the actual (RP) choice, and 0 otherwise. These coefficients are aimed at capturing habit and similar effects. The inertia coefficients are positive in sign if the respondents are reluctant to change their type of parking.

Since there were two different ways of expressing parking costs, we used the model with an artificial tree structure. The structural coefficient, θ, shows that the variance of error term is greater for those observations where parking cost was for a single trip. This is in line with the more general finding that it is easier to model choices related to work trips than choices related to trips made infrequently.

According to the goodness-of-fit measure most often used, ρ^2, the model is not very good. The low value 0.1421 indicates that some kind of segmentation is needed. However, more than two segments and, hence, two separate models, are not possible, since there would not be enough observations. Different ways of segmentation were tested. The best one is based on the purpose of the trip, with separate models for the trips between home and place of work and for other trips. The models presented in Table 8.3 are somewhat less complicated than the model without segmentation (for instance, monthly parking cost is divided by 22).

Table 8.3 Model for choice of parking location, home-based work trips and other trips

	Home-based work trips		Other trips	
Variable	Coefficient	t-value	Coefficient	t-value
Walking distance (m)				
part between 0 and 500 m	–0.002521	–4.3	–0.004756	–10.3
part over 500 m	–0.0001224	–0.1	–0.003424	–5.5
Parking cost (FIM)				
paid by the respondent	–0.1047	–8.9	–0.08519	–7.2
paid by the employer	–0.03011	–4.0	–0.03803	–4.0
Type of parking (dummies)				
facility	0.8629	8.9	–0.08859	–1.2
on-street (reference)	0	—	0	—
parking area	0.1061	0.6	0.2275	1.8
private	–0.09524	–0.5	–0.2065	–0.6
same as before (inertia)	0.8046	8.5	0.3316	4.9
Time needed to find a free parking place (min)	–0.2682	–7.3	–0.3610	–11.9
	$\rho^2 = 0.2048$ (N=1,378)		$\rho^2 = 0.1099$ (N=1,609)	

Not surprisingly, it seems that walking distance is less important in the case of work trips than for other trips. It can also be concluded from the models that parking facilities are more important (if everything else is equal) for work trips than for other trips. Nevertheless, it is quite likely that some of the respondents did not fully understand the idea of a stated preference exercise and chose on-street parking because 'you do not have to walk so much'. On the other hand, the inertia coefficient for work trips is unrealistically high, which weakens the validity of the results obtained.

Choice of Travel Mode

The second part of the SP survey concerned mode choice. Table 8.4 presents two mode choice models that have been estimated with the choice data of 341 respondents, and not the full data set, with 379 respondents. Observations were screened out if the travel time by car was more than one hour or if it was clear that the respondent's choice was not based on level-of-service variables but on some exogenous factors. Parking cost is divided by 2, since other cost and all time components are for a one-way trip.

Table 8.4 Mode choice models without segmentation

	Simple model		Structured model	
Variable	Coefficient	t-value	Coefficient	t-value
Car time (min, up to 60 min)	–0.04873	–2.8	–0.05101	–2.8
Car fuel cost (FIM)	–0.02974	–2.9	–0.04024	–3.6
Car parking cost (FIM):				
paid by the respondent	–0.07160	–13.3	–0.07229	–12.5
paid by the employer etc.	–0.02117	–7.8	–0.02750	–7.9
per month (FIM/month)			–0.0005645	–1.7
Car walking distance (m)	–0.001876	–9.1	–0.002036	–8.9
Type of parking (dummies):				
facility (reference)	0	—	0	—
on-street	–0.5586	–4.7	–0.5346	–4.1
parking area	–0.9435	–6.5	–0.9186	–6.1
private	–1.283	–9.4	–1.322	–8.7
Car dummy (reference)	0	—	0	—
Public transport dummy	–0.6882	–2.7	–0.6273	–2.2
PT in-vehicle time (min)	–0.03923	–4.9	–0.03536	–4.1
PT waiting time (min, first 5 min):	–0.1016	–2.9	–0.09386	–2.5
PT walking distance (m)				
part below 400 m	–0.002205	–6.2	–0.002411	–6.3
part above 400 m	–0.0008424	–4.4	–0.0009569	–4.4
PT travel cost (FIM):	–0.03359	–2.8	–0.06033	–4.1
per month (FIM/month)			–0.004307	–3.0
PT number of transfers	–0.1592	–1.0	–0.2807	–1.6
Structural coefficient (θ)			0.5648	–3.6 θ=1
2,702 observations	$\rho^2 = 0.1540$		$\rho^2 = 0.1605$	

1 euro is 5.94573 Finnish marks (FIM)

The second model in Table 8.4 includes an artificial tree structure, in the same way as in the parking location choice model. The first model is simpler in structure in such a way that monthly parking cost has been divided by 22 and the price of a one-month season ticket for public transport has been divided by 44. The second-model estimate of the coefficient for monthly parking cost is quite close to zero (–0.0005645×0.5648), which can be explained by the fact that a number of respondents chose to travel by car in spite of the high monthly parking cost, usually paid by the employer. Therefore, the ratio of the coefficients for parking costs is much higher than 22: (–0.02750)/(–0.0005645×0.5648) = 86.3. On the other hand, the equivalent ratio for public transport travel cost is less than 44: (–0.06033)/(–0.004307×0.5648) = 24.8.

Table 8.5 Mode choice model, home-based work trips and other trips

	Home-based work trips		Other trips	
Variable	Coefficient	t-value	Coefficient	t-value
Car time (min, up to 60 min)	–0.02185	–1.0	–0.06390	–2.2
Car fuel cost (FIM)	–0.04081	–3.2	–0.01849	–1.0
Car parking cost (FIM):				
paid by the respondent	–0.08960	–10.3	–0.08402	–9.5
paid by the employer etc.	–0.01402	–4.5	–0.07445	–9.0
Car walking distance (m)	–0.001413	–4.6	–0.002522	–8.4
Type of parking (dummies):				
facility (reference)	0	—	0	—
on-street	–0.6515	–3.1	–0.5815	–3.7
parking area	–0.9825	–3.6	–1.072	–5.6
private	–1.022	–5.7	–2.059	–5.0
Car dummy (reference)	0	—	0	—
Public transport dummy	–0.7066	–1.7	–0.5997	–1.6
PT in-vehicle time (min)	–0.02836	–2.8	–0.05735	–4.2
PT waiting time (min, first 5 min)	–0.09883	–1.9	–0.08194	–1.6
PT walking distance (m):				
part below 400 m	–0.001617	–3.0	–0.002880	–5.5
part above 400 m	–0.0006576	–2.6	–0.001137	–3.5
PT travel cost (FIM)	–0.02445	–1.5	–0.05348	–2.4
PT number of transfers	–0.2331	–1.0	–0.3617	–1.3
	$\rho^2 = 0.1667$ (N=1,266)		$\rho^2 = 0.1969$ (N=1,436)	

The mode choice models are not very realistic in terms of the trade-offs that can be derived from the parameters. In this respect, the second model, with separate coefficients for monthly cost, is superior to the first model. Nevertheless, the value of car travel time is quite high, (–0.05101)/(–0.04024) = 1.3 FIM/min = 13 euros/hour, while the value of public transport in-vehicle time is about 6 euros per hour. According to the second model, one transfer from one public transport vehicle to another equals 8 minutes' time in the vehicle ((–0.2807)/(–0.03536)), on average. The coefficient for the number of transfers was not very stable, and the trade-off varied between 4 and 17 minutes, depending on the type of model. These figures reflect car users' valuations, since all car parkers are also car drivers. Several studies carried out in the Helsinki metropolitan area have indicated that public transport users are not so afraid of transfers; the same trade-off is typically between 4 and 8 minutes of in-vehicle time, depending on the type of transfer.

It appears that car parkers do not like to walk. According to the models estimated, the value of car walking time is 24 euros per hour ((–0.002036)/(–0.07229) = 0.03 FIM/m = 24 euros/h, 5 km/h walking speed assumed), and the value of public transport walking time is as high as 34 euros per hour. The value of public transport waiting time is about 16 euros per hour, which may be too high. Nevertheless, the ratio of the coefficient of waiting time to the coefficient of in-vehicle time is (–0.09386)/(–0.03536) = 2.7, which is within a reasonable range.

The value of goodness-of-fit measure, ρ^2, calculated with respect to alternative specific constants (car and public transport dummies), is 0.1605 (or 0.1540). The values would be acceptable if the models were based on revealed preference data (i.e. actual choices), but with stated preference data we should expect a slightly higher value. There were also some modelling difficulties that were probably caused by the general heterogeneity of the observations, in particular by the great variation in parking cost, as well as trip-chaining and other situational constraints (e.g. need for the car during the day).

Table 8.5 presents the coefficients of separate models for work and other trips. Again, it was not possible to estimate more than two separate models for two segments. Moreover, it was not possible to obtain reliable estimates of all the parameters of the second model in Table 8.4. Therefore, the first, simplified, non-structural model was used. Surprisingly, the model for work trips has a lower ρ^2 than the model for other trips. Furthermore, it is difficult to find any pattern that could explain the differences between the coefficients of the two models. Perhaps the sub-samples are too small (159 and 182 respondents, respectively), after all. Another possible explanation may be the fact that the correlation between travel time and travel cost is quite high, at least as far as stated preference surveys are concerned.

8.3.2 Results on Choice of Parking Location

The walking distance from the parking place to the destination is certainly one of the most important factors when choosing parking location or travel mode. The present stated preference survey supports this assumption, although the results are not very convincing in terms of statistical reliability. According to the models explaining the choices of parking location and travel mode, a 100-metre walk equals about 1 euro in parking cost, about 4 minutes of travel time, or one and a half minutes of the time needed to find an unoccupied parking place. In other words, car drivers are willing to pay 1 euro more in order to get 100 metres closer to their destination, or are ready for a three-minute search for a parking place in order to avoid a 200-metre walk (some other studies have indicated that car drivers are willing to spend even more time to find a place as close as possible to their destination).

According to the models, a parking facility is preferred to on-street parking, if everything else is equal. Nevertheless, habit and previous experience often determine where to park. Besides, real-time information about empty parking places, which is available for some parking facilities, may help to influence the making of the decision. Unfortunately, the present stated preference exercise does not provide any information about this kind of dynamic decision making. Last but not least, it should be noted that, in real life, car drivers often have incomplete information about the different parking options available for them, which is an important problem for modelling and impact analysis.

8.3.3 Modal Effects

In order to study the effect of parking cost on modal split, parking cost was increased by 30 percent over that in the original data gathered and the mode choice models were applied to the modified data. It was required, however, that the parking charge should be at least 3 Finnish marks (FIM) (0.50 euro) per hour for the first ten hours. This increase in parking cost resulted in an 8 to 14 percent decrease in car share; that is to say, 8 to 14 percent of present car users would switch to public transport. The change is somewhat bigger in the case of trips between the home and the workplace than in the segment for other trips. On the other hand, it is not possible to estimate changes in trip distribution (choice of trip destination) with the mode choice model only. For instance, how many shopping trips would be made elsewhere, or would not be made at all, if parking costs were increased as described above? Nevertheless, tests were conducted to investigate what would happen (concerning modal split) if the cost of parking were doubled or increased to a minimum cost of 6 FIM (1 euro) per hour. This would decrease car share by 21 to 38 percent.

In some cases, a private car can be a cheaper, as well as a faster, alternative to public transport. If the cost of parking were always at least the same as the public transport fare for regional trips (trips up to about 40 km in the Helsinki metropolitan area; in 2000 a 30-day ticket cost 340 FIM or about 57 euros), car share would decrease by 8–14 percent. The change is quite small, because 340 FIM divided by 22 days per month or about 15 FIM (about 2.60 euros) per day is quite a small amount of money for parking, at least for most of the respondents, particularly if the employer pays. If public transport fares were to be decreased by 30 percent, car share would decrease by only about 1–3 percent. On the other hand, the flat fare system with two zones (trips within one city mean one zone, trips between cities two zones) makes short trips that cross a border quite expensive, and therefore tests were conducted to investigate what would happen if the public transport fare were always the same as the cost of fuel for the corresponding car trip, or less. This change in pricing would decrease car share by 0 to 2 percent.

According to the models estimated, relatively small changes in the cost of parking are not a very efficient way to discourage car drivers from using their cars. A more efficient approach would be to regulate the locations of parking facilities, parking areas, and on-street parking so that car users would have to walk more. The problem is that the cities have almost no control over parking in private parking places, often provided by employers for their employees. Nevertheless, if the walking distance from parking place to destination were always 400 metres, car share would decrease by 8–11 percent (depending on the model). If the walking distance were the same as that needed with the public transport alternative, car share would decrease by 13 to 15 percent. This distance is often more than 400 metres, because it has not been measured from the closest stop or station but from the stop or station that is included in the optimum public transport route, bearing in mind that car drivers do not like transfers.

If public transport in-vehicle time were decreased by 15 percent, car share would decrease by about 2 to 3 percent. If all connections that at present include transfers were replaced by direct routes without transfer waiting and walking times, 10 to 12 percent of those car users who now have no direct public transport alternative (less than one-fifth of all respondents) would use public transport. This means a 1 to 4 percent decrease in car share.

The modelling is based on the assumption that all the respondents are homogeneous with respect to their preferences, which is, of course, not true. It would be better to estimate separate models for each person. This is not possible, however, since statistical models such as a logit model require hundreds of observations in order for the estimation of the model to be successful. To take account of different preferences ('taste variations'), one could study how various socio-economic factors (age, sex, income, etc.) affect choices. In the present survey, however, the questionnaire did not include any socio-economic questions, since it was considered important that the questionnaire should be easy to fill in. Nevertheless, gender could be identified because the first names of the respondents were known. No statistically significant difference could be found in a comparison between men and women.

8.3.4 Validity of Results

The value of travel time (i.e. the trade-off between time and money) and several other trade-offs between two variables can be derived from the coefficients of the models estimated. It should be noted, however, that the role of these trade-offs is twofold in modelling; on the one hand, they provide new information about people's preferences, but, on the other hand, the 'reasonableness' of the trade-offs is an important criterion for the selection of the best models and specifications. For instance, a model that produces very small or very high values of time will probably be rejected even if goodness-of-fit and other statistical measures show that the model is good. The selection of model specification is based on experience and earlier studies, and it is often more of an art than a science. Therefore, any final conclusions should not be drawn on the basis of a single study alone.

Table 8.6 Mode choice model, respondent-specific constants

	Model with respondent-specific constants		Model without respondent-specific constants	
Variable	Coefficient	t-value	Coefficient	t-value
Car time (min)	–0.08064	–2.7	–0.04350	–2.7
Cost (FIM):				
paid by the respondent	–0.1198	–14.4	–0.05641	–11.6
paid by the employer	–0.04123	–9.0	–0.02095	–7.3
Car walking distance (m)	–0.002137	–6.5	–0.001563	–7.3
Type of parking (dummies):				
facility (reference)	0	—	0	—
on-street	–1.103	–4.4	–0.4675	–3.5
parking area	–3.175	–9.5	–1.013	–6.7
private	–1.930	–5.0	–1.050	–7.3
Car dummy (reference)	0	—	0	—
Additional dummies for:...				
women	–0.3705	–1.8	–0.03907	–0.4
respondent-specific	(176 coefficients)		(no coefficients)	
Public transport dummy	–1.486	–2.6	–0.5197	–1.9
Additional dummies for				
business trips	1.189	4.6	0.6774	5.3
trips to Tapiola	–0.8843	–2.4	0.01489	0.1
PT in-vehicle time (min)	–0.1303	–7.1	–0.05229	–5.2
PT waiting time (min)	–0.02733	–1.2	–0.02511	–1.9
PT walking distance (m):				
part below 400 m	–0.003885	–6.4	–0.002333	–6.1
part above 400 m	–0.0005875	–1.8	–0.0006729	–3.0
PT number of transfers	–0.05995	–0.2	–0.2350	–1.6
	$\rho^2 = 0.3172$ (N=2,436)		$\rho^2 = 0.1321$ (N=2,436)	

One of the drawbacks of stated preference data is the fact that the observations are not entirely independent, since there are typically about ten (here eight) choices from each respondent. This contrasts with the key assumption of a logit model, which states that the random utilities associated with each observation and each alternative are identically and independently distributed. It is widely recognized that this dependence results in downward-biased standard errors of parameter estimates. However, possible effects on parameter estimates are still often neglected in practice. Bates and Terzis (1997) have shown that this can have quite severe practical consequences in cases where stated preference data with repeated observations are used to estimate logit models of mode choice. Nevertheless, the problem can be partly avoided if observation-specific constants (e.g. car dummies) are used. Unfortunately, it is not possible to estimate a respondent-specific constant for a particular alternative if the respondent has not chosen that alternative at least once. This is something that should be taken into account when designing the experiment. In the present stated preference survey, the experimental design failed in this respect; in a test model (Table 8.6) only 60 percent (176 of 291) respondent-specific constants could be estimated. Due to different scales, the coefficients of the model with, and the model without, respondent-specific constants are not directly comparable. Nevertheless, the ratios of the coefficients show that the key parameters do not differ significantly and therefore it seems that the data do not suffer very much from this kind of violation of the key assumptions of a logit model.

8.4 Discussion

From a methodological point of view, the two-stage survey worked quite well. The response rate to the first stage was quite low, but this could have been expected in the light of the fact that questionnaires were put between the windscreen and the windscreen wiper in the same way and place as traffic wardens put parking tickets. Of course, there are several things that can be criticized with hindsight. For instance, the respondents were not asked who paid their fuel costs or if they had a company car. In addition, more detailed information about different kinds of 'free' and subsidized parking (employee parking, customer parking, etc.) was apparently needed in modelling. Nevertheless, some compromises had to be made in order to keep the first-stage questionnaire as short as possible.

The modelling results obtained are, with a few exceptions, quite reasonable. It must be admitted, however, that some of the models that were estimated with the data gathered are not very satisfactory in terms of statistical reliability. For instance, there were problems with the correct signs of the coefficients, which is not typical of stated preference surveys. Some of the modelling difficulties can be attributed to the experimental design, in particular to the fact that car travel time and the number of transfers were varied across individuals but not across the eight choices presented to each respondent. Furthermore, the correlation between travel time and travel cost is quite high in some model types, depending on whether the cost of parking is included in total travel cost or whether there are separate parameters for parking and fuel costs. The reason for the relatively high correlations is the fact that the variables were customized to actual trips and that they were varied only to such a degree that they could still be regarded as realistic. Nevertheless, many of the problems stem from the general heterogeneity of the respondents and of the trips they were making at the time of recruitment. In addition, trip chaining and other situational constraints could not be properly taken into account in the survey and modelling.

Acknowledgements

The author wishes to thank Helsinki Metropolitan Area Council (YTV) and the Ministry of Transport and Communication for financing the project as a subcontractor to YTV in the COST 342 project titled 'Parking policy measures and their effects on mobility and the economy', as well as a part of LYYLI, the Finnish research and development programme for an environmentally-friendly urban form and transport system, 1997–2001.

References

Axhausen K.W. and Polak J.W. 1991: Choice of Parking: Stated Preference Approach, Transportation, Vol. 18, 1, 59–81

Bates J.J. and Terzis G. 1997: Stated Preferences and the 'Ecological Fallacy', in European Transport Forum 1997, Transportation Planning Methods, Proceedings of Seminar F, Volume P415, PTRC Education and Research Services Ltd., London, 155–171

Ben-Akiva M.E. and Lerman S.R. 1985: Discrete Choice Analysis: Theory and Applications to Travel Demand, MIT Press, Cambridge, MA

Ben-Akiva M.E. and Morikawa T. 1990: Estimation of Travel Demand Models from Multiple Data Sources, in Proceedings of the 11th International Symposium on Transportation and Traffic Theory, Yokohama, July 1990

Bradley M.A. and Daly A.J. 1991: Estimation of Logit Choice Models using Mixed Stated-preference and Revealed-preference Information, Paper presented at the 6th International Conference on Travel Behaviour Research, Quebec, May 1991, also in Stopher P.R. and Lee-Gosselin M. (eds.) 1997: Understanding Travel Behaviour in an Era of Change, Pergamon, Elsevier, Oxford, 209–231

Hensher D.A. and King J. 2001: Parking Demand and Responsiveness to Supply, Pricing and Location in the Sydney Central Business District, Transportation Research Part A: Policy and Practice, Vol. 35, 3, 177–196

Kurri J. and Laakso J.-M. 2002: Parking Policy Measures and their Effects in the Helsinki Metropolitan Area (partly in Finnish, partly in English), Helsinki Metropolitan Area Publication Series C 202:18

Louviere J.J., Hensher D.A. and Swait J.D. 2000: Stated Choice Methods: Analysis and Applications, Cambridge University Press, Cambridge, UK

Ortúzar J. de D. and Willumsen L.G. 1994: Modelling Transport (second edition), John Wiley & Sons, Chichester, West Sussex, UK

Pearmain D., Swanson J., Kroes E. and Bradley M. 1991: Stated Preference Techniques: a Guide to Practice (second edition), Steer Davies Gleave and Hague Consulting Group

9 Acceptability of Marginal Cost-based Pricing in Urban Transport

Jens Schade

Dresden University of Technology, Dresden, Germany

9.1 Introduction

For a variety of reasons, economists strongly recommend the implementation of differentiated pricing strategies in transport as well as in other domains of every day life. Technological research has already solved many of the practical requirements for the introduction of flexible urban transport pricing schemes. However, apart from technological and legal problems, the lack of public acceptability is recognized as one of the main obstacles to the implementation of road pricing measures. Likewise, the acceptability of systems is assumed to be a major factor influencing effectiveness of the implementation of a system (Van der Laan 1998). Numerous studies indicate the rejection of pricing measures by a majority of car drivers (e.g. Bartley 1995; Jones 1991a, 1991b; Luk and Chung 1997). For instance, results of the TransPrice-study[1] show that – with only little variation between cities – car drivers most accept public transport improvements, followed by the provision of more park and ride facilities and, surprisingly, then by access restrictions for inner cities (Schlag and Schade 2000). Reducing parking space has much lower acceptability, but least accepted generally are fees concerning both the stationary traffic (parking fees) and, most strongly refused, driving on urban roads (distance-based pricing), even if it is only imposed at times with high traffic density (congestion pricing/peak-load pricing). These findings raise two questions: a) How high is the current acceptability level of different road pricing strategies?; and b) Which factors influence the degree of acceptability? The chapter is organized as follows: In Section 9.2, a brief overview about previous acceptability research is given. The theoretical framework for acceptability analysis is outlined in Section 9.3 where the main factors influencing the acceptability of pricing strategies are reported. Section 9.4 presents the method used for the public acceptability survey carried out in Athens, Como, Dresden and Oslo. Then, the marginal cost-based pricing strategies which were to be evaluated by the respondents and the examined sample are described. The results are reported in Section 9.5. The descriptive results mainly deal with the present acceptability level of the different pricing strategies. The question of which factors influence the degree of acceptability is investigated by using multivariate statistical methods. Finally, in Section 9.6 the results are summarized and discussed.

9.2 Previous Research

Since the late 1980s, questions concerning the acceptability of road pricing measures have become increasingly relevant. Giuliano (1994) accredits this to the renewed interest in implementing pricing measures in urban areas. At the same time, there are still significant barriers to road pricing. A variety of public concerns regarding electronic road pricing systems have already been identified: for example, misunderstanding (complexity), equity/fairness issues, privacy issues, or tax resistance (see e.g. Giuliano 1994; Jones 1998; May 1992; Goodwin 1989). More recently, public opinions about road pricing have been studied with the help of attitude surveys, which identified possible additional factors (Jones 1991a, see below). But so far, in particular, statistical analyses of the interdependence of the investigated factors with acceptability have been missing.

Bartley (1995) presents the results of the EU project MIRO which, on the one hand, aimed to measure users' awareness of urban traffic problems and, on the other hand, explored their awareness, perceived effectiveness and acceptability of various policy measures. The main conclusion is that user awareness is loosely related to user acceptance (subject to belief in the effectiveness of the relevant scheme), while only few systematic interrelationships were found between socio-economic, location-based or mobility-related variables, on the one hand, and acceptability of travel demand management (TDM) measures, on the other. A similar study was conducted by Rienstra et al. (1999), who found a relation between problem perception, perceived effectiveness and acceptability. Jakobsson et al. (2000) found an income effect on acceptability, but emphasize the pervasive influence of psychological factors on the acceptance of coerced reforms. In summary, current research has identified a range of criteria and conditions which may influence road pricing acceptability. However, the few studies which have tried to determine acceptability conditions empirically are lacking a sufficient theoretical basis to satisfactorily describe and explain the phenomena.

In other traffic-related fields, the acceptability construct has been researched more closely, especially for the implementation of new technical systems, such as automatic violation registration-systems or speed control-systems (e.g. Rothengatter et al. 1991; Van der Laan 1998) as well as for technological innovations in general (e.g. Davis 1986; Rogers 1995). Here, the acceptability of the respective systems has been seen to be primarily determined by attitudes and influenced by additional system-specific characteristics. In many cases, the social psychological attitude-theory of planned behavior (Ajzen 1991) which describes the relation between attitudes and behaviour, has been used as a theoretical basis.

Although, at present, a general acceptance theory does not exist, it is undisputed that attitudes are of great relevance for agreeing or disagreeing with something.

9.3 Theoretical Background

This section reviews the main factors which contribute to the overall low acceptability of pricing measures in transport and, hence, hinder its introduction. The presentation follows the heuristic model of Schlag (Schlag and Teubel 1997; Schade and Schlag 2000) which defines the following essential issues determining acceptability:

1. Problem perception,
2. Important aims to reach,
3. Mobility-related social norms,
4. Knowledge about options,
5. Perceived effectiveness of the proposed measures,
6. Equity (Personal outcome expectation),
7. Attribution of responsibility,
8. Socio-economic factors.

9.3.1 Problem Perception

The perception of traffic-related problems is a necessary precondition for regarding problem-solving measures as important (Steg and Vlek 1997). It is assumed that high problem awareness will lead to increased willingness to accept solutions for the perceived problems. A multitude of studies have shown that perception of mobility-related problems is particularly high in densely populated regions (e.g. Jones 1991a). Yet empirical findings on the influence of problem perception on acceptability are inconsistent. Although several studies found a relationship between problem perception and the acceptability of various pricing measures (Rienstra et al. 1999), other results show that it is precisely the groups who perceive traffic congestion as one of the biggest problems reject road pricing more strongly than groups perceiving mainly environmental problems (e.g. Harsman et al. 2000; Schade 1999). Therefore a distinction is made between traffic-related problem perception (like congestion, etc.) and more environment-related problem perception (e.g. air pollution, noise, etc.).

9.3.2 Important Aims to Reach

There are different general aims connected to road pricing measures (e.g. financial aims, ecological aims, demand management, etc.). On the one hand, motorists as well as others pursue certain mobility aims. The potential conflict between these perhaps different aims is crucial for the question of acceptability. The problem is that, in principle, a large number of competitive aims and interests is imaginable. Derived from the concept of 'social dilemma' social vs. personal aims are distinguished (Dawes 1980). It is assumed that more socially-orientated aims, such as preference for bus and railway or air improvements, are in accord with those of road pricing. On the other hand, personal aims of gain maximization, like 'I would like to use my car, whenever I want', compete with those of road pricing. It is to be expected that a higher valuation of common social aims will be positively related to the acceptability of road pricing, while pursuing personal and gain maximizing aims may lead to a refusal of road pricing, because of the threatened restriction of personally important aims (cf. Schade and Schlag 2000; Jaensirisak 2001).

9.3.3 Mobility-related Social Norms

Perceived social norms and perceived social pressure refer to perceived opinions of significant others (family, friends) multiplied by the importance of others' opinions for the individual (cf. Ajzen 1991). More precisely, social norms refer to the respondent's assumption about whether his significant other would think that he should accept the strategy. These normative beliefs are concerned with the likelihood that important referent individuals or groups approve of performing a given behaviour. Thus, the more favourable the perceived social norm (pressure) is with respect to a presented pricing strategy, the stronger should be an individual's acceptability of the strategy (cf. Jakobsson et al. 2000; Kitamura et al. 1999).

9.3.4 Knowledge of Options

Environmental awareness research has established that knowledge about the right action is a necessary but not sufficient prerequisite for conservation-conscious behaviour (Bell et al. 1990). So, any new measure that is to change the demand for traffic depends on good user information. Among other issues, the background of the problem, the aims of the measure, and its concrete realization have to be clear (Schlag 1998). Although this causal connection has not been settled yet, previous studies show that well-known demand management measures meet with a higher rate of acceptability than unknown measures (MIRO 1995; Schlag and Schade 2000). It is usually hypothesized, that a higher effectiveness evaluation depends on how well-known the measure is, and that this effectiveness-judgement has an influence on the acceptability. But findings are inconsistent. Steg and Vlek (1997) found that knowledge has a negative effect, with a lot of information leading to a higher assessment of effectiveness but, compared with a less-informed control group, to a significantly lower acceptability of restrictive measures. In psychological terms, a distinction must be made between whether a person feels well or badly informed or whether he or she actually is well- or badly-informed.

Hence, a differentiation is made between what is called *objective* knowledge and the *subjective* assessment of the individual's own knowledge.

9.3.5 Perceived Effectiveness

If someone recognizes traffic problems and their consequences (problem perception), and identifies at least in part the aims of changing these problems (reducing traffic congestion, declining environmental damage etc.), he or she has to answer the crucial question of whether the proposed measures are of appropriate effectiveness and efficiency. *Effectiveness* refers to the degree to which the aims of the measure can be reached. *Efficiency*, on the other hand, means the cost-benefit-relation of a concrete measure (e.g. road pricing) compared with other possible measures (e.g. access restriction). Because the efficiency criterion – due to its complexity – is difficult to investigate or to communicate, up to now, it is mostly perceived effectiveness that has been examined (MIRO 1995; Steg and Vlek 1997). Findings show a lower degree of perceived effectiveness (e.g. to reduce traffic congestion) being ascribed to road pricing than to restrictive measures, such as closing inner cities for car traffic, or to measures which improve public transport (Luk and Chung 1997). Here, lower scores in perceived effectiveness usually correlate with lower acceptability of the particular measure, and vice versa. But so far, facts defining the direction of this relationship are missing (cf. Bartley 1995). Rienstra et al. (1999, p. 190) mention 'that strategic responses on perceived effectiveness may occur when respondents try to justify their rejection of painful policy measures by claiming that they perceive them as ineffective'.

9.3.6 Equity

When discussing the introduction of road pricing measures, arguments often arise, which question the equity or fairness of such a system (Teubel 2000). Here, equity primarily refers to the distribution of costs and benefits (Giuliano 1994). From a psychological point of view perceived justice – among other things – is of major concern as a basic requirement for acceptability. Justice, as people perceive it, may differ from objective distribution of costs and benefits but certainly depends on it as one major parameter influencing personal perceptions. And, as with most personally-mediated perceptions, it differs not only between different situations (intra-individual variance) but also between people in the same situation, and even between people with comparable objective costs and benefits (inter-individual variance). Therefore, besides rational cost-benefit-calculations, additional variables, which also influence the personal cost-benefit-ratio, must be taken into account. However, in the present study equity is tentatively operationalized as 'personal outcome expectations'. It is expected that the more people/respondents perceive advantages following the introduction of road pricing, the more they will be willing to accept it.

9.3.7 Attribution of Responsibility

The construct of responsibility attribution stems from the norm activation theory of Schwartz (1970) and represents a central component for explaining altruistic behaviour. In previous years, this model has been used in studies which attempt to explain environmental behaviour (Homburg and Matthies 1998). It is assumed that environment-preserving behaviour becomes more likely if persons perceive damaging consequences of their own actions on the environment and on others, and at the same time ascribe the responsibility for the consequences to themselves. In the context of acceptability of road pricing, attribution of responsibility appears to be a precondition for perceiving problems caused by traffic (problem awareness). It is possible to distinguish between the (attribution of) responsibility for problem causation, on the one hand, and the responsibility for problem solving on the other. It may also be questioned to whom the responsibility to solve the problems is attributed. On the one hand, it can be ascribed to the individual, i.e. internal. But responsibility may also be attributed externally (e.g. to the state). If the individual is considered as at least partly responsible for solving the problems, this should lead to increased agreement with measures raising the price, or restricting the use of the car. But, if only external instances are regarded as responsible for the solution of traffic problems, a negative effect on the support for road pricing measures can be expected. Because many traffic problems, such as traffic congestion, are often integrated in a complex system of interdependent relations and effects, it is to be assumed, that responsibility attribution can depend on both internal and external factors simultaneously. This means that someone who is aware of his own responsibility, may equally consider other (external) instances as responsible. Steg and Vlek (1997) report first results showing a positive connection between responsibility attribution and TDM-acceptability.

9.3.8 Socio-economic Impacts

It may be assumed that the acceptability of road pricing also depends on socio-economic status. Following economic theory, it is to be expected that low income groups should be more opposed to road pricing because of their higher marginal utility of money, and their decreased willingness to pay to reduce externalities (Rienstra et al. 1999). Conversely, road pricing is expected to receive more support from higher income groups because their value of time is higher and their marginal value of income generally lower. Overall, it is assumed that the advantages of road pricing will be mainly of benefit for the better-off (Banister 1994). Thus, higher acceptability levels are to be expected here. Strikingly, in their empirical analysis, Rienstra et al. (1999) found that it was the lowest income group who perceived pricing measures as most effective, and, also, that income level had no significant impact on the support for pricing measures. Further, they examined the relation between other socio-economic variables and model variables like 'problem awareness', 'perceived effectiveness' and 'support of policies'. They summarized "that the support for policy measures is influenced to a lesser extent by the personal features of respondents" (Rienstra et al. 1999, p.198). All in all, contrary to economic theory, the effects of income and related socio-economic variables on the acceptability of road pricing are expected to be rather low.

9.4 Method

9.4.1 Questionnaire

The survey was conducted in Athens (Greece), Como (Italy), Dresden (Germany) and Oslo (Norway). The questionnaire reflects the different variables outlined in the acceptability model. Earlier experiences with questionnaires in the EU projects MIRO (Vierth and Göthlin 1995) and TransPrice (Schlag and Schade 2000) have been integrated. For a more detailed description of the background of the questionnaire see Schade and Schlag (2000). All variables mentioned in the previous section were operationalized by a four-graded scale from 1 (e.g. 'know nothing at all', 'absolutely unacceptable') to 4 ('know a lot', 'totally acceptable') with the exception of: personal outcome expectations (equity), which has a three-graded scale, i.e. -1 ('expected disadvantages'), 0 ('no change') and +1 ('expected advantages').

9.4.2 Sample

The sample consisted exclusively of *motorists*. A quota sampling was used which consider demographic criteria with regard to age, gender, occupation and place of residence (see Table 9.1 below). Respondents were selected at random within the constraints of the quota.

In each of the four cities, the samples were drawn from the local telephone directory. Subjects were contacted by phone and asked whether they possessed a valid driving licence for private vehicles and were willing to take part in the survey. If this was the case, a questionnaire was sent to respondents. After completing the questionnaires, respondents were invited to return them. Surveys were carried out from December 1998 to January 1999. Due to specific local conditions, the procedures for selecting the sample in the four different cities varied slightly. In Athens, Como and Dresden, surveys were conducted by the local AFFORD partner. In Oslo the Norwegian Gallup Institute carried out, the survey. In Athens, personal interviews were carried out for cultural reasons. The total sample of the four cities (Athens, Como, Dresden, Oslo) is not representative. However, the sex distribution of the sample is a good approximation of the ratio of active car drivers of women and men (but not their respective population share).

Table 9.1 Sample characteristics

	Total	Athens	Como	Dresden	Oslo
N	954	150	238	281	285
Sex (male)	61.9%	61.7	64.9	64.5	56.9
Age (years)	M= 44.3	36.5	45.6	45.4	46.3
	SD = 14.6	11.5	13.6	15.5	14.7
Employment status					
Self-employed	12.8%	26.7	19.1	5.3	7.7
Employed	57.2%	54.0	48.7	54.1	69.2
Student/trainee	6.2%	8.7	3.8	7.5	5.6
Pensioner	14.6%	2.0	17.4	20.6	13.0
Unemployed	2.6%	1.3	0.4	7.5	0.4
Other	6.6%	7.3	10.6	5.0	4.1
Household's gross income per month					
<1000 EUR	15.7%	34.7	11.7	20.3	5.0
1001-3000	49.9%	48.0	60.2	58.0	34.4
3001-6000	19.9%	10.0	11.0	13.9	38.2
>6000	8.6%	6.0	5.0	2.1	19.3
Missing	5.9%	1.3	12.3	5.7	3.2
Usual travel mode to work					
Car	67.1%	60.7	68.2	64.8	52.6
Public transport	20.9%	30.7	11.0	13.5	25.3
Bicycle	6.3%	2.7	4.2	5.7	8.8
Walk	5.6%	6.0	7.2	2.8	5.3

9.4.3 The Road Pricing Strategies

The detailed pricing strategies (policy packages) presented in the public acceptability surveys have been developed by Milne et al. (2000, pp.71-79), who discuss the economic rationale for designing marginal cost based policy packaging, the general categories of policy packages and who give examples of actual packages. Two packages (strategy A and B) were applied in all cities, as described in Tables 9.2 and 9.3.

A. 'best practice second best', the 'strong' package,
B. 'acceptable', the '*weak*' package.

Strategy (*A*), the so-names 'strong' package comprises time-differentiated cordon pricing and an increase in parking charges and fuel taxes as well as some way of revenue redistribution. Strategy (*B*) the 'weak', but acceptable package also – but to a lesser extent – comprises cordon pricing as well as an increase in parking charges and fuel taxes, and revenue redistribution. There were minor differences in the packages between Oslo and the other cities. This is due to the current toll ring in Oslo.

Table 9.2 'Best practice second best' – Pricing strategy A

"Charge motorists toll cordon with charges of 2 EUR during the morning peak (7.00 - 9.00 a.m.) and 0.5 EUR thereafter, parking charges increased by 0.5 EUR/hr, fuel taxes increased by 0.5 EUR/litre, and use the revenues: two-thirds to lower labour taxes, one-third to invest in capacity expansion of known road traffic bottlenecks."

Table 9.3 'Acceptable' – Pricing strategy B

"Charge motorists toll cordon charges of 1 EUR at all times (including nights and weekends), parking charges increased by 0.25 EUR/hr, fuel taxes increased by 0.125 EUR/litre, and use the revenues: one-third to lower fixed vehicle taxes, one-third to invest in capacity expansion of known road traffic bottlenecks and/or to improve parking facilities, one-third to improve the quality of public transport."

9.5 Results

9.5.1 Descriptive

Table 9.4 summarizes the descriptive results (mean values) for the evaluation of the two pricing strategies by survey respondents. As can be seen for the total sample, the subjective *knowledge* levels for both strategies are very low. This is not surprising because the strategies are new. However, there are differences between the sites. There is very little knowledge in Dresden and Oslo and rather little in Athens and Como. This gap could be due to the fact that: a) in Oslo, the question referred to the objective knowledge because of an already-existing pricing system, and b) in Dresden, as part of a former socialist country the knowledge about *pricing* instruments is generally low.

Table 9.4 Overall evaluations of strategies A and B

	Strategy	Knowledge	Perceived effectiveness	Personal outcome expectations (equity)	Acceptability
Total sample	A	1.50	2.39	-.21	1.80**
	B	1.48	2.34	-.16	2.22
Athens	A	1.70	2.51	.07*	1.96**
	B	1.69	2.56	.30	2.29
Como	A	1.92**	2.23*	-.39*	1.80**
	B	1.72	2.38	-.28	2.17
Dresden	A	1.32	2.37	-.60	1.65**
	B	1.39	2.37	-.37	2.07
Oslo	A	1.23	2.50**	.16**	1.85**
	B	1.27	2.15	-.11	2.38

All mean values can vary from 1 (e.g. know nothing at all, absolutely unacceptable) to 4 (know a lot, totally acceptable) with the exception of personal outcome expectations (equity) where values can vary from -1 (expected disadvantages) to +1 (expected advantages).
Wilcoxon Signed Ranks Test:
* Difference between strategy A and B is significant at the 0.05 level.
** Difference between strategy A and B is significant at the 0.01 level.

The Athens and Como respondents – on the other hand – seem to feel better-informed about the strategies. The *effectiveness* evaluation of the two strategies with respect to the reduction of inner city traffic is much higher than the knowledge level. Thus, respondents may believe that demand management is to some extent capable of successfully addressing current transport problems, and, accordingly, that the public is prepared to trust these measures even if they are new and unknown (Schlag and Teubel 1997). In Athens and Como, the perceived effectiveness of strategy B is higher than that of strategy A (but only in Como slightly significant); in Dresden, it is the same; and in Oslo, the stronger strategy A is perceived as more effective than strategy B. Again, this finding could be due to the fact that the respondents of the Oslo sample have experience with an existing pricing scheme, allowing a more realistic evaluation of the effects than in the other cities, where the evaluation may be very subjective.

Concerning *personal outcome expectations* (equity), for the total sample there are no significant differences in the expectations about personal (dis-) advantages between the two packages. All in all, the majority of respondents tend to expect disadvantages from the introduction of the two strategies. However, there are differences between the sites. In Athens, Como and Dresden, respondents expect more advantages from strategy B than of strategy A. In Athens, personal outcome expectations generally tend be rather positive, with even a majority expecting advantages of the 'acceptable' strategy. Only the Oslo respondents expect more personal advantages of strategy A than of strategy B.

In general, both strategies are hardly *acceptable* for the respondents (see Table 9.5). As expected, rejection is stronger for strategy A. But there is a significant increase in support from strategy A to strategy B. Nevertheless, even the 'acceptable' strategy B is still not accepted by a majority. In Dresden, rejection of both strategies is strongest. In Como, attitudes towards both pricing strategies are also very negative. In Oslo, there is a strong refusal of strategy A but strategy B is more accepted. In Athens, attitudes towards both strategies are in general, less negative.

Table 9.5 Acceptability (% who rated the strategy as rather or totally acceptable)

Strategies	Support in %				
	Total	Athens	Oslo	Dresden	Como
A	20	25	24	17	15
		Oslo	Athens	Como	Dresden
B	39	48	43	34	31

Thus, even if a package solution with transparent revenue hypothecation receives stronger support than single pricing measures (Jones 1991b; Schlag and Schade 2000), the packages tested here find no majority among the motorists. However, compared with the 'strong' strategy A, the acceptability of the 'weak' strategy B has nearly doubled. On the basis of the descriptive figures above, the increase of acceptability from strategy A to strategy B cannot be explained. Variables do not vary significantly between the two strategies overall. However, there are some rather unsystematic differences on the city level. To analyze the relations between the different variables further statistical investigations on a multivariate basis are necessary.

9.5.2 Multivariate Results

A two-step procedure was adopted. First, the extensive data was reduced to an appropriate minimum by a factor analysis. Variables such as 'problem perception', 'important aims to reach' and 'the internal attribution of responsibility' are based on a multitude of items, which cannot all be examined with regard to their relation with acceptability. The results of the factor analyses were satisfying, and are presented and discussed in greater detail in Schade and Schlag (2000).

In a second step, regression analyses were applied to investigate which variables contribute to the explanation of acceptability. In more detail, it has to be ascertained:

- whether the acceptability of road pricing strategies is related to the perception of problems resulting from traffic. The hypothesis is that a high problem perception leads to increased dealing with options for solving the perceived problems;
- whether a higher valuation of common social aims leads to a higher level of acceptability of road pricing and whether pursuing personal and gain maximizing aims results in a refusal of road pricing, because of a threatened restriction of personally important aims;
- whether a higher level of subjective knowledge leads to an increase of acceptability for pricing strategies;
- whether lower scores in perceived effectiveness correlate with lower acceptability for a particular measure and vice versa. Or, does a strategic response on perceived effectiveness occur because respondents try to justify their rejection of painful policies by claiming that they perceive them as ineffective?

Table 9.6 Stepwise multiple regression analysis of the acceptability of strategy A

Predictor Variables	R^2	R_A	β
Social norm	.215	.355**	.354
Perceived effectiveness	.262	.189**	.207
Personal outcome expectations	.299	.227**	.203
Knowledge	.303	.074*	.066
Constant		.152	
F total = 100.829**; df = 4/926			
* .01 < p <.05; **p < .01			

Note: R^2: Coefficient of determination (fit), R: Regression coefficient (Estimate of the change in the dependent variable that can be attributed to a change of one unit in the independent variable); β (beta): Standardized regression coefficient.

In this study, linear stepwise analyses were carried out, which terminate when no new variable is able to improve the prediction of the dependent variable. First the predictive power of the independent variables which were examined in direct relation to the two pricing strategies were analysed. The independent variables are 'level of knowledge', 'perceived effectiveness', 'personal outcome expectations' and 'perceived social pressure' to accept the strategy (social norm). The dependent variable is *acceptability*, i.e. the degree of approval or disapproval of the respective strategy (A or B). As can be seen in Table 9.6, three significant variables qualified as accounting for nearly 30 percent of the criterion variance. A high acceptability of strategy A is to be expected: if social pressure to accept such a strategy is regarded as high; if the impact of strategy A to reduce inner city traffic is perceived as effective; and, if personal advantages following the introduction of strategy A are expected. The subjective 'level of knowledge' does not improve the equation, although it is significant. The surprising result is that social norm has the most predictive value of all variables. Besides the influence of 'social norm' only cost-benefit evaluations predict acceptability of the strategy.

The regression analysis for strategy B as shown in Table 9.7, confirms the findings above. In the case of strategy B, the three variables account for 38 percent of the criterion variance, which is a slight improvement. Again, 'social norm', 'perceived effectiveness' and 'personal outcome expectations' were significant, with 'social norm' having the strongest predictive power. The subjective level of knowledge is no longer significant.

Table 9.7 Stepwise multiple regression analysis of the acceptability of strategy B

Predictor Variables	R^2	R_B	β
Social norm	.301	.364**	.378
Perceived effectiveness	.347	.206**	.207
Personal outcome expectations	.381	.224**	.214
Constant		.522**	
F total = 189.816**; df = 3/924			

* .01 < p <.05; **p < .01

Lastly, the remaining background variables, 'problem perception', 'important mobility aims' and 'responsibility-attribution' were integrated – together with the above mentioned variables ('social norm', 'perceived effectiveness', 'personal outcome expectation', 'knowledge') – into a general model. An arithmetic mean for all variables was calculated combining the two strategies A and B. All analyses were carried out for the total sample. The results are shown in Table 9.8. The variables examined in direct relation to the two pricing strategies ('social norm', 'perceived effectiveness', 'personal outcome expectations', etc.) reveal again the highest predictive power. 'Social norm', i.e. the perceived social pressure to accept or not to accept the measure, is the most influential predictor. Besides 'social norm', especially 'personal outcome expectations' prove to be an influential predictor of acceptability: respondents who expect definite advantages show a significantly higher acceptability for the specific measures. Subjects expecting only slight advantages and subjects expecting disadvantages from the above-mentioned strategies were compared in a socio-economic check. This reveals that it is mainly subjects who drive less kilometres on average a year (r= –.24, p<.01), who do not belong to an automobile association (r= –.13, p<.01), and who prefer not to take the car to work (r= –.24, p=.01), who expect advantages from road pricing for themselves.

Table 9.8 Multiple stepwise regression analysis for the acceptability of strategy A and B

Predictor variables	R^2	R_{A-B}	β
Social norm	.236	.269**	.274
Personal outcome expectations	.317	.296**	.270
Perceived effectiveness	.343	.164**	.170
Traffic-related problem perception	.359	-.179*	-.123
General (societal) important aims to reach	.369	.113**	.083
Internal attribution of responsibility	.376	.078**	.088
Knowledge	.380	.068*	.065
Constant		.320	
F total =80.34**; df =7/918			

* .01 < p <.05; **p < .01

The effectiveness evaluation of a strategy also appears to be a variable with high predictive power. This can be related to Rienstra et al. (1999, p.190), who, as mentioned in Section 9.3.6, suggest that '*strategic responses on perceived effectiveness may occur when respondents try to justify their rejection of painful policy measures by claiming that they perceive them as ineffective.*' A tentative test of this hypothesis may be carried out when the variable 'personal outcome expectations' is included. The assumption is that persons who expect mainly disadvantages evaluate the measure as being ineffective to justify their disapproval of road pricing. Although there is a relation between the expectation of disadvantages and effectiveness evaluated as low (r = .18**), this correlation is considerably lower than the correlation between outcome expectation and acceptability (r = .47**) and between effectiveness and acceptability (r = .32**). If a partial correlation is carried out, with the impact of acceptability excluded, then, for the correlation between outcome expectations and effectiveness, a coefficient of r = .05 (n.s.) is obtained. That is, the relation between the expectation of advantages or disadvantages from road pricing and the acceptability of the strategies is not statistically related to the effectiveness evaluation of the strategies. Thus, these findings indicate that, in this study, strategic responses are not likely.

Regarding the background variables, besides 'problem awareness', especially the 'general important aims to reach' qualifies as predictor for acceptability. This means that respondents who perceive socially important aims, such as rights for cyclists or improvements in urban living conditions, significantly more clearly approve of the strategies. Concerning problem awareness, a perception of traffic-related problems like traffic congestion or lack of parking space actually appears to hinder the acceptability of road pricing.

9.5.3 Socio-economic Impacts

The following socio-economic variables were included in the stepwise regression analyses: 'sex', 'age', 'persons living in the household', 'persons under 17 living in the household', 'employment status', 'household income',[2] 'yearly driven kilometres by car', 'membership of an automobile association' and 'usual mode of transport to work' (e.g. car or bike etc.).[3] The socio-economic impact on all major variables was examined (see Appendix). The results reveal clearly that perceptions, attitudes and evaluations towards pricing strategies are influenced only to a very low extent by respondents' socio-economic status. This confirms the findings of Odeck and Brathen (1997), Rienstra et al. (1999) and Harrington et al. (2000) which all failed to find income effects. Regarding the acceptability of pricing strategies only two dummies of the 'usual travel mode to work' variable qualified. Respondents who mainly use public transport or bike accept the strategies more often than people who mainly go by car. All other socio-economic variables have *no explanatory value* concerning the acceptability variable. The explained criterion variance is very low (2%). Nevertheless, there is an interesting additional finding. Although the hypothesis that higher income groups support pricing strategies could not be confirmed, it could be shown that these groups are more likely to expect advantages from the strategies, whereas lower income groups tend to expect disadvantages.

9.6 Summary and Discussion

Several studies and research projects all over the world have considered transport pricing measures and policies – including various kinds of road and parking pricing – as promising attempts to solve the urgent traffic problems in urban areas. One important precondition for the successful implementation of pricing measures is *public acceptability*. Empirical findings have shown that the acceptability of such measures is very low. But only a few studies have tried to empirically investigate the variables influencing the acceptability of pricing measures.

As expected, it has been found that, at present, road pricing is incapable of winning major support among motorists. In general, the stated *acceptability* of both strategies: the 'best practice second-best' strategy and the strategy assumed to be quite 'acceptable', is fairly low. Predictably, rejection is stronger regarding the 'strong' strategy A. But there is a significant increase of support from strategy A to strategy B. Nevertheless, the majority of respondents do not accept the 'acceptable' strategy B either. But, there are still differences between the four sites. In Dresden, rejection of both strategies is very strong, which can be at least partly explained by its socialist past. Although this finding should be further examined, it has to be considered when introducing road pricing in the future EU-member states of Eastern Europe. Refusal is also very strong in Como. In Oslo, there is strong rejection of strategy A, but strategy B is more acceptable. In Athens, the attitudes towards both strategies are generally less negative. Thus, even if a package solution with transparent revenue hypothecation receives stronger support than single pricing measures, the packages tested here find no majority among the motorists. However, compared with the 'strong' strategy A, the acceptability of the 'weak' strategy B pricing has nearly doubled and reaches almost 40 percent, which is not too far from a majority. Furthermore, in this study only motorists have been surveyed. Thus, it is assumed that this sample reflects people with rather conservative attitudes towards the investigated issues (Schade and Schlag 2001).

Multivariate analyses were applied to investigate which variables contribute to the explanation of acceptability. The analyses showed that, in particular, the variables 'social norm', 'personal outcome expectations' and 'perceived effectiveness' are positively connected with acceptability of pricing strategies. On the other hand, the results reveal that socio-economic characteristics (e.g. income) influence the perceptions, attitudes and evaluations of respondents concerning the pricing strategies only to a minor extent. That means that the more social pressure there is to accept the respective strategy is perceived, then the more personal advantages are expected following the introduction of the measure; and that the more pricing strategies are evaluated as effective by respondents, the higher is the acceptability of a pricing strategy.

One surprising result is that 'social norm' has the most predictive power of all variables. It is assumed that, if there is a lack of experience and knowledge as a basis for the evaluation of the presented strategies, then an approach to an individual's tentative ideas and valuations and especially to perceived social valuations is to be expected. In an ill-defined but maybe relevant field, a kind of striving for concordance between own preferences and the those of important others has emerged. Most people strive for social integration and consonance (Festinger 1957). This in turn clarifies the importance of social norms. If social norms can be changed in a favourable way towards road pricing, a respective alignment of personal attitudes could be expected. The pressure towards conformity exercised by relevant others is one of the strongest factors influencing personal opinions, feelings and behavioural intentions, especially in a situation with a rather uncertain physical basis for judgement. But these influences may go in both directions; favouring or opposing pricing strategies. Today, the latter tendency is more probable. Thus, the results support the social pressure hypothesis (cf. Jakobsson et al. 2000).

Besides the social norm, 'personal outcome expectations' prove to be an influential predictor of acceptability. This is not surprising, but suggests, that the advantages of the pricing packages have to be communicated in a very credible and convincing manner. In addition, the evaluation of the measures' 'effectiveness' also predicts the acceptability of pricing strategies. But, no indications were found that respondents justify their rejection of the policies by claiming that they perceive them as ineffective. Thus, the hypothesis of a 'strategic response' on perceived effectiveness could not be confirmed in this survey (Rienstra et al. 1999). However, it is recommended that this assumption should be tested more explicitly in future studies. All other variables like 'traffic related problem perception', 'societal aims to reach', 'internal attribution of responsibility' and 'subjective knowledge' only contribute to the explanation of acceptability to a negligible extent.

The present study has not examined all factors which might influence the acceptability of transport pricing. Details of the acceptance object such as fee level, purpose, types of charging and revenue distribution have been applied by using only two different types of exemplary pricing schemes. Besides this, the results suggest that further impacts on acceptability, such as emotional variables (e.g. annoyance, reactance, price aversion) or the trust-concept from acceptability research of genetic applications (e.g. Siegrist 2000) might also contribute to an improvement of the acceptance prediction for innovative TDM. However, the results show that psychological results have a surprisingly dominating influence on the acceptability of road user fees.

Acknowledgement

The author gratefully acknowledges the contribution of all AFFORD partners whose work is reflected in this article: the University of Leeds (ITS), United Kingdom; the Institute of Transport Economics (TOI), Norway; the Politecnico di Milano (MIP), Italy; the Free University of Amsterdam, the Netherlands; the Centre for Interdisciplinary Systems Research (C.I.S.R.), Greece. The project was coordinated by the Government Institute for Economic Research (VATT), Finland.

References

Ajzen I. 1991: The Theory of Planned Behavior, Organizational Behavior and Human Decison Processes, Vol. 50, 179–211

Banister D. 1994: Equity and Acceptability in Internalising the Social Costs of Transport, in OECD/EMCT (eds.) Internalising the Social Costs of Transport. Paris, 153–175

Bartley B. 1995: Mobility Impacts, Reactions and Opinions. Traffic Demand Management Options in Europe: The MIRO Project, Traffic Engineering and Control, Vol. 36, 596–603

Bell P.A., Fisher J.D., Baum, A. and Greene, T.C. 1990: Environmental Psychology, Harcourt Brace Jovanovich College Publishers, Fort Worth

Davis F. D. 1986: A Technology Acceptance Model for Empirically Testing New End-user Information Systems: Theory and Results, Sloan School of Management, Massachusetts Institute of Technology, Cambridge, MA

Dawes R.M. 1980: Social Dilemmas, Annual Review of Psychology, Vol. 31, 169–193

Festinger L. 1957: A Theory of Cognitive Dissonance, Stanford University Press, Stanford, CA

Goodwin P.B. 1989: The 'Rule of Three': A Possible Solution to the Political Problem of Competing Objectives for Road Pricing, Traffic Engineering + Control, Vol. 39, 495–497

Giuliano G. 1994: Equity and Fairness Considerations of Congestion Pricing, in Curbing Gridlock, Vol. 2, National Research Council. Special report 242, Washington DC, 250–279

Harrington W., Krupnick A. and Alberini A. 2001: Overcoming Public Aversion to Congestion Pricing, Transportation Research Part A, Vol. 35 , 93–111

Harsman B., Pädam S. and Wijkmark, B. 2000: Ways and Means to Increase the Acceptance of Urban Rp, Deliverable D4. EU-project PRIMA, funded by the European Commission, 4th Framework Transport RTD

Homburg A. and Matthies E. 1998: Umweltpsychologie: Umweltkrise, Gesellschaft und Individuum, Juventa, Weinheim

Jakobsson C., Fujii S. and Gärling T. 2000:. Determinants of Private Car Users' Acceptance of Road Pricing, Transport Policy, Vol, 7, 153–158

Jaensirisak S. 2001. Public Valuation of Road User Charging: Selfish and Social Perspectives, Traffic Engineering + Control, 42,7, 221–225

Jones P.M. 1991a: UK Public Attitudes to Urban Traffic Problems and Possible Countermeasures: A Poll of Polls, Environment and Planning C: Government and Policy Vol., 9, 245–256

Jones P.M. 1991b: Gaining Public Support for Road Pricing through a Package Approach, Traffic Engineering + Control, Vol. 4, 194–196

Jones, P.M. 1995: Road Pricing: The Public Viewpoint, in B. Johansson and Mattson L.G. (eds), Road Pricing: Theory, Empirical Assessment and Policy, Kluwer, Boston, 159–180

Jones P.M. 1998: Urban Road Pricing: Public Acceptability and Barriers to Implementation, in K.J. Button and Verhoef E.T. (eds) Road Pricing, Traffic Congestion And The Environment. Issues of Efficiency and Social Feasibility, Edward Elgar Publishing, Cheltenham, Glos, UK, 263–284

Kitamura R., Nakayama S. and Yamamoto T. 1999: Self-Reinforcing Motorization: Can Travel Demand Management Take Us Out of the Social Trap?, Transport Policy, Vol. 6, 135–145

Luk J. and Chung E. 1997: Public Acceptance and Technologies for Road Pricing, ARRB Transport Research Ltd., Research Report No. 307, Vermont South

May A.D. 1992: Road Pricing: An International Perspective, Transportation, Vol. 19, 313–333

Milne D., Niskanen E. and Verhoef E. 2000: Operationalisation of Marginal Cost Pricing within Urban Transport, VATT Research Reports 63, Helsinki

MIRO 1995: MIRO Final Report, Deliverable 8 to Commission of the European Communities, DG XIII DRIVE Programme, Brussels

Odeck J. and Brathen S. 1997: On Public Attitudes Toward Implementation of Toll Roads – the Case of Oslo Toll Ring, Transport Policy, 4, 73–87

Rienstra S. A., Rietveld P. and Verhoef E. T. 1999: The Social Support for Policy Measures in Passenger Transport. A Statistical Analysis for the Netherlands, Transportation Research Part D, Vol. 4, 181–200

Rogers E. M. 1995: Diffusion of Innovations, Free Press, New York

Rothengatter D., Waard D. de, Slotegraf G., Carbonell Vaya E. and Muskaug R. 1991: Social Acceptance of Automatic Policing and Information Systems, Deliverable Report 1033/D7, DRIVE Project V1033

Schade J. 1999: Individuelle Akzeptanz von Straßenbenutzungsentgelten, in Schlag B. (eds), Empirische Verkehrspsychologie, Pabst Science Publishers, Lengerich, 227–244

Schade J. and Schlag B. 2000: Acceptability of Urban Transport Pricing, VATT Research Reports 72, Helsinki

Schade J. and Schlag B. 2001: Akzeptierbarkeit von Nachfragemanagement- und Preismaßnahmen in europäischen Städten, Internationales Verkehrswesen, Vol. 53, 72–77

Schlag B. 1998: Zur Akzeptanz von Straßenbenutzungsgebühren, Internationales Verkehrswesen, Vol. 50, 308–312

Schlag B. and Schade J. 2000: Public Acceptability of Traffic Demand Management in Europe, Traffic Engineering + Control, Vol. 41, 314–318

Schlag B. and Teubel U. 1997: Public Acceptability of Transport Pricing, IATSS Research, 21, 134–142

Schwartz S.H. 1970: Moral Decision Making and Behavior, in Macaulay J. and Berkowitz L. (eds.) Altruism and Helping Behavior, Academic Press, New York, 221–279

Siegrist M. 2000: The Influence of Trust and Perceptions of Risks and Benefits on the Acceptance of Gene Technology, Risk Analyses, Vol. 20, 195–203

Steg L. and Vlek C. 1997: The Role of Problem Awareness in Willingness-to-Change Car-Use and in Evaluating Relevant Policy Measures, in Rothengatter T. and Vaya E.C. (eds.), Traffic and Transport Psychology, Pergamon, Amsterdam, 465–475

Teubel U. 2000: The Welfare Effects and Distributional Impacts of Road User Charges on Commuters – An Empirical Analysis of Dresden, International Journal of Transport Economics, Vol. 27, 231–255

Van der Laan J.D. 1998: Acceptance of Automatic Violation-Registration Systems, PhD Thesis, University of Groningen, The Netherlands

Vierth I. and Göthlin L. 1995: MIRO – DRIVE II: Attitutes to Demand Mangement Strategies. Local Results for Göteborg, TFK – Transport Research Institute, Stockholm

Endnotes

[1] EU Project TransPrice: Trans Modal Integrated Urban Transport Pricing for Optimum Modal Split.

[2] It has to be noted that the highest income group consists mainly of respondents from Oslo because of the higher average income level of income in Norway.

[3] All categorical data were dummy-coded.

PART C: Freight Transport Innovations

10 Transport and Logistics as Network Competencies in a Localized Industrial Cluster

Leif Gjesing Hansen

Roskilde University, Roskilde, Denmark

10.1 Introduction

It is hardly surprising that space matters in terms of industrial organization, not only in terms of mere physical distance in relation to accessibility to supplies and markets (Weber 1909), but also in terms of accessibility to general socio-economic resources (Dicken 1992). Analyses of the spatial organization of industries and networks of firms often focus on production and distribution systems formed by the material flows of production and distribution units. However, the significance of transport operations and activities, connecting the individual production units together in a network of material flows, is often underestimated, and transport seen only as a derived demand activity.

This chapter argues that transport is more than infrastructure and transport capacity for the processes governing industrial organization. It is an integrated part of the competitive advantage for single firms' production systems, and for local networks of industrial organization. This perspective has not been sufficiently reflected in the recent renewed interest in the dynamics of agglomeration economies, which generates local clusters of industrial networks based on scale and external economies (Krugman 1991, 1995).

Theoretical developments in recent years on the theory of firms' networks and in regional economics have provided new perspectives on the dynamics of regional growth and advantages. In order to develop this argument, the area being analysed needs to be made clear. Most economic investigations use the country or the individual firm as a unit of analysis. In contrast, this chapter focuses on groups of firms, in particular on clusters of firms. A cluster is defined as a geographical and sectoral concentration of firms. Such clustering allows for efficiency gains that isolated firms can rarely attain. These gains are captured in the concept of collective efficiency, defined as the competitive advantage derived from local external economies and coordinated action (Foss 1996; Amin and Cohendet 2000; Schmitz 1999). These localized external economies and forms of joint action, through coordination of resources and distribution of competencies among firms, go well beyond functional economic input-output relations. The scope of this chapter is to present an analytical framework to explore the meaning and working of their collective efficiency.[1]

Concepts such as 'industrial districts' and 'learning regions' have been proposed to frame the localization of external economies in terms of input-output systems and processes of learning and innovation (Piore and Sabel 1984; Morgan 1997). In these approaches, the local cluster is conceptualized as a platform for the generation and coordination of resources and competencies (Lawson 1999). These analytical frameworks have predominantly been applied to empirical studies of production firms and production systems, with an explicit focus on the innovation process of different kinds of products, procurement, marketing, etc. The transport links between the individual production units within a network or production system have only been included in this analytical framework to a very limited extent. Generally speaking, transport has been treated as a means of overcoming the friction of distance by focusing on transport costs. Rarely has the focus been on the innovation and learning processes which result from interaction between actors of production and transport systems. By applying this analytical framework to a case study of the interaction between two types of sectors – furniture production and transport services in a local setting – this chapter shows that it is necessary to shift the analytical focus from generic to specific resources and assets in analyses of the role of transport in regional development. Thus, this chapter addresses the following research questions:

1. How are competencies of transport and logistics developed and reproduced in networks of production and transport firms?
2. Where are the competencies organizing the material flows and transport activities located geographically?
3. In what way do these networks of competencies depend on geographical clustering among firms?

This chapter presents results from a study of the network organization of furniture and transport firms in a territorial setting and the capabilities that are generated and sustained across industrial sectors in a localized socio-economic context. The organization of transport in relation to the furniture industry is determined by a complex relationship of the product and its transformation in a value chain, the geographical localization of suppliers, producers, retailers and transport operators, and the governance structures among the individual actors. This exploratory study identifies specific forms of transport networks, of material flows and organization of transport activities, that form local clusters of relations between relatively small firms. They contribute to a 'local transport competence' to the advantage of the furniture producers, which benefit from a specialized competence in handling furniture goods and direct transport chains from the factories to the stores or final consumers. The importance of this local transport and logistical competence seems to be greater for the competitive advantage of the local furniture industry than further improvements in the quality of the regional transport infrastructure.

In Sections 10.2 and 10.3, the theoretical perspectives on, respectively, location and territorial embeddedness will be presented to illustrate the analytical focus shift in theories and conceptions of regional development from generic to specific resources and assets. This shift should also be analytically developed in order to devote more attention to the organizational aspects of transport networks instead of merely focusing on physical infrastructure networks, as generators of economic development in specific regions. In Section 10.4, an analytical perspective will be proposed in order to highlight the close relationship between material flows, transport activities, infrastructures and governance structures in inter-firm networks. The empirical findings of the case study will then be presented in Section 10.5 and followed by a synthesis of the findings and theoretical reflections in the final Section 10.6.

10.2 Freight Transport as a Residual of Firms' Location and Trading Links

Location theory, inspired by neoclassical economics, has been an approach that tried to construct predictive models of optimal locations for individual firms in relation to their suppliers of raw material and intermediate products, on the one hand, and markets for intermediate and final product, on the other hand. The classical theories of, for example, Alfred Weber (1909) and Walter Isard (1956) developed 'space equilibrium models' based on assumptions of economic rationality. Weber (1909) assumed in his analysis that factories would be positioned with respect to raw material sources and markets at points that minimized transport costs.

The implications were that there would be a simple linear flow of products from the raw material source to the final point of sale via a single processing plant. Deductions for locational behaviour made from these premises and assumptions included the claim that a rational location for a new plant would be one that: 1) maximized proximity to raw materials where there was substantial weight loss in production, and 2) maximized proximity to the market where production resulted in a gain in weight – hence minimizing transport costs. Weber's hypothetical factories were embedded in very simple supply chains comprising one tier of suppliers upstream and one tier of customers downstream. This model-based perception of the rational locational behaviour of firms thereby oversimplifies the pattern of freight movement in many contemporary industrial sectors where factories are entwined in complex networks linking them to many tiers of suppliers, subcontractors, distributors and transport operators.

In Braunerhjelm and Svensson (1998), a review of recent theoretical developments within model-based location theory is presented, which also includes the factors of knowledge and R&D processes as determinants of firms' locational pattern. According to these authors, the major argument in recent location and trade theory is: "(...) assuming that firms implement identical technologies, location is shown to depend on the interaction of costs of production and costs of market access. It is through this link that trade liberalization, either globally or regionally, affects the location of firms and the extent to which production is geographically dispersed or concentrated" (Braunerhjelm and Svensson 1998, p. 102). This simplification is to a large degree caused by the model-based approach, which necessitates assumptions and demarcations from the complexities in the 'real' world: "(...) neoclassical location theory is not merely a branch of neoclassical economics, although it shares some common features, notably its normative approach. Normative theory is concerned with what should happen, given certain assumptions, rather than with what may actually happen in reality. To that extent, it is an 'artificial' approach based on a set of rigid assumptions" (Dicken and Lloyd 1990, p. 10). The critique of the classical location theories should, therefore, not target the theory as such, but focus on its limited understanding of the dynamics at work within more complex systems of production.

In Venables (1996), cost and demand linkages are similarly perceived as important driving forces for the agglomeration of vertically-linked firms, but the author also argues that relocation of firms is hampered by a number of immobile factors of production, and the location of final markets. These opposite working forces are termed, respectively, 'centripetal' and 'centrifugal', referring to their capability to either cause dispersion or concentration of economic activity among and within regions. Thereby, Venables (1996) modifies somewhat the more or less implicit assumption in classical location theory that improvement of transportation and communication technologies is a primary force in organizing economic activities, since production factors and processes are identical at every location. Thus, Burmeister and Colletis-Wahl (1998) underline that models of neoclassical location theory take production as "(...) a combination of generic production factors, exclusively procured through market transactions" and space as "(...) a reservoir of generic resources that can be transferred or are accessible at a certain cost" (Burmeister and Colletis-Wahl 1998, pp. 203-204). Indeed, these models do not include place-specific or time-specific factors and processes.

Without rejecting Weber's classical location model for firms, in the next section we will attempt, to extend his approach by elaborating theories and concepts that incorporate a resource and socio-economic perspective on firms' location pattern and embeddedness in specific territories. By defining transport and logistics as a part of the concept of 'production', this work seeks to widen the significance of transport for single firms and production systems as more than mere transport costs in relation to the distance and location of supply and markets.

10.3 Transport Systems as a Resource for Production Systems

Theoretical developments in recent years within theories of firm networks and in regional geography have provided a better understanding of firms' locational patterns as a condition for regional development or as a consequence of the socio-economic development within specific regional territories. Therefore, space is not only interpreted as being a neutral plane, but also represented as a social construction by economic actors located in different places. This approach has been developed within a long tradition of multidisciplinary research focusing on different aspects of local territories, ranging from industrial districts and growth poles (Marshall 1919; Peroux 1950) to local systems or networks of innovation and production (Crevoisier and Maillat 1993; Camagni 1993).

In Storper (1995), the territorial dimension, conceptualized as regional spaces, is seen as a vital element in the formal and non-formal social relationships among individuals, firms, institutions, etc. Storper's point is that the non-formal social relationships – termed *untraded interdependencies* – are not transferable to a market, but generate region-specific assets in production.[2] Such untraded interdependencies rely on various types of knowledge and practices. The firms involved are tied to various kinds of networks with other firms, both through formal and informal exchanges, through labour markets, public institutions, and local or national rules of action, customs, understandings, and values. By identifying these untraded interdependencies, this approach has seriously challenged the orthodoxy of the comparative advantage in location analyses in regional science (Camagni 1993a; Amin and Thrift 1994; Conti, Malecki and Oinas 1995).

Storper (1995, p. 206) has summarized the argument as follows:

"1. Technological change is *path dependent;*
2. It is path dependent because it involves interdependencies between choices made over time – choices are sequenced in time, not simultaneous, and often irreversible.
3. These choices have a spatial dimension, which is closely tied to their temporal uncertainty and interdependence. Some inter-organizational dependencies within the division of labour, that is, input-output or network relations, involve some degree of territorialization. But in *all* cases where organizations cluster together in territorial space in order to travel along a technological trajectory, they have interdependencies which are *untraded*, including labour markets and conventions, or common languages and rules for developing, communicating and interpreting knowledge (however, direct input-output relations may also play a role here)."

Storper's approach, among others, gives a holistic understanding of the spatial dynamics, which affects the location of individual firms and, in a broader sense, the embedding of firms within place-specific socio-economic environments. This approach goes beyond the causality implied by traditional location theory that the main driving forces behind locational patterns of firms could primarily be explained by prices in production factors or transport costs.

Burmeister and Colletis-Wahl (1998) have drawn a parallel between the concept of untraded interdependencies and the concepts of *specific assets* and *resources* that are not traded directly on market terms. By reference to the work of Colletis and Pecquer (referred to in Burmeister and Colletis-Wahl 1998), they distinguish between *resources* as potential production factors, and *assets* as "production factors that are in use in a production process". The characteristics of these assets and resources can also be differentiated in terms of their specificity and linkages to specific processes, relations, actors, territories, etc. through the concepts of *generic* and *specific* resources and assets (Burmeister and Colletis-Wahl 1998).

Generic resources and assets are not tied up to a specific firm and its activities, and can therefore be transferred among firms and activities. Specific assets and resources are, on the contrary, related to specific production processes and contexts. Assets of this type cannot be transferred and reproduced easily in another context. Specific resources could be exemplified by the collective and non-formal knowledge on products, markets, technology etc. that characterize clusters of firms within the same industry or market. This knowledge is not made explicit or codified, but is embedded in the practice and experience of individuals and firms, and which has the potential of being brought into use – as a resource – by other individuals and firms affiliated to this cluster. These specific assets and resources are therefore critical factors in the way firms differentiate and specialize their activities. They also reflect how firms are tied into various forms of networks of inter-firm relations that affect their development trajectory.

Burmeister and Colletis-Wahl (1998) translate the inclusion of specific resources and assets in traditional location analysis into the following two assumptions:

- "Regional advantage is mainly related to the path-dependent process of the creation of specific resources.
- Transport infrastructure is, in most cases, a generic resource, which can be used in different production processes."
  (Burmeister and Colletis-Wahl 1998, p. 205)

These assumptions challenge the often-stated perception of transport infrastructure as an essential element for regional and industrial development. However, one should go further to state the significance of transport and logistics as an activity that affects, and is affected by, the organization and localization of firms. A third assumption can be proposed:

- The significance of transport systems for regional advantage is primarily in the specific resources embedded in organizational networks between production and transport firms.

From these three assumptions, an analytical framework has been developed and applied in the empirical analysis of the role of transport firms for the coordination of resources and the development of capabilities in the furniture industry of Salling. It is therefore assumed that the regional advantage is less directly related to the investments in terms of generic traffic infrastructure than to the organizational network between furniture and transport firms.

The network – in the case of geographical proximity between the firms – could be conceptualized as a localized governance structure in transport networks. A framework to analyse the interaction between transport systems and production systems is developed in the next section.

10.4 The Interactions between Systems of Production and Transportation: A Network Framework

In this section, transport is presented as an integrated part of the competitive advantage for single firms production and distribution systems and localized networks of industrial organization.

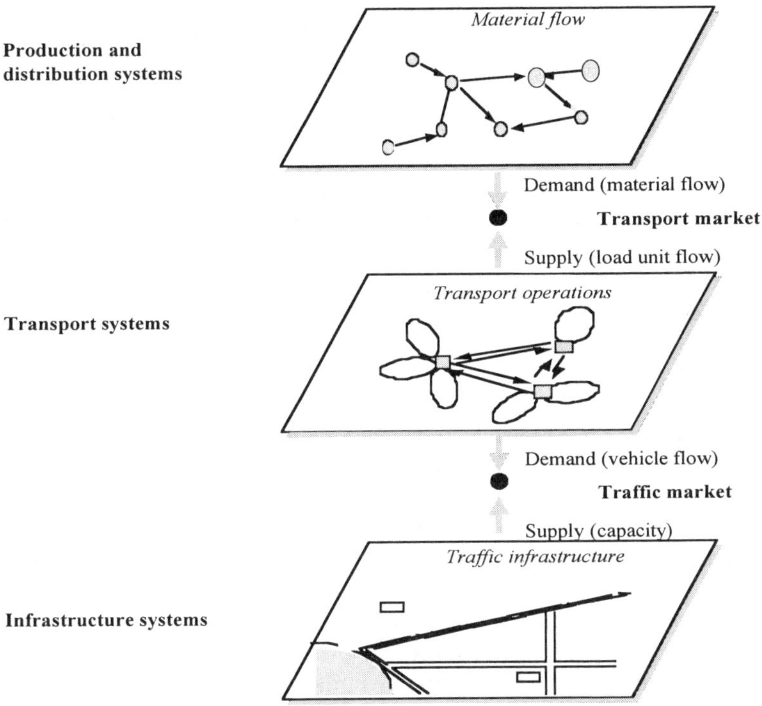

Figure 10.1 A three-layer system approach of functional relations between systems of production and distribution, transport and infrastructure (revised figure from Wandel and Ruijlk 1993)

The perspective in Figure 10.1 centres on the functional aspects of road-based freight transport in terms of interdependent systems. The first layer represents the production logistics activities of manufacturing or trading firms in terms of nodes that are interconnected via links of *material flows*. The second layer involves the logistic activities of transport service companies and reflects the flow of load units and vehicles between nodes as *transport operations*. The infrastructure system represents the third layer and it creates supply opportunities for vehicle movements in the form of *traffic infrastructure*: for example, as traffic supply in terms of space and time. Actors embedded in all three types of systems therefore govern the flow of goods, transport operations, and traffic infrastructure. The actors involved have different capabilities for affecting the design of parts of the total system. The following discussion will concentrate primarily on the systems of production and distribution, and transport.

Types of decisions affecting the design of logistics and transport systems can be manifold – from corporate management's decisions to open or close down production facilities, to the truck driver deciding which way to drive his lorry. In McKinnon (1998), an analytical framework is proposed to differentiate between different levels of decision making within the field of logistics:

1. *Logistic structures*: i.e. numbers, locations and capacity of factories, warehouses and terminals.
2. *Pattern of trading links*: created by commercial decisions on sourcing, sub-contracting and distribution, thereby constituting a material flow linking a company's premises to those of its trading partners.
3. *Scheduling of product flow*: the programming of production and distribution operations translates trading relationships into discrete freight flows. Adherence to a just-in-time (JIT) regime, for example, usually requires frequent delivery of small orders.
4. *Management of transport resources*: within the framework defined by decisions at the previous three levels, transport managers still have discretion over the use of transport resources.

Levels 1–2 represent the physical and geographical layout of production and distribution systems, which can be termed as *input-output networks*. These levels characterize a traditional geographical perspective on transport, focusing on location and distance between interacting production units, thereby determining the significance of transport in a production and distribution system solely as costs.

Levels 3–4 concern the organization of material flows and management of transport operations. These levels represent a *coordination network* of the activities performed by the actors within similar systems or across different systems (see Figure 10.1 above). The coordination dimension is likely to be of particular importance in production and distribution systems displaying a high degree of specialization and social division of production processes.

By considering all four levels, transport costs and geographical distance appear to be only some of the many factors that define the interrelationship between production and distribution systems, on the one hand, and transport systems, on the other hand. Beyond geographical distance, and the organization of input and output linkages a coordination network is necessary to establish a kind of organizational proximity among actors in production and transport systems.

As summed up at the end of Section 10.3 – in the form of three assumptions – the significance of transport systems as a driving force for regional advantage is as much a question of organizational as of physical networks of flows and infrastructures. On this basis, an analytical approach has been used in the empirical enquiry in order to integrate the functional dimensions of material and vehicle flow (input-output networks) with the organizational dimension of governance structures embedded in inter-organizational relationships (coordination networks).

The furniture industry of Salling in Denmark and its cooperation with local transport firms has been selected as a case for this study. The Danish furniture industry is relatively concentrated in clusters and consists mainly of small- and medium-sized firms, especially firms in Salling's industrial district (Lorenzen 1999). Although the firms are modest in size, their links to suppliers and customers are geographically extensive, and therefore necessitate a complex of transport links facilitating the material flow of supply and end products. In this organization of material flows, local transport firms play an important role in facilitating transportation and logistical networks.

10.5 The Furniture Industry of Salling: An Example of Regional Transport Networks and Capabilities

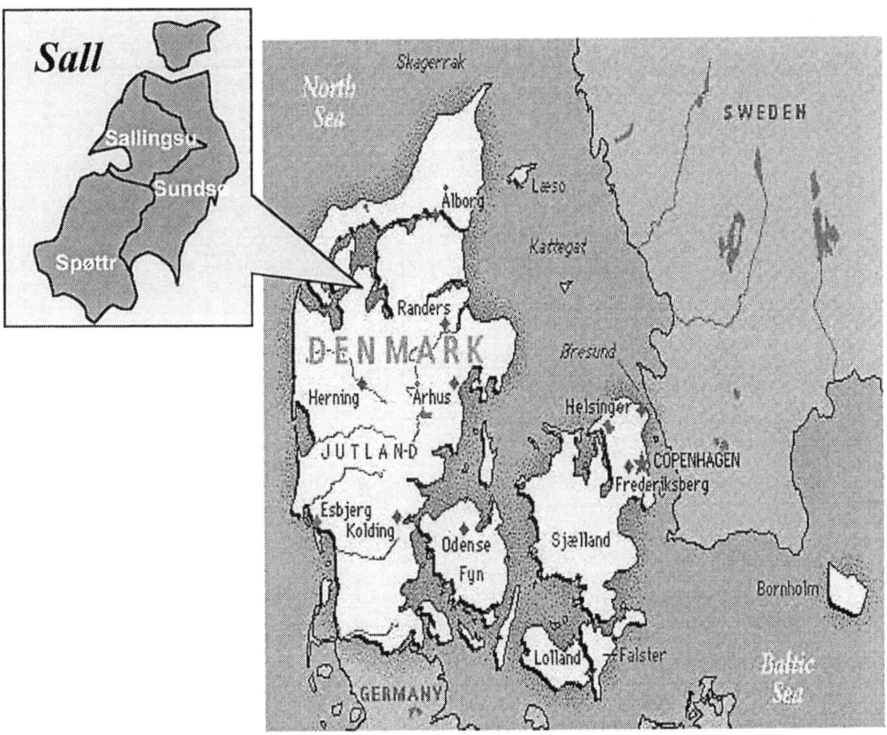

Figure 10.2 The region of Salling defined as the three municipalities of Sallingsund, Sundsoere and Spoettrup within the County of Viborg in the north-west of Denmark

The Danish furniture industry, in general, exhibits two interesting features in terms of the transport chain approach: wood is a dominant input in the manufacturing process and the industry has a large number of firms geographically concentrated in specific areas. Several studies have pointed to a particularly localized and specialized network among local furniture firms in the area of Salling (Nielsen and Storgaard 1981; Kristensen 1992; Maskell 1997; Lorenzen 1999) – a peninsula in the north-western part of Jutland and located in the county of Viborg, as illustrated in Figure 10.2 above.

In Section 10.4 above, Figure 10.1, which displays the material flow in the furniture industry as a value chain, illustrates the challenging complexity of the industrial flows from raw material through processing and distribution to consumption. As indicated in Figure 10.3 below, the direct input of the furniture producers is a manifold mixture of wood, metal, chemicals and agricultural goods (Maskell 1997). Other inputs such as energy, packing, machinery etc. are not included even though these goods also require transport.

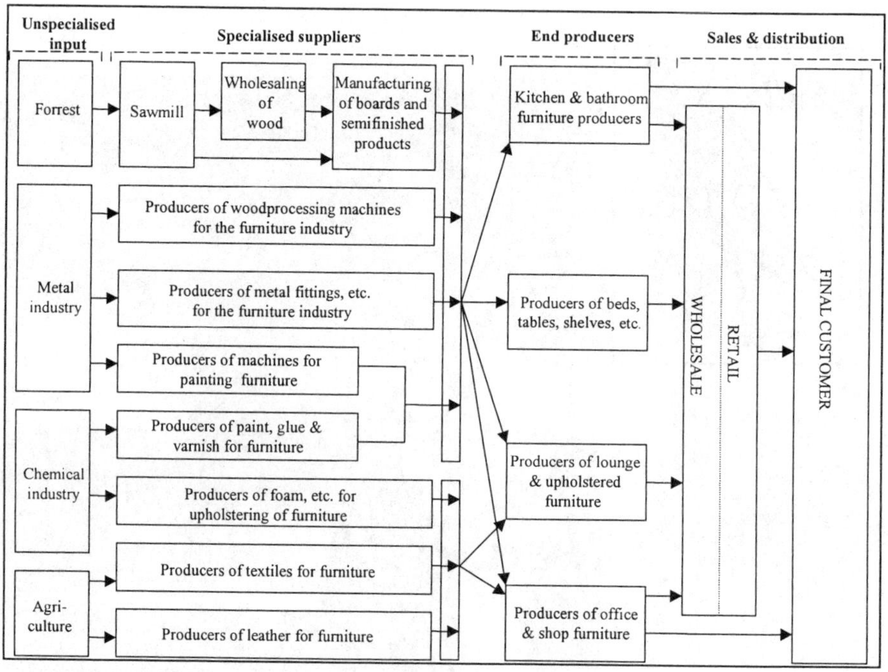

Figure 10.3 A model of the value chain of the furniture industry (adapted from Maskell 1997, p. 153).

The concentration of the furniture industry in Salling has no straightforward explanation in terms of traditional location factors. There is no significant forest to supply the furniture industry, and the major share of the furniture goods is exported to foreign markets. The development of the furniture industry seems to have been closely tied in with a local entrepreneurial spirit. Until the 1970s, only a small number of furniture firms were located in Salling, as well as a similarly small number of carpentry firms related to the construction industry. The general economic recession in the middle of the 1970s caused a crisis in the construction industry, but instead of closures, the carpenters used their knowledge from working with wooden materials for the manufacturing of wooden furniture. The knowledge of furniture production and markets inherited from the small number of existing furniture firms was shared with the newcomers and developed as a common asset and resource of the region in the following years (Lorenzen 1999).

For that reason, the competitiveness of the local furniture industry has not only been the result of product quality and design content (characterized by the original furniture producers of Salling). It has also been driven by qualitative flexibility in terms of quick response to heterogeneous demand, customization and continuous product innovations allowing for constant reorientation to customers' needs in different markets with various style preferences. In addition, production flexibility plays a central role in the way that the majority of local furniture producers are both end-producers and sub-contractors. Therefore, if there is lower demand for its own end products, the single firm minimizes the idle time of its production capacity by offering sub-contracting work to other local firms.

The furniture producers in Salling are mainly end-producers and therefore highly dependent upon the coordination of the material flow through the value chain – see Figure 10.3 above. Most of the wood used to make furniture is pinewood imported from Sweden and Finland, and the primary markets for pinewood furniture are Germany, Denmark, Norway, and Sweden. By analyzing the transport network from the specialized suppliers through the end-producers to the distribution of the value chain, the present study has identified where the capability of governance in the transport network is located organizationally and territorially.

10.5.1 Design of the Empirical Study and Analytical Approach

The study was carried out from 1999 to 2000 and was based on a questionnaire followed by telephone interviews involving 56 furniture firms from different regions of Denmark and from different product segments of the furniture industry. The purpose was to identify general patterns of organizing transport and logistics in relation to the actors' organizational and geographical position along supply chains in this industry. The telephone interviews based on the themes and categories in the questionnaire made it possible to clarify ambiguities experienced by the respondents.

In the questionnaire, the firms were asked to describe their: 1) logistical structure (location of production, warehouses, supply and market regions); 2) trading links (relations to suppliers and customers); 3) organization of product flows (production on-demand or to stock, just in time, etc.), and 4) management of transport (own transport, external freight firms, local or non-local transport firms).

The findings were elaborated on through a series of in-depth interviews in selected firms related to the wholesaling of unprocessed wood and transport services for the purpose of analyzing discrete examples of transport chains in the furniture industry. The analytical interest was limited to input flows of wood, processing flows of wooden furniture components and output flows of wooden furniture by furniture producers. Even though the products of the furniture industry consist of multiple types of inputs, wood is the dominant component in terms of transported volume.

10.5.2 Results of the Survey

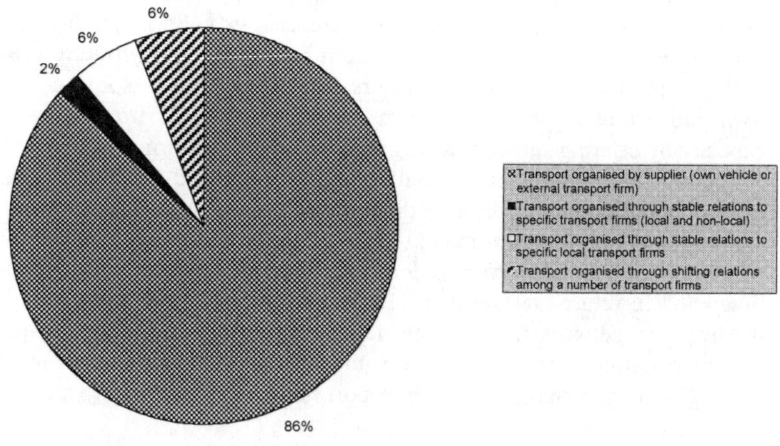

Figure 10.4 Organization of inward-bound transport of unprocessed wood in Denmark

Results from the questionnaire show a marked difference in the way various modes of organizing transport by furniture firms are brought into action. On the one hand, the transport flows of unprocessed wood to furniture producing firms is usually organized by the wholesalers or the sawmills that buy transport services and therefore choose the transport mode and the forwarder or transport operator – (see Figure 10.4 above). For the furniture producing firms of Salling, this transport flow of unprocessed wood is furthermore concentrated in specific transport corridors from sawmills in Sweden and Finland. In the specific case of furniture producing firms located in Salling, this indicates that the governance of the *interregional input* transport chain is located organizationally and territorially outside Salling.

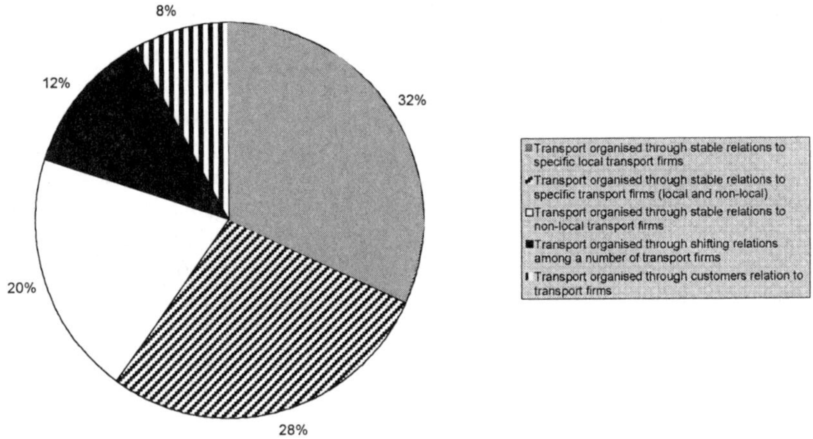

Figure 10.5 Modes of organizing outward-bound transport of furniture in Denmark

On the other hand, the study indicates that the furniture producing firms do usually possess the competence to organize the transport of furniture to their customers. The furniture producing firms tend to use transport firms, which have organized their outward-bound transport flows for longer periods of time – (see Figure 10.5 above). The Salling furniture firms' use of primarily local transport firms reveals a highly localized structure of decision making concerning the organization and management of the transport chain related to *intra-regional distribution* and *interregional output*.

Table 10.1 Location of transport firms used by furniture-producing firms located in Salling and Viborg County, the remaining part of the western part of Denmark, and the eastern part of Denmark

	Location of furniture firms		
Location of transport firms	Salling and Viborg County	Western Part of Denmark	Eastern Part of Denmark
Salling and Viborg	80 pct.	62,5 pct.	0 pct.
Western Part	30 pct.	37,5 pct.	40 pct.
Eastern Part	0 pct.	0 pct.	40 pct.
Multiple locations in Denmark	35 pct.	25 pct.	80 pct.

Source: Gjesing Hansen 2002.

The study has not only confirmed a localized specialization in furniture production, but also identified a highly localized specialization within furniture transportation in Salling and the surrounding County of Viborg. This specialization in furniture transportation is reflected in the way in which furniture-producing firms located in Salling and the surrounding county of Viborg use locally based freight haulage firms. However, furniture producers in other parts of Denmark do, to a large extent, also use freight haulage and forwarding firms located in Salling and Viborg County (see Table 10.1 above). These results from the questionnaire and in-depth interviews indicate that a number of transportation firms in Salling and the County of Viborg seem to have benefited from the geographical clustering of furniture production in terms of developing a nation-wide reputation of being skilled and competent in handling and organizing transportation and logistics of furniture.

In relation to the furniture producing firms, localized in Salling, their use of primarily local transport firms indicates a highly localized governance structure of the *intraregional distribution* and *interregional output* of the total transport chain of the furniture industry in Salling as a whole (illustrated in Figure 10.6).

This localized capability seems to be embedded in the inter-organizational relations between the local furniture producers and transport firms that have evolved since the growth of furniture production in Salling from the mid-1970s onwards.

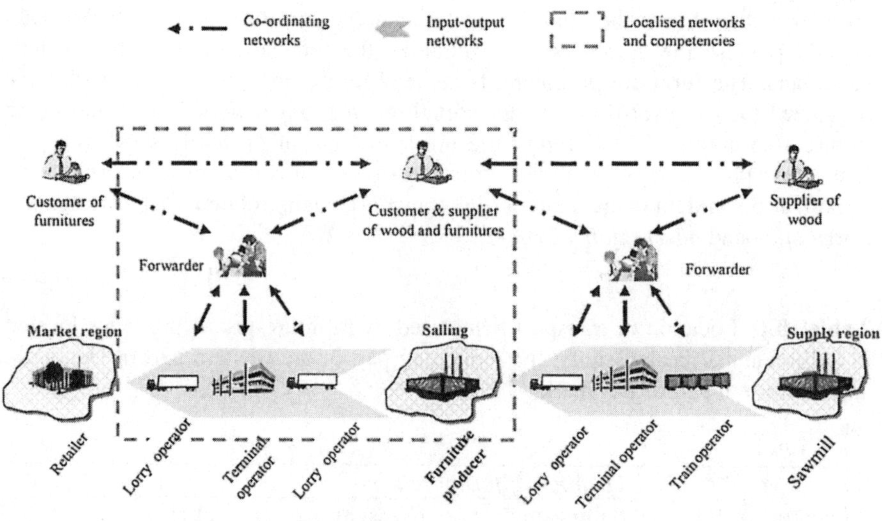

Figure 10.6 A demarcation of the localized competence in the network of actors involved in organizing the transport and logistical chains of the furniture industry in Salling

A major difference between input and output of the producers in Salling is the variety of load sizes and frequencies of delivery. Supply of non-processed pinewood is normally organized as weekly deliveries directly from the Swedish or Finish sawmills, or by Danish wholesalers. The furniture producers very often do not have a large inventory, but instead rely on just-in-time deliveries. The sawmills or the wholesalers manage the inventory of non-processed wood. By using contracts of 3-6 months with the furniture producers, the wholesalers and sawmills have the capability of organizing full load transport by ship or train in a frequent – but not necessarily fast – transport schedule. The full loads are divided into part loads for different customers at the loading terminals in Denmark (ports or railway terminals). However, the cluster of furniture firms using pinewood, particularly in Salling, makes co-distribution to several customers in one load possible.

A somewhat different procedure is used when the finished furniture goods are to be distributed to the final markets. The major difference is that individual producers can make consignments of varying sizes, from a single chair to a truckload of goods. The furniture industry in Salling includes firms that produce small batch orders of, say, a single sofa, as well as firms that mass-produce 'knock-down' furniture to large retail chains. As in the case of non-processed wood, the furniture producers often do not have storage capacity for the finished furniture goods. To cater for this need, the transport companies used by the producers have established dedicated storage facilities. These companies were originally small providers of all-round transport services and were often chosen as transport operators by the local furniture firms because of their proximity – probably not just spatially but also culturally. The growth of the local furniture industry during the 1970s and up till the 1990s also brought about a parallel growth in the market for furniture transport by the local transport firms. Most of them decided not to provide all-round transport services and specialized solely in furniture transport during the 1980s.

Out of this closely knit network, between a small number of local transport firms and a major part of the local furniture firms in Salling, a regional network has been developed based on co-distribution. Some of the major challenges for the transport companies have been the unpredictable shifts in consignment sizes and the ability to consolidate the transport of furniture goods to particular export markets. This has required the development of fixed routes for collecting furniture to specific markets on specific days of the week – for example, the collection of furniture to Germany every Monday and Wednesday. Alternatively, the transport companies collect all the furniture goods for different markets from one producer and bring it to the storage facility of the transport company, where it can be split and consolidated within a week before shipment to the final markets. The furniture producers and their customers seem to accept up to a week of distribution time to export markets in Europe, which gives the transport companies time to consolidate furniture goods – as far as possible – in full loads.

10.6 Conclusions

The findings suggest that geographical proximity, both among the local furniture firms and between these and local transport firms represent an important condition for generating external economies in terms of a regional transport and logistics system. The study indicates that a kind of localized transport and logistical competence has been developed to the advantage of small- and medium-sized furniture producers, and that relatively small transport operators have gained a specialized competence in handling furniture goods and organizing the logistical management of distribution in the out-bound parts of the product and transport chains. Two forms of capabilities seem to be important in generating specific and localized competencies in the field of transport and logistics of the furniture industry cluster in Salling:

1. economies of scale and scope through networking, and
2. resources of specific knowledge in production and transport that are transformed through socio-economic networks into a common competence across the sectorally-defined boundaries of these industries.

First, the large number and concentration of furniture firms has made it possible for a small number of highly specialized transport firms to sustain a dense and frequent distribution and pick-up network. This locational pattern of the local furniture industry gives the transport firms the ability to gain scale economies by consolidation through the pick-up of furniture goods via multiple stops and through the system of storage and distribution at the warehouses of the transport firms.

Closely related to the scale economies of the local transport system, the dense distribution and pick-up network also contributes to economies of scale for many of the small- and medium-sized furniture firms in terms of markets and shipment size. A common feature of these firms' activities is their ability to serve different types of customers in different markets with great variations in consignment sizes. Many of the firms send consignments of both single items of furniture and, part and full loads of furniture.

A second important feature of the relationship between furniture producer and transport operator is the capability to coordinate and organize the material flow in an efficient way. This efficiency is dependent on the knowledge of transport demand in terms of quality and quantity within an extremely short time horizon, knowledge which the local transport firms accumulate over time via long-standing relationships with specific furniture firms. This knowledge thus represents a kind of untraded interdependency among local furniture and transport firms. In particular, the involved actors have developed common geographical markets of furniture and transport; for example, the transport firms have specialized their route network in accordance with the developed markets for furniture products. Therefore, the knowledge from each involved firm is being transformed into a common competence, which generates a collective efficiency in the organization of input and output flows.

This common competence is, however, not easy to describe explicitly, since its character is often embedded in the ongoing social interactions between individuals and organizations. In Section 10.2, this was referred to as an untraded interdependency. Our study indicates that this interdependency is to some extent territorially embedded within a dominant relationship between local furniture firms and transport firms.

The study, on which this chapter is based, confirms findings in other studies of industrial clusters that the competitiveness of a local cluster cannot be explained solely by focusing on individual firms or production systems. The clustering of the furniture firms is essential to their success in the way it generates external economies through specialization, division of labour, and accumulation of shared knowledge. This study has thus enlarged the scope of the dynamic forces that condition the development of a local industrial cluster by including the role of transport and logistics.

References

Amin A. and Thrift N. (eds) 1994: Globalization, Institutions and Regional Development in Europe, Oxford University Press, Oxford

Amin A. and Cohendet P. 2000: How to Govern Simultaneously Transactions and Commences: Should Organizational Learning Rely on Isolated Properties or Embedded Practices?, Paper presented at DRUID's Summer 2000 Conference, Rebild in Denmark, 15th-17th June: The Learning Economy – Firms, Regions and Nation Specific Institutions

Braunerhjelm P. and Svensson R. 1998: Agglomeration in the Geographical Location of Swedish MNFs, in Braunerhjelm P. and Ekholm K. (eds.), The Geography of Multinational Firms, Kluwer Academic Publishers, Boston, 99–115

Burmeister A. and Colletis-Wahl K. 1998: Proximity in Networks: The Role of Transport Infrastructure, in Reggiani A. (ed.), Accessibility, Trade and Locational Behaviour, Ashgate, Aldershot, 201–218

Camagni R. 1993b: Local 'Milieu', Uncertainty and Innovation Networks: Towards a New Dynamic Theory of Economic Space, in Camagni R. (ed.): Innovation Networks: Spatial Perspectives, Belhaven Press, London 121–144

Conti S.; Malecki E.J. and Oinas P. (eds.) 1995: The Industrial Enterprise and Its Environment: Spatial Perspectives, Avebury, Aldershot

Crevoisier O. and Maillat D. 1993: Technological Innovation and Territorial Development: An Approach in Terms of Networks and Milieux, in Camagni, R. (ed), Innovation Networks: Spatial Perspectives, Belhaven Press, London, 35–54

Dicken P. 1992: Global Shift: The Internationalization of Economic Activity, Paul Chapman, London

Dicken P. and Lloyd P.E. 1990: Location in Space, Harper Collins Publishers, New York

Drewes Nielsen L., Gjesing Hansen L., Kornum N., Nedergaard K. and Aastrup J. 1999: Godstransport I Et Kædeperspektiv – Erhverv, Miljø og Planlægning (Freight transport in a chain perspective – industry, environment and planning). Working Paper for the Ministry of Environment and Energy, Ministry of Environment and Energy, Copenhagen

Foss N.J. 1996: Higher–Order Industrial Capabilities and Competitive Advantage, Journal of Industry Studies, Vol. 3, 1, 1–20

Gjesing Hansen L. 2002: Logistik– og transportkæder i et regionalt perspektiv (Logistical Chains and Transport in a Regional Perspective), Report for the Danish Transport Council, Copenhagen (forthcoming)

Isard W. 1956: Location and Space-Economy, John Wiley, New York

Kristensen P.H. 1992: Industrial Districts in West Jutland, Denmark, in Pyke F. and Sengenberger W. (eds.), Industrial Districts and Local Economic Regeneration, ILO Publications, Geneva, 122–173

Krugman P. 1991: Geography and Trade, The MIT Press, Cambridge, MA

Krugman P. 1995: Development, Geography, and Economic Theory, The MIT Press, Cambridge, MA

Lawson C. 1999: Towards a Competence Theory of the Region, Cambridge Journal of Economics, Vol. 23, 2, 151–166

Lorenzen M. 1999: Localized Learning and Community Capabilities, Copenhagen Business School, Copenhagen

Marshall A. 1919: Industry and Trade, London

Maskell P. 1997: Localized Low-Tech Learning in the Furniture Industry, in Eskeleinen H. (ed.), Regional Specialisation and Local Environment – Learning and Competitiveness, NordREFO, 3, 145–170

McKinnon A. C. 1998: Logistical Restructuring, Freight Traffic Growth and the Environment, in Banister D. (ed), Transport Policy and the Environment, Routledge, London, 97–109

Morgan K. 1997: The Learning Region: Institutions, Innovation and Regional Renewal, Regional Studies, Vol. 31, 5, 491–531

Nielsen L.D. and Storgaard K. 1981: Teknologi og regional brancheudvikling i træ- og møbelindustrien (Technology and Regional Industrial Development within the Wood and Furniture Industry), DTI, Høje Taastrup

OECD 1996: Integrated Advanced Logistics for Freight Transport, OECD, Paris

Perroux F. 1950: Economic Space: Theory and Applications, Quarterly Journal of Economics, Vol. 64, 1, 89–104

Piore M. and Sabel C.F. 1984: The Second Industrial Divide, Basic Books, New York

Schmitz H. 1999: Collective Efficiency and Increasing Returns, Cambridge Journal of Economics, Vol. 23, 465–483

Storper M. 1995: The Resurgence of Regional Economics, Ten Years Later: The Region as a Nexus of Untraded Interdependencies, European Urban and Regional Studies, Vol. 2, 3, 191–221

Venables A. 1996: Equilibrium Locations of Vertically Linked Industries, International Economic Review, Vol. 37, 2, 341–359

Wandel S. and Ruijgrok C. 1993: Innovation and Structural Changes in Logistics: A Theoretical Framework, in Giannopoulos G. and Gillespie A. (eds.), Transport and Communications Innovation in Europe, Belhaven Press, London, 233–258

Weber A. 1909: Theory of the Location of Industries, University of Chicago Press, Chicago

Endnotes

[1] External economies exist when private costs or benefits do not equal social costs or benefits. When social costs are higher than private costs, one can speak of external diseconomies. When social benefits are higher than private benefits, one can speak of external economies.

[2] This concept draws on the work of Francois Perroux in the 1950s, who noted that an economy consists of 'spaces', in part having to do with the density of common types of knowledge or similar types of machines among firms, or a common knowledge of how to work similar types of basic materials or inputs (Peroux 1950). From the point of view of economists, the concepts of external economies or higher order capabilities are perhaps more familiar terms than untraded interdependencies (Foss 1996; Schmitz 1999). The concepts imply a kind of co-operation or interaction among firms, which generates a collective efficiency for the involved actors. This efficiency is embedded in, and a result of, an ongoing interaction process in a particular location in time and space. Therefore, the particular collective efficiency gains and its institutionalisation of involved actors behaviour cannot be transferred or traded from one situation of interaction to another.

11 A Methodology to Evaluate Potential Locations for Intermodal Barge Terminals: A Policy Decision Support Tool

Cathy Macharis

Vrije Universiteit, Brussels, Belgium

11.1 Introduction

Intermodal transport is the transport of unit loads by the combination of at least two modes of transport in a single transport chain. Goods are transported with most of the route travelled by rail, an inland waterway system or via an ocean-going vessel, with the shortest possible initial and final journeys by road. Due to environmental and congestion pressures this type of transport has received much attention over the last few years. Intermodal terminals, necessary for the transhipment of unit loads from one mode to another, are very important in this transport chain. In recent years, several new terminal projects have started up in Belgium (see Macharis and Verbeke 1999).

The recent tendency towards using containers in maritime traffic has put stronger pressure on the hinterland networks of the ports. The economies of scale that were attained by the use of larger maritime vessels have partially compensated for this. It is, however, becoming more and more difficult to transport goods to their end destination due to congestion in the port and on the roads. In the meantime, due to the growth of freight, the minimum economic volume is attained due to the start up of new shuttles for inland shipping and rail transport. Over the last few years, there have been various initiatives to implement new inland terminals. Between 1990 and 1999, the number of existing terminals in the inland waterway/road sector has grown from 2 to 5, and 6 new initiatives are expected in the near future. This indicates a clearly beneficial trend that stimulates further growth in the intermodal transport sector and the ports.

The Belgian government has, until now, not followed a clear location policy. This has resulted in an excess of new initiatives. This evolution will inevitably lead to inefficiencies (e.g. insufficient capacity utilization and a decrease in the resulting profitability). On the other hand, the intermodal transport market may require active government support to compete effectively with the road networks and to reinforce the competitive position of the seaports.

In this chapter a methodology is proposed to evaluate new terminal projects. It is primarily designed for inland waterway/road terminals, but the methodology can easily be extended to other combinations. This methodology can become the base of an active location policy. This is followed by a stakeholders' approach, outlining the problems concerning the decision making process taking into consideration the points of view of all of those involved.

Next, in Section 11.2, we will discuss existing evaluation models, while Section 11.3 proposes a new methodology. In a preliminary phase, described in Section 11.4, the terminals are evaluated with respect to their traffic potential and their influence on the market area of the existing terminals. The result of applying this methodology to the Belgian situation is given in Section 11.5. The multi-criteria analysis method is described in Section 11.6 which enables a comprehensive evaluation of confirmed future terminal projects.

11.2 Overview of the Literature

As *intermodalism* is a rather new topic in transportation research, only a few models have been developed to assist policy makers in solving complex decision problems in this field (Bontekoning, Macharis and Trip 2002). This section provides a short overview of models developed to evaluate new possible locations of intermodal terminals.

A good starting point is a study conducted for the European Commission, 'Cost-benefit and multi-criteria analysis for nodal centres for goods' (1996). In this study, an overview is given of the evaluation methods and criteria used in investment decisions for intermodal terminals in various EU countries. It proposes an integrated framework for the evaluation of nodal centres, in which a financial analysis and a social cost/benefit analysis are combined. In both evaluation methods, all the effects have to be expressed in monetary values, which makes them difficult to use. Criteria such as complementarity with ports or existing terminals are very difficult to measure in monetary units. A qualitative criterion is more appropriate in such cases.

In the Neptune project of the European Union (Maggi 1998), the decision criteria for the location of logistics nodes are determined. The location factors are divided into three main categories: namely, economic-financial factors,[1] qualitative factors[2] and geographic-infrastructure factors.[3] A logistics node can contain an intermodal terminal but is usually much larger. No formal evaluation method has yet been developed in this project.

Ashayeri and Rongen (1997) propose a framework to select a site for a European Distribution Centre (EDC). An EDC is a site in Europe where goods are collected, stored, grouped and transported and can comprise an intermodal terminal (although the latter is not a necessity). Ashayeri and Rongen propose a 7-step procedure to develop new structures for a company's distribution. A model is proposed to select the optimal sites according to the cost minimization criterion (cost of transport and location). The number of sites resulting from the model determines the type of distribution model (centralized versus decentralized distribution). If this is not possible, due to a lack of data, a grid model can be used.

Taking other evaluation criteria into account, such as transport and location costs, turnaround time and a qualitative/quantitative location factor index, makes it possible to refine this outcome in a further step. This final evaluation is carried out by the ELECTRE method.

The studies of Maggi (1998) and Ashayeri and Rongen (1997) are not really applicable to the location of intermodal terminals due to the different kind of criteria that have to be considered when the size of the project is much larger.

Several papers also address the problem of finding the *optimal* location of intermodal terminals (Rutten 1995; Arnold and Thomas 1999; Groothedde and Tavasszy 1999; Declercq and Verbeke 1997; Van Duin and Van Ham 2001; Macharis 2000). In order to do so, network models are set up. And, in combination with a linear programming model or other techniques, such as the simulated annealing technique in Groothedde and Tavasszy (1999), the optimal location is found. In this chapter, however, it is not the optimal locations which are sought, but an evaluation procedure for the intermodal terminal projects that are proposed by the private sector.

As indicated in the discussion of the existing models, an evaluation model for intermodal terminals does not yet exist. In order to support the decision making of the Belgian government, we have developed the 'LAMBIT' model (Location Analysis Model for Belgian Intermodal Terminals). An overview of the methodology for this model will be given in the next section.

11.3 Overview of the Methodology Adopted

Figure 11.1 gives a complete overview of the proposed evaluation methodology. The preliminary phase is the determination of the traffic potential of the terminal projects. In order to have a sustainable terminal, the traffic potential in the surrounding area of the terminal must be large enough to support it. Furthermore, the impact of the new projects on the market area of the existing terminals must be studied. A network model developed in Section 11.4 will determine this traffic potential and the impact on the existing terminals.

Furthermore, the proposed locations will have to be screened according to set standards (large enough for an intermodal terminal, grants by the local district, ability to get permissions, etc.).

In the next phase, a more comprehensive evaluation of the terminal projects is executed. The criteria used in this evaluation represent the aims of the parties involved: namely, the users of the terminal, the operators/investors and the community as a whole. For each of these groups a multi-criteria analysis will be executed. A multi-criteria analysis provides insight into the strengths and weaknesses of several alternatives according to several criteria. Ultimately, a ranking of the alternatives is given. Some hundreds of methods exist (see, e.g., the overview in Vincke 1992). Here, the PROMETHEE method (Brans 1982; Brans and Mareschal 1994; Macharis et al. 1998) is used. The results of the analysis of the affected parties are brought together in order to get a global ranking of the projects. A sensitivity analysis closes the procedure.

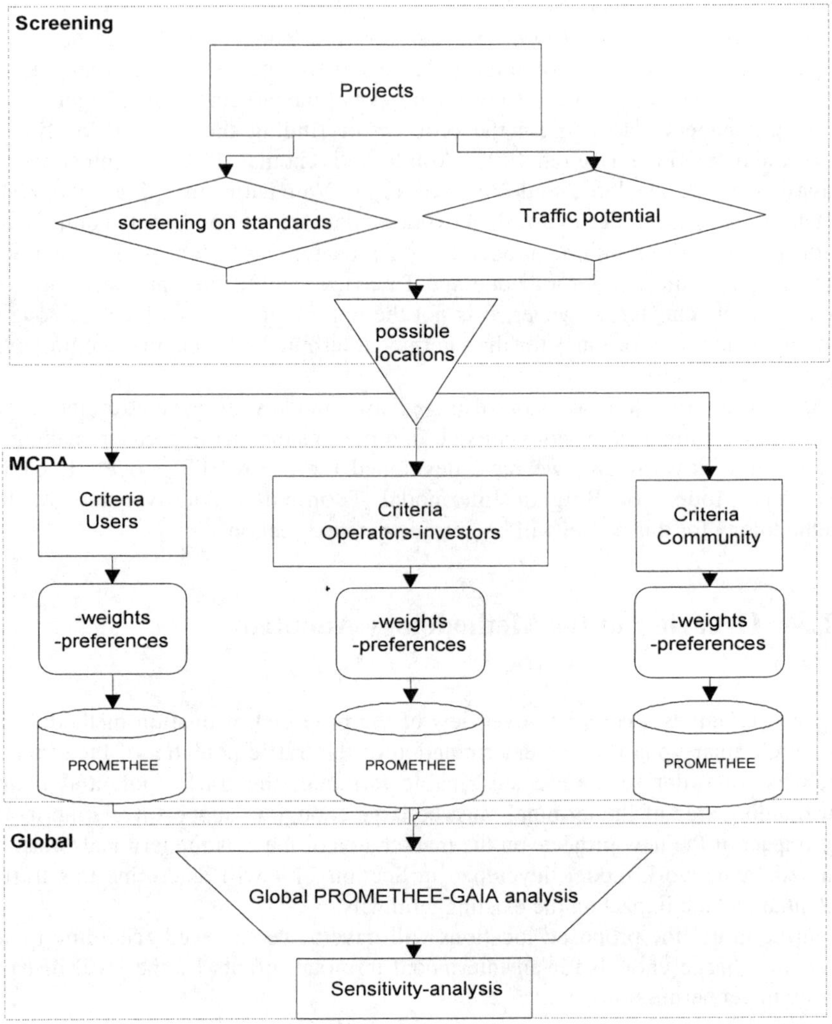

Figure 11.1 Overview of the methodology
Source: Macharis 2000.

It is important to note that the results of the preliminary study by the network model are particularly relevant for the evaluation of some criteria. For example, the profitability of the terminal (a criterion of the investors) will be directly linked with the traffic potential. Also the complementarity with other terminals can be valued according to the results of the network model.

11.4 Network Model

11.4.1 Data

The reliability of the results of the traffic potential analysis depends on the reliability of the data used in the model. Every year, the National Institute of Statistics (NIS) collects data concerning freight transport from Belgium and abroad (*Statistique du commerce intérieur et des transports*). During the data collection – a survey among some 60,000 companies – the detailed data at district level is obtained for Belgium and data covering wider regions is obtained for surrounding countries. This basic data is aggregated to produce the data for publication.

For the purpose of the present analysis, a selection was taken from the database. Only the origin to destination matrices for containers were selected, and only those to and from the port of Antwerp. In this way, we were able to get a view of the potential shift from road to intermodal road/inland waterway for the maritime container movements. The current container movements that had already shifted from road to rail were thus not taken into account. The authorities are primarily interested in a modal shift from road to inland waterway and not between more environmentally-friendly modes already being used, such as between rail and inland waterway. Therefore, this approach to the analysis is the most logical.[4]

11.4.2 Location Analysis for Belgian Intermodal Terminals

The first step was for the network to be set up. The road network, the inland waterway network, and the network of the final haulage by road from the terminal to the end destination were stored in three different levels (on the base of the maps of TELE ATLAS). The second step was to set up the generalized cost function. According to the real price structure, the variable prices are allocated to each of the following: road, inland waterway segment and the fixed prices of the terminals.[5] These fixed prices include the transhipment costs and the fixed parts of the transportation prices.

Using a shortest path algorithm,[6] the shortest path and corresponding cost are determined for each municipality for the two alternative modes, i.e. the route over the road network and the combined route via one of the inland waterway terminals. Using EXCEL software to calculate the formula, the cheapest option is chosen. The market area of each terminal can easily be visualized by highlighting the municipalities for which a certain terminal provides an alternative for road transport in terms of price. The calculation of the potential of the terminal is done by aggregating the numbers of containers that are transported to and from these municipalities (making use of the NIS data). Only a small percentage of this potential will effectively shift after implementing the terminal as not only the price but other criteria are important for the modal choice. Besides the determination of the possible traffic potential of the new projects, the model can also assist in determining the impact of the new terminals on the market area of the already existing terminals. This can be done by inserting the terminals in time phases in the model, according to the planned start of the new projects. As noted in Table 11.1, four phases were identified according to the implementation planning of new terminal initiatives in Belgium.

Table 11.1 Implementation phases of new terminals

Year	Terminals
1998	Avelgem Container Terminal (AVCT) in Avelgem, Water Container Terminal (WCT) in Meerhout and Luik (Renory)
End 1999	+ Trimodal Container Terminal (TCT) in Willebroek, Puurs Inland Terminal Zeekanaal (PITZ) in Puurs and Container Terminal (CT) Gent
End 2000	+ Hessenatie in Vilvoorde, Port of Brussels, Genk and Wielsbeke
Future?	+ La Louvière, Roeselare and Brugge

Source: Macharis (2000).

11.5 Result of the Analysis

In Figure 11.2, the situation of the Belgian terminal landscape at the end of 1999 is given. At that time, six terminals were active in Belgium: namely, WCT in Meerhout, AVCT in Avelgem, CT Gent in Gent, PITZ in Puurs, TCT in Willebroek, and the terminal of Luik (Renory). The municipalities where intermodal road/inland waterway transport is cheaper than road transport alone are highlighted on the map. The highlighted areas thus represent the market area of the terminals.

The terminals in Meerhout and Avelgem can reach several municipalities. The new terminals in Willebroek, Puurs and Gent are too close to the port of Antwerp and cannot offer a competitive price. As previously explained, the final potential of the terminals can be determined by summing up the container traffic of the concerned municipalities. In Table 11.2, the potential of each market area is indicated.

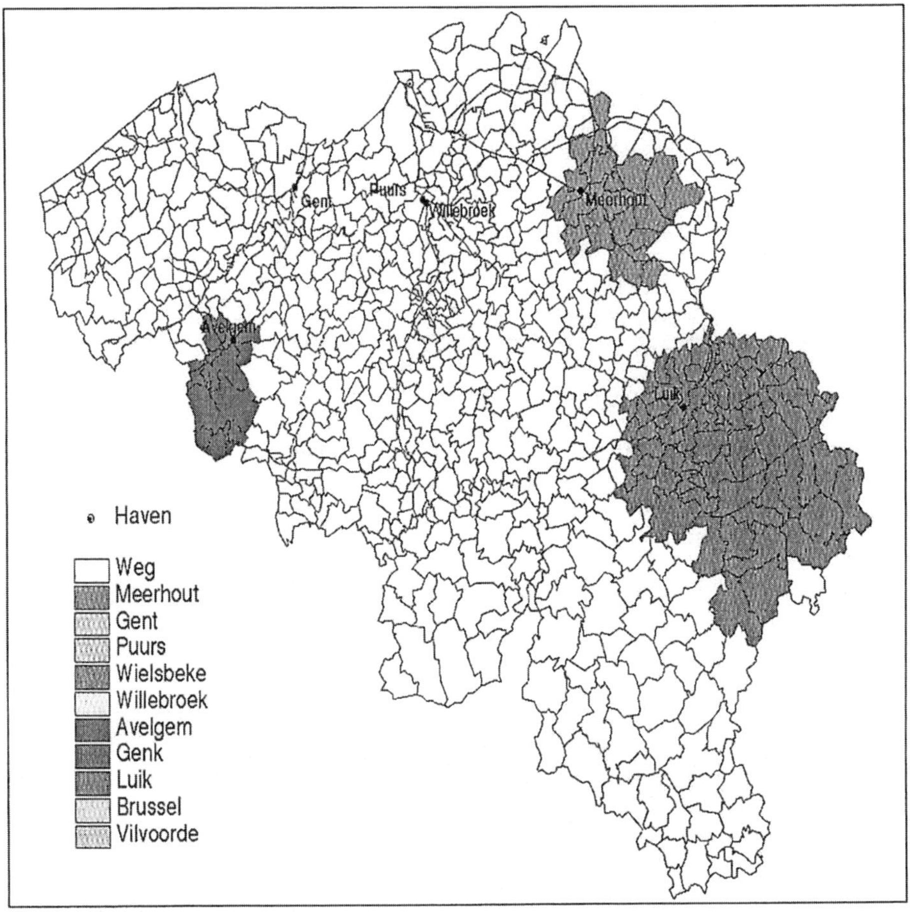

Figure 11.2 Market areas for intermodal barge terminals (end 1999)
Note: Weg=Municipalities where road transport alone is the cheapest choice, Haven=Port
Source: Own set up in ARCINFO and ARCVIEW, Data 1999.

The potential volume only indicates how many tonnes (or TEUs[7]) are actually transported via the road to the municipalities concerned. Only a small percentage of this potential will be effectively shifted once the terminal is built as other mode choice variables are also important (for a good overview of modal choice variables, (see Vannieuwenhuyse, 1999).

Table 11.2 Potential of the terminals (end 1999)

	Volume(tonnes)	Volume(TEUs)	Number of municipalities
Road	18,464,898	1,709,713	490
Meerhout	1,596,856	147,857	18
Gent	0	0	0
Puurs	0	0	0
Willebroek	0	0	0
Avelgem	638,818	59,150	12
Luik	1,058,346	97,995	69

Source: Macharis 2000.

In Figure 11.3 the situation as of mid-2001 is presented. After completion of the works, terminals will also be active in the Port of Brussels, Vilvoorde, Genk and Wielsbeke.

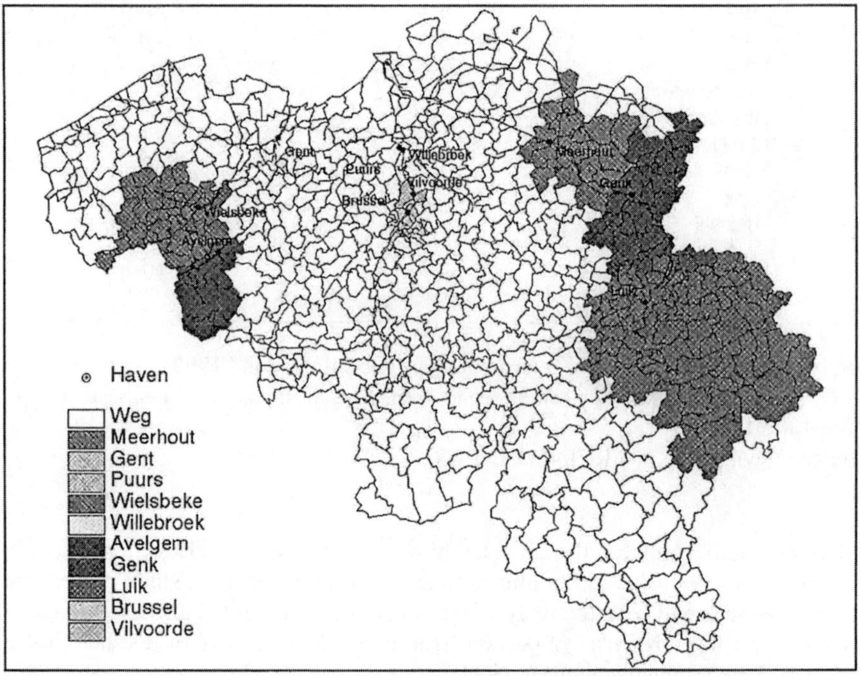

Figure 11.3 Market areas for intermodal barge terminals (situation mid 2001)
Note: Weg = Municipalities where road transport alone is the cheapest option; Haven = Port.
Source: Own set-up in ARCINFO and ARCVIEW, Data 1999.

This comparison with the previous figure indicates clearly that Wielsbeke will absorb some of the market area of Avelgem. The new terminal in Genk will absorb the market areas of both Meerhout and, to a lesser extent, Luik. In Table 11.3, the potential of each terminal is expressed in tonnes, TEUs and the number of municipalities which represent the market area. A comparison with Table 11.2 indicates that Meerhout only loses 1,667 TEUs (or 1%) of potential and Avelgem 26,539 TEUs (or 45% of the original volume).

Table 11.3 Potential (situation mid-2001)

	Volume (tonnes)	Volume (TEUs)	Number of municipalities
Road	14,302,218	1,324,279	430
Meerhout	1,578,856	146,190	16
Gent	0	0	0
Puurs	0	0	0
Wielsbeke	2,599,050	240,653	28
Willebroek	0	0	0
Avelgem	352,198	32,611	11
Genk	1,052,148	97,421	25
Luik	1,040,346	96,328	66
Brussel	261,224	24,187	11
Vilvoorde	572,878	53,044	2

Source: Macharis 2000.

In Figure 11.4, the terminals in La Louvière and Roeselare are also added. These terminals can, with current market prices, cover a large market area.

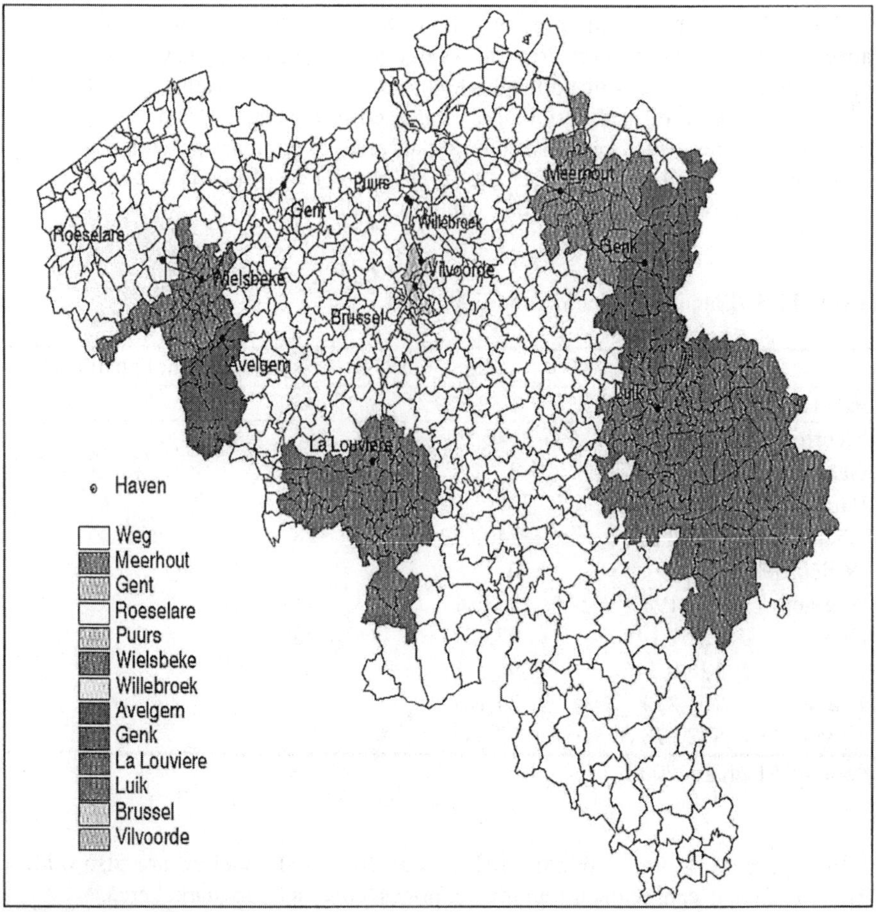

Figure 11.4 Market areas for intermodal barge terminals (possible future situation)
Note: Weg = Municipalities where road transport alone in the cheapest option.
Source: Own set-up in ARCINFO and ARCVIEW, Data 1999.

In Table 11.4, the volumes complying with the criteria are shown.

Table 11.4 Potential (future)

	Volume (tonnes)	Volume (TEUs)	Number of municipalities.
Road	13,171,309	1,219,565	390
Meerhout	1,578,856	146,190	16
Gent	0	0	0
Roeselare	688,989	63,795	18
Puurs	0	0	0
Wielsbeke	2,167,481	200,692	20
Willebroek	0	0	0
Avelgem	352,198	32,610	11
Genk	1,052,148	97,421	25
La Louvière	873,489	80,878	31
Luik	1,040,346	96,328	63
Brussel	261,224	24,187	12
Vilvoorde	572,878	53,044	3

Source: Macharis 2000.

11.6 Multi-criteria Analysis

In the first phase, the problem is seen from the viewpoint of the three groups of stakeholders. The specification of the criteria (Section 11.6) follows this stakeholders' approach. The criteria obtained are further defined using measurable indicators. The importance of each criterion (the weights) is discussed in Section 11.2. The applied multi-criteria approach is explained in Section 11.3. In the second phase, the points of view of the three stakeholders are brought together in a global analysis (Section 11.4). An application of this approach is illustrated in the methodology in Section 11.5.

11.6.1 Identifying the Criteria

All relevant criteria were determined according to the aims and goals of the parties involved. These groups are (European Commission 1996):

1. The users: transport operators, forwarders, consignors and consignees;
2. The terminal owners and/or operators;
3. The community as a whole.

Previous empirical research by Macharis and Verbeke (1999) determined the goals and aims of the different stakeholders. The criteria of the involved groups for the location of an intermodal terminal can be conflicting. The *users* look initially to the logistical desirability of the project. They want to minimize the cost of transport, which can be seen as the sum of the excess cost and the value of the transportation time. Even more important than transportation time in maritime container transport is the reliability and the frequency of the service. The services offered by the terminal and the connection with other transport modes are two additional criteria for the user.

The terminal *owners/operators* are more concerned with the financial side of the project. They need to be sure that the terminal is financially viable. Furthermore, the possibility to expand and the capacity of the waterway are important decision variables for this group.

The goals of the *community* as a whole concern minimizing the effects of building the new intermodal terminal on the environment and congestion and maximizing these on employment. The complementarity of the new terminal with the ports and the existing terminals is also taken into account.

It was necessary to translate these criteria into measurable indicators with a specific measuring method. In Tables 11.5 to 11.7, the criteria and their weights, and the indicators and their measuring methods are synthesized.

11.6.2 Determining the Preferences of the Decision Maker

In the previous steps, the possible projects (alternatives) and the criteria to evaluate them were determined. Also the methods used to evaluate the alternatives (a_1, …a_n) with respect to the criteria were discussed. An evaluation table can now be set up (see Figure 11.5).

	$f_1(.)$	$f_2(.)$	…	$f_j(.)$	…	$f_k(.)$
a_1	$f_1(a_1)$	$f_2(a_1)$	…	$f_j(a_1)$	…	$f_k(a_1)$
a_2	$f_1(a_2)$	$f_2(a_2)$	…	$f_j(a_2)$	…	$f_k(a_2)$
…	…	…	…	…	…	…
a_i	$f_1(a_i)$	$f_2(a_i)$	…	$f_j(a_i)$	…	$f_k(a_i)$
…	…	…	…	…	…	…
a_n	$f_1(a_n)$	$f_2(a_n)$	…	$f_j(a_n)$	…	$f_k(a_n)$

Figure 11.5 Evaluation table

This evaluation table can be analyzed in several ways (for an overview of the different multi-criteria analysis methods, see Vincke 1992; Pomerol and Romero 1993; De Brucker et al. 1998). Depending on the method, additional information relating to the preference structure of the decision maker is needed in order to further analyze the table.

We chose to use the PROMETHEE method because its principles can be easily explained to the decision makers, it is well adapted to the treatment of both quantitative and qualitative criteria, and the availability of the Decision Lab software makes it easy to analyze data and to perform sensitivity analyses. In the PROMETHEE method, two additional types of information are required, namely:

- The relative importance of the criteria (the weights);
- The way the decision makers compare actions according to a criterion (the preference functions).

This additional information defines the preference structure of the decision maker.

- The weights

Several methods for determining the weights were developed (for an overview of these methods, see Nijkamp, Rietveld and Voogd 1990; Eckenrode 1965). The PROMETHEE method does not provide guidelines for determining the weights, assuming that the decision maker is able to give weights to the criteria when the number of criteria is not too large. In order to structure this weighing process, the dual comparison procedure suggested by the AHP method was followed. This multi-criteria method was developed by Saaty (1982, 1988, 1995) and is supported by the EXPERT CHOICE software.

The modal choice variables considered important by the users when choosing between several transportation modes were determined using a study by Vannieuwenhuyse (1999). For the other two stakeholders, the weights of the criteria were elicited by means of a workshop with key individuals from the parties concerned.

- The preference function

The preference function translates the deviation of the evaluations of two actions on a particular criterion in a preference degree from 0 to 1.

Let:

$$P_j(a,b) = G_j\left[f_j(a) - f_j(b)\right] \tag{11.1}$$

$$0 \leq P_j(a,b) \leq 1 \tag{11.2}$$

be the preference function associated with criterion $f_j(.)$, where G_j is a non decreasing function of the observed deviation between $f_j(a)$ and $f_j(b)$.

In order to facilitate the selection of the preference function, six basic types have been proposed (see Brans 1982; Brans and Vincke 1985; Brans and Mareschal 1994).

For the LAMBIT model, the best suitable preference function is already selected according to the type of scale that is used (numeric, ordinal, ...) and the order of magnitude of the evaluations. The user of the LAMBIT model is, of course, free to change these preference functions.

11.6.3 Multi-criteria Decision Analysis

The PROMETHEE method computes the following quantities for each stakeholder r (r=1, ...R) and actions a and b:

$$\begin{cases} \pi^r(a,b) = \sum_{j=1}^{k} P_j(a,b) w_j^r \\ \phi^{+r}(a) = \sum_{x \in A} \pi(a,x) \\ \phi^{-r}(a) = \sum_{x \in A} \pi(x,a) \\ \phi^r(a) = \phi^{+r}(a) - \phi^{-r}(a) \end{cases} \tag{11.3}$$

For each *a* belonging to the set A of alternatives, $\pi(a,b)$ is an overall preference index of *a* over *b* taking into account all the criteria, with $\phi^{+r}(a)$ and $\phi^{-r}(a)$ measuring, respectively, the power and the weakness of *a* with regard to the other alternatives, and $\phi^r(a)$ is a value function, such that the higher it is, the better *a*. We call $\phi^r(a)$ the *net flow* of alternative *a* for stakeholder r (in the LAMBIT-model : r=1,..3).

For each stakeholder three main PROMETHEE tools can be used to analyze the problem :

- the PROMETHEE I partial ranking;
- the PROMETHEE II complete ranking;
- the GAIA plane.

The PROMETHEE I partial ranking ranks the alternatives, but in some cases alternatives are incomparable. This happens usually when the first alternative is good with respect to the particular criteria on which the second is weak, and vice versa. PROMETHEE I invites the decision maker to further investigate the problem.

PROMETHEE II provides a complete ranking of the alternatives from the best to the worst. The net flow is used here to rank the alternatives.

The GAIA plane (Geometrical Analysis for Interactive Aid) graphically displays the relative position of the alternatives with regard to the criteria and the conflicts between the criteria (according to the principal component analysis).

Additional tools such as the 'Walking Weights' and the decision marker can be used to further analyze the sensitivity of the results according to the weights (Brans and Mareschal 1994).

The application will further clarify the use of these tools in the Decision Lab software.

11.6.4 Global Ranking

In this phase, the points of view of the different affected parties are taken together (see Macharis, Brans and Mareschal 1998). The same PROMETHEE tools are available to the decision maker. In the GAIA plan, the conflicts between the points of view of the affected parties become very clear. This gives a good view of the trade-offs that will need to be made.

The global net flow ϕ^G is calculated as a weighted average of the individual net flows:

$$\phi^G(a_i) = \sum_{r=1}^{R} \sum_{j=1}^{k} \phi_j^{\,r}(a_i).w_j.\omega_r, \qquad i = 1,2,\ldots n, \tag{4}$$

where ω_r represents the relative importance of the stakeholder r.

11.6.5 Application

The Flemish government has commissioned an evaluation of two inland waterway terminals, planned relatively close to each other (Roeselare and Wielsbeke). The government wanted to know whether it should support terminals, whether these terminals would be viable, and whether they would benefit the community at large if they were built. Two reference terminals (Avelgem and Meerhout) were included in the analysis in order to have benchmarks. This section reports on this evaluation exercise. In order to keep the data confidential, the four terminals have been labelled as A, B, C and D.

An evaluation table was established for each stakeholder (user, operator/investor, and community). In these tables, the evaluation of the four terminals are given on the respective criteria which represent the goals of the stakeholders. An analysis can then be performed for every individual group of stakeholders.

Results of the User Model

In Figure 11.6, the PROMETHEE II ranking is shown for the user model. The net flows indicate the overall attractiveness of an alternative. Terminal C is ranked first, as its net flow ϕ is 0.16 followed immediately by terminal A, with a net flow of 0.14.

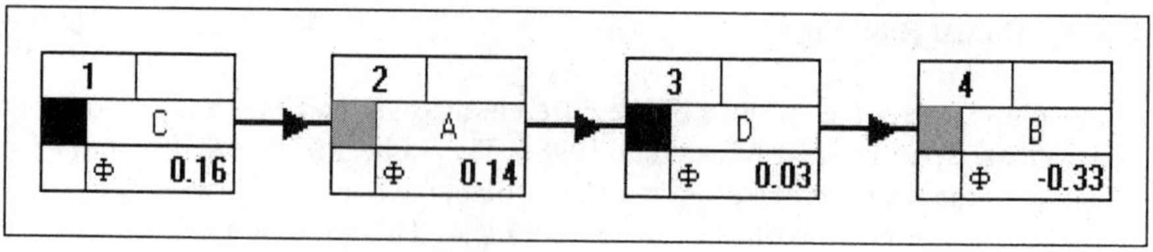

Figure 11.6 PROMETHEE II ranking of the user model
Source: Own set-up.

The GAIA plane graphically displays the relative position of the alternatives with regard to the criteria and the conflicts between criteria. In Figure 11.7, a clear visual representation of the decision problem is obtained.

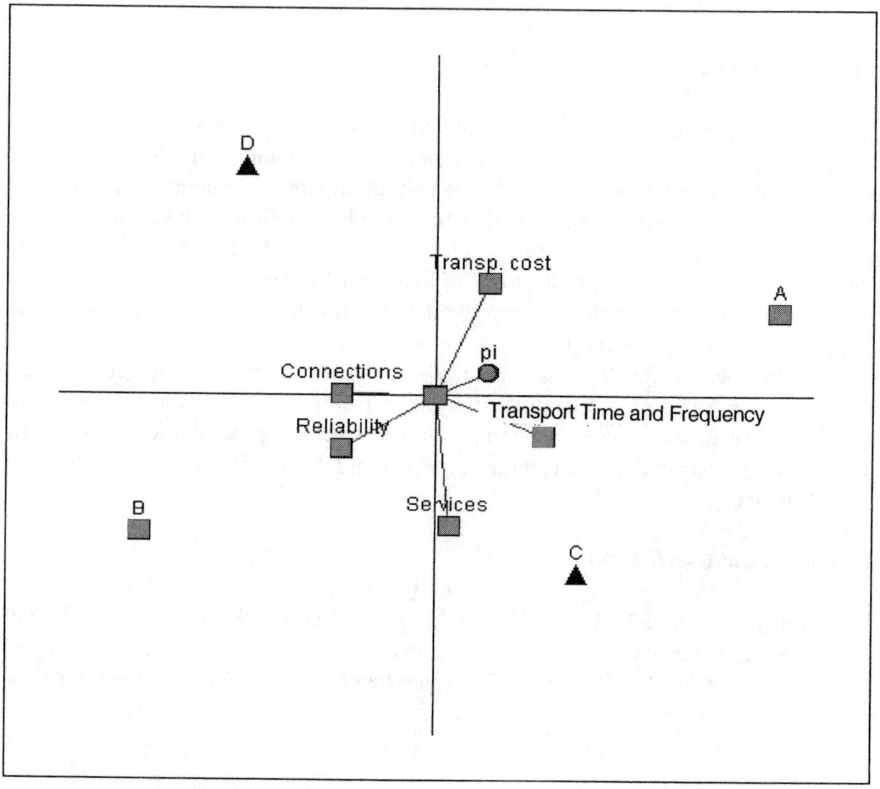

Figure 11.7 GAIA plane of the user model
Note 1: See text below for an explanation of superimposition of 'Transport time' and 'frequency'.
Source: Own set-up.

The points on the graph represent the six criteria. The direction of the lines indicate in which area of the plane the best actions will be located. For example, B obtained good scores on the criteria 'connections' and 'reliability', but poor scores on 'transportation cost', 'transportation time' and 'frequency' (these two last criteria are subject to an identical projection, which makes the figure less clear). A distinction is made between the reference terminals (A and B, denoted by a square) and the terminal projects (C and D, denoted by a triangle). It is also possible to establish the conflicting or non-conflicting nature of the criteria. The criteria 'transportation time' and 'frequency' coincide, suggesting that these two criteria are positively related to each other. The pi-decision marker points to the project which obtains high scores from an overall perspective according to the weights of the criteria. In this application, terminal A seems to be the most preferred according to all the criteria of the users and the relative importance they attach to it.

Results of the Investors/Operators Model

In the investors/operators model (Figure 11.8 and Figure 11.9), the profitability of the project received the largest weight as it is considered the most important objective of this group of stakeholders. The pi-decision marker is thus very close to the profitability marker. Terminal B is clearly the most preferred and there is no strong conflict between the criteria.

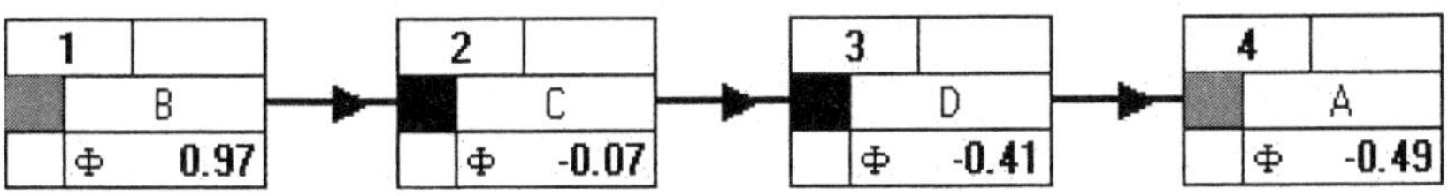

Figure 11.8 PROMETHEE II ranking of the investors/operators model
Source: Own set-up.

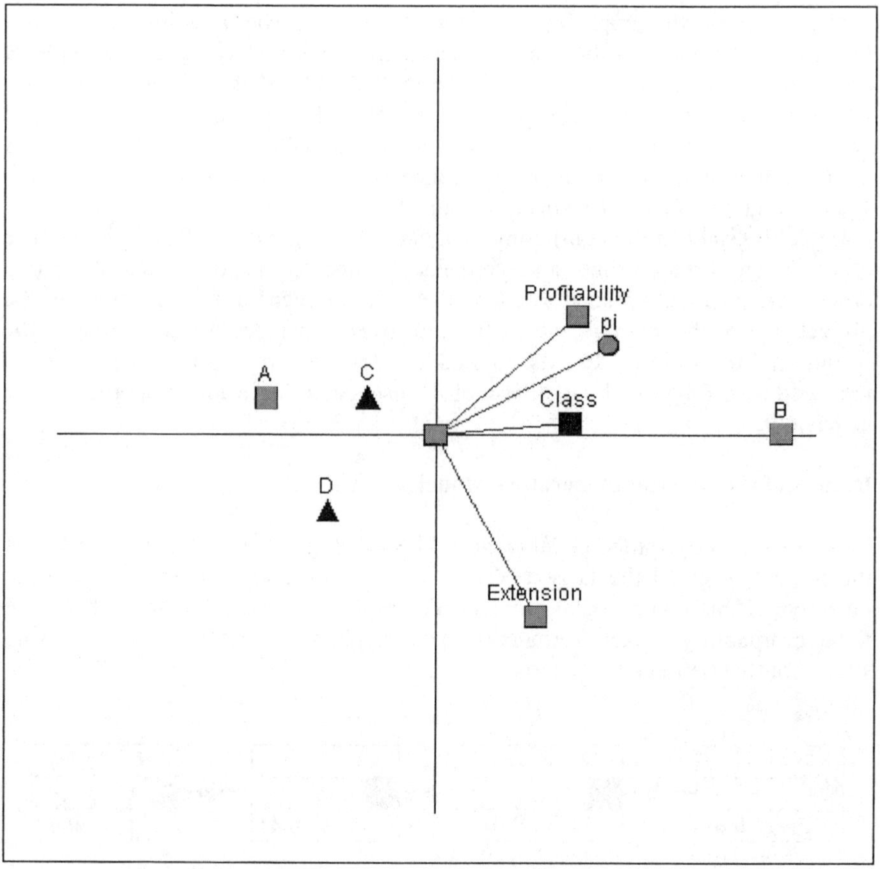

Figure 11.9 GAIA plane of the investors/operators model
Source: Own set-up.

Results of the Community Model

In the Community model, ten sub-criteria were considered. They can be synthesized in four categories, namely:

- Complementarity (network efficiency): Complementarity of the terminal with the ports and the existing terminals.
- Congestion: The local and global influence.
- Employment (regional development): Employment within the terminal and in the companies attracted by the new terminal.
- Environment: Reduction of emissions, energy use, land use and visual intrusion.

In Decision Lab, it is possible to group the criteria in such a way as to have a better view in the GAIA plane. To clarify the GAIA representation in this case only the four category criteria are represented. A more detailed view including the ten criteria is also possible.

In Figure 11.10 and Figure 11.11, terminal B again appears to be favoured above the others, but terminal C is more strongly differentiated from terminals A and D than in the previous stakeholders' model. This can be explained by their opposite profiles on employment and environment.

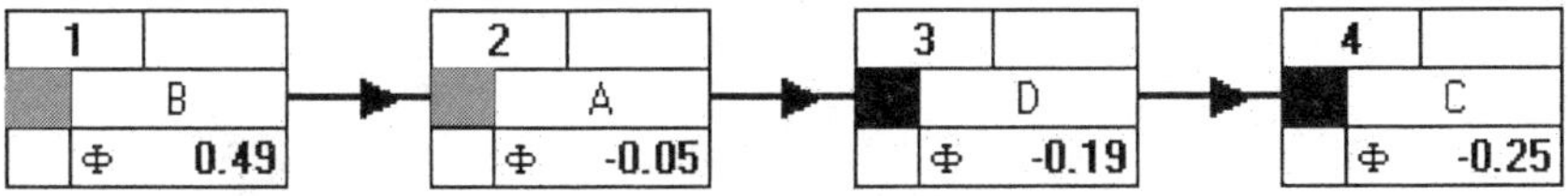

Figure 11.10 PROMETHEE II ranking of the community model
Source: Own set-up.

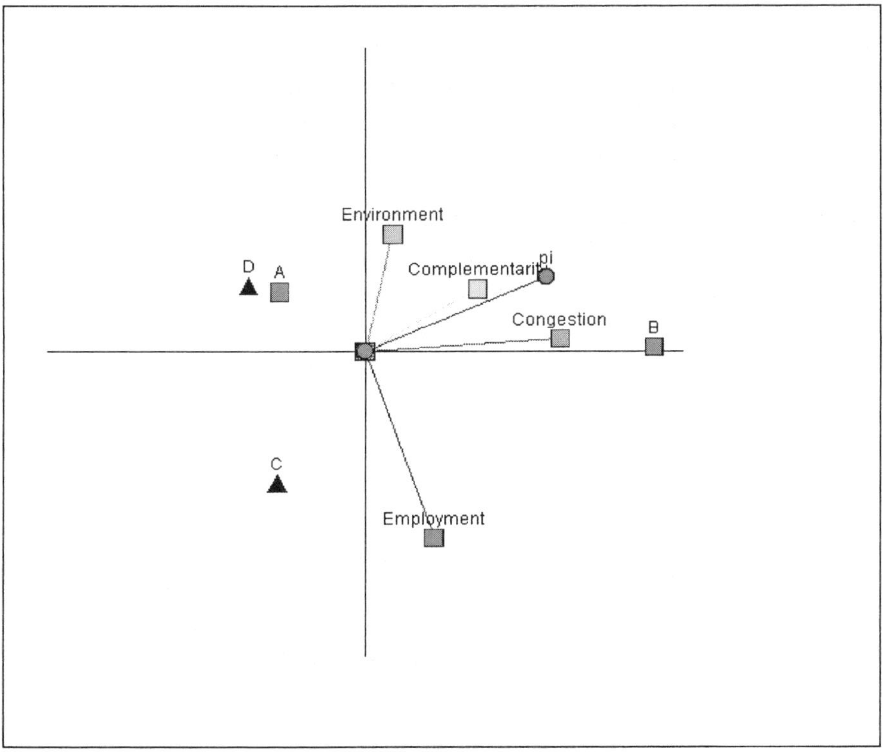

Figure 11.11 The GAIA plane of the community model
Source: Own set-up.

Globalization of the Results

In this phase, the points of view of the stakeholders were pooled together in order to obtain a global ranking of the projects. Here, the different stakeholders were given the same weight. It is, however, possible to change these weights to observe how this affects the results.

The global PROMETHEE II ranking procedure positions project B as the preferred terminal (see Figure 11.12).

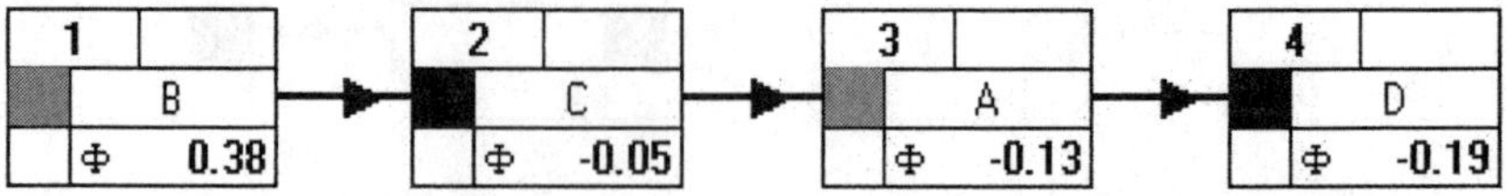

Figure 11.12 Global ranking
Source: Macharis, with the aid of Decision Lab (2000).

In the GAIA plane (Figure 11.13), the conflict between the stakeholders' points of view is now visualized. Terminal B obtains favourable scores on the goals of the investors/operators and on the goals of the community. The goals of the user conflict with the goals of the community (the two criteria fall on the other side of the axis). Project C obtains favourable scores on the goals of the users. The pi-decision marker, i.e. the overall decision, points to project B.

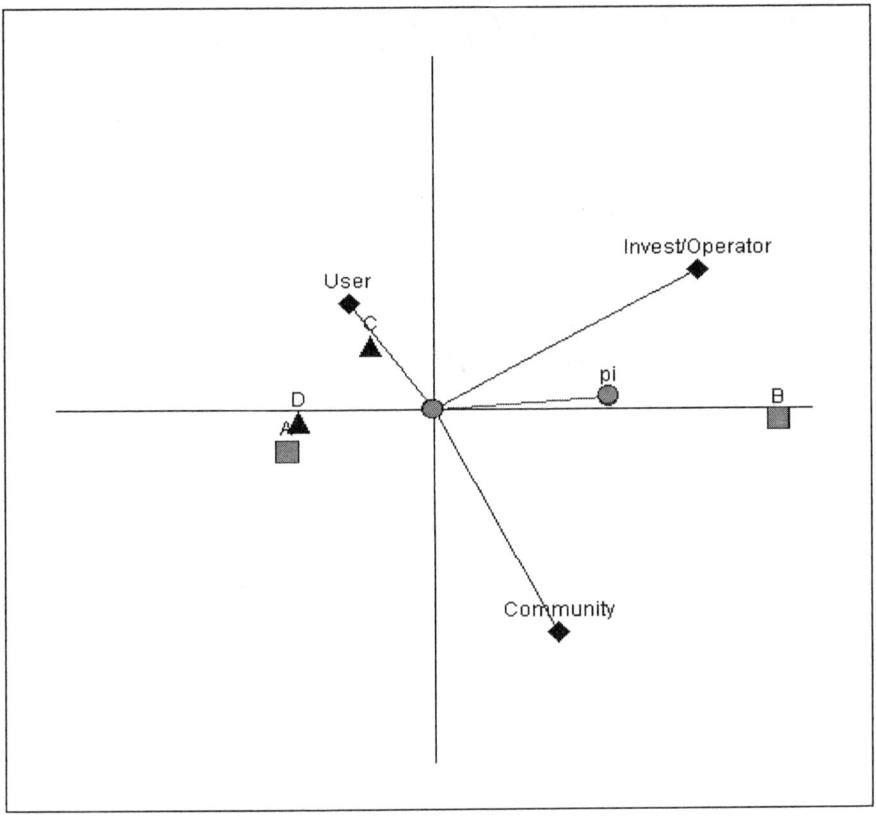

Figure 11.13 GAIA plane of the global problem
Source: Own set-up.

Sensitivity Analysis

A sensitivity analysis in a multi-criteria framework aims at determining the effect of changing the weights on the ranking of the alternatives.

A very interesting sensitivity tool is the use of 'stability intervals' (Mareschal 1988). These intervals give the allowable percentage changes in weights that do not affect the ranking (see Figure 11.14). If the weights, which were initially equal for each group of stakeholders (column 1), are changed by more than the allowable interval (columns 2 and 3), a change in the ranking will be generated. The last three columns of Figure 11.14 show the same information expressed in percentages.

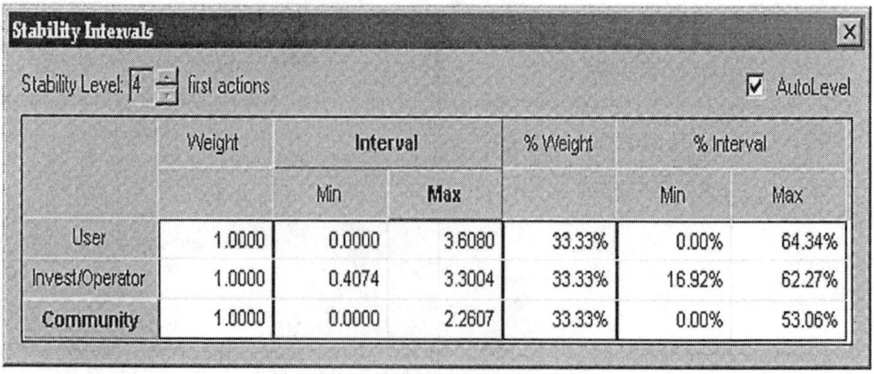

	Weight	Interval		% Weight	% Interval	
		Min	Max		Min	Max
User	1.0000	0.0000	3.6080	33.33%	0.00%	64.34%
Invest/Operator	1.0000	0.4074	3.3004	33.33%	16.92%	62.27%
Community	1.0000	0.0000	2.2607	33.33%	0.00%	53.06%

Figure 11.14 Stability intervals for the global problem
Source: Own set-up.

11.7 Conclusion

The choice of an optimal location for an intermodal terminal is a very important decision for the private investor as well as for the government. The investor has to gain a realistic idea of the traffic potential before he will invest any considerable amount of money. The policy makers need instruments that can help them to analyze the effects of new intermodal terminals on the terminal landscape and to make a comparison between several terminal projects.

The LAMBIT methodology allows a comparative evaluation of a set of terminal projects according to several criteria. These criteria were set up according to the points of view of several stakeholders. This enables a multi-party analysis that clearly identifies any areas where possible conflicts may arise. The strengths and weaknesses of the projects are exposed.

The most important criterion is the traffic potential of the new terminal, which was determined in a preliminary phase. A network model was set up, so as to obtain a realistic estimation. On the basis of the real market prices and the freight volumes that are currently transported on the road, the model makes it possible to estimate this traffic potential. It further enables us to see the competitive effects of new terminals on existing ones and to simulate several policy options.

References

Arnold, P. and Thomas I. 1999: Localisation des Centres de Transbordement dans un Système Multi-reseaux: Essai de Formalisation, L'Espace Géographique, Vol. 3, 193–204

Ashayeri J. and Rongen J.M.J. 1997: Central Distribution in Europe: A Multi-Criteria Approach to Location Selection, The International Journal of Logistics Management, Vol. 8, 1, 97–109

Bontekoning Y.M., Macharis C. and Trip J.J. 2002: Is a New Applied Transportation Research Field Emerging? – A Review of Intermodal Rail-truck Freight Transport Literature, Transportation Research Part A

Brans J.P. 1982: L'Ingéniérie de la Décision. Elaboration d'Instruments d'Aide à la Décision. Méthode PROMETHEE, in Nadeau R. and Landry M. (eds.), L'Aide à la Décision: Nature, Instruments et Perspectives d'Avenir, Presses de l'Université Laval, Québec, Canada, 183–214

Brans J.P. and Mareschal B. 1994: The PROMCALC & GAIA Decision Support System for MCDA, Decision Support Systems, Vol. 12, 297–310

Brans J.P. and Vincke Ph. 1985: A Preference Ranking Organization Method: the PROMETHEE Method, Management Science, Vol. 31, 647–656

Daskin M.S. 1995: Network and Discrete Location. Models, Algorithms and Applications, Wiley, New York

De Brucker K., Verbeke A. and Winkelmans W. 1998: Sociaal-economische Evaluatie van Overheidsinvesteringen in Transportinfrastructuur, Garant, Leuven

Declercq E. and Verbeke A. 1997: Locatieanalyse en Beleidsrelevantie met een Toepassing voor de Binnenvaartsector in Vlaanderen, Tijdschrift Vervoerswetenschap, Vol. 33, 3, 195–211

Eckenrode R.T. 1965: Weighting Multiple Criteria, Management Science, Vol. 12, 3, 180–192

European Commission 1996: Transport Research APAS, Cost-benefit and Multi-criteria Analysis for Nodal Centres for Goods, Office for Official Publications of the European Community, Luxemburg

Groothedde B. and Tavasszy L.A. 1999: Optimalisatie van Terminallocaties in een Multimodaal Netwerk met Simulated Annealing, Proceedings of Vervoerlogistieke Werkdagen 1999, Connekt, Delft 1999

Macharis C. 2000: Strategische Modellering voor Intermodale Terminals. Socio-economische Evaluatie van de Locatie van Binnenvaart/weg Terminals in Vlaanderen, PhD Thesis, Vrije Universiteit Brussel, Brussel

Macharis C. and Verbeke A. 1999: Intermodaal Vervoer. Economische en Strategische Aspecten van het Intermodaal Vervoer in Vlaanderen, Garant, Leuven

Macharis C., Brans J.P. and Mareschal B. 1998: The GDSS Promethee procedure, Journal of Decision Systems, Vol. 7, 283–307

Maggi E. 1998: The Location of Logistics Nodes, in Neptune Supporting Co-operation, Study on the Location of Logistic Nodes, ISTIEE, Trieste, 1–29

Mareschal B. 1988: Weight Stability Intervals in Multicriteria Decision Aid, European Journal of Operational Research, Vol. 33, 54–64

Nijkamp P., Rietveld P. and Voogd H. 1990: Multicriteria Evaluation in Physical Planning, Elsevier Science Publishers, Amsterdam, 219

Pomerol, J.C. and Barba-Romero S. 1993: Choix Multicritères dans l'entreprise. Principe et Pratique, Hermes, Paris, 391

Rutten B.J.C.M. 1995: On Medium Distance Intermodal Rail Transport, PhD Thesis, Delft
Saaty T.L. 1982: Decision Making of Leaders, Lifetime Learning Publications, Wadtsworth, Belmont
Saaty, T.L. 1988, The Analytic Hierarchy Process, McGraw Hill, New York
Saaty T.L. 1995: Decision Making For Leaders, the Analytic Hierarchy Process for Decisions in a Complex World, RWS Publications, Pittsburgh
Van Duin R. and Van Ham, H. 2001: A Three-Stage Modeling Approach for the Design and Organization of Intermodal Transportation Services, Proceedings of the IEEE International Conference on Systems, Man and Cybernetics, Part 4, 11-14 Oct., San Diego, CAL., USA
Vannieuwenhuyse B. 1999: Gedetailleerde Bespreking en Voorstelling van Enquêteresultaten. Internal report, KUL, Industrieel beleid, Leuven
Vincke Ph. 1992: Multicriteria Decision-Aid, John Wiley & Sons, New York
Visual Decision Inc. 1999: Decision Lab 2000 – Getting Started Guide, Montreal, Canada

---

Endnotes

[1] Main costs: primary transport costs, secondary transport costs, inventory and capital investments, storage costs, costs of buildings, land costs, labour costs and source of skilled labour, energy costs. A reduction of costs can be obtained by the economies of agglomeration.

[2] Examples are standard of services, fiscal legislation of a particular country, trade union/political situation, influence on the environment.

[3] Examples are standard of services, fiscal legislation of a particular country, trade union/political situation, influence on the environment.

[4] A further refinement of the model could, however, also take account of these flows and of the existing road/rail terminals.

[5] The real market prices, in force in December 1999, for the unimodal road transport and the intermodal inland waterway/road transport were collected from among experts for all relevant origin-destination relations that are considered in this chapter. Given the confidentiality of these data they could not be added in an annex.

[6] In ARCINFO the algorithm of Dijkstra is used (for a description of this algorithm, see e.g. Daskin (1995)).

[7] To convert ton to TEUs the following ratio is used: 1 TEU=10.8 tonnes.

12 Rail-rail Hub-terminals as an Alternative for Shunting: An Explorative Comparative Case Study

Yvonne Bontekoning and Jan Jacob Trip

Delft University of Technology/OTB Research Institute for Housing, Urban and Mobility Studies, Delft, The Netherlands

12.1 Introduction

Shunting yards are important exchange nodes for those intermodal freight flows that are too small for frequent direct point-to-point shuttle connections. To maintain a daily or more frequent shuttle service, trains need to have a certain minimum length and loading degree to be economical. However, by bundling such small freight volumes (consolidation), it is still possible to achieve a frequent connection. Such bundling networks could be line networks, hub-and-spoke networks or collection-and-distribution networks (Figure 12.1). At the start and end points of these networks, trucks take care of drayage to and from the shipper or receiver.

Bundling networks require nodes where load units or wagons can be exchanged between trains. It is very common to apply the shunting technique at such nodes. But shunting has some serious disadvantages that negatively influence the competitiveness of intermodal transport. First, shunting is a rather time-consuming operation that affects the transit time of the train, which is one of the main mode decision determinants for shippers (see, e.g., Oum 1979). Especially in Europe, where the common intermodal distance lies between 450 and 1200 kilometres, shunting operations may take between 10 to 50 percent of the chain lead-time of a train. Second, the time-consuming shunting operations also affect the cycle time of rail wagons. The more time shunting takes, the lower the productivity of the rail wagons. Third, despite the fact that many shunting yards have modern equipment to precisely brake the speed of the shunted wagon, shippers also fear the risk of damage to freight due to shunting. Fourth, shunting yards are often situated close to expanding urban areas that suffer noise pollution from the shunting operations. Finally, shunting yards are land consuming.

In this chapter we address the problem of shunting in hub-and-spoke networks. Recently, terminal manufacturers have proposed alternative solutions for shunting in the form of (new-generation) hub-terminals and elaborated these concepts into detailed designs. At these hub-terminals, standardized load units (such as containers, swap bodies and sometimes semi-trailers) are transhipped from one train to another, instead of shunting rail wagons. Hub-terminal operations strongly rely on the advanced possibilities of information and communication technology (ICT). As a result of ICT, more complex and more simultaneous operations can be carried out. In various innovative terminal designs, robots replace crane and terminal truck drivers. None of these hub-terminals have been implemented yet, for three main reasons. First, these hub-terminals are relatively expensive compared with shunting yards that are often already depreciated. Next, these terminals apply new and sometimes very sophisticated technology. And finally, hub-terminal operational strategies are very different from shunting.

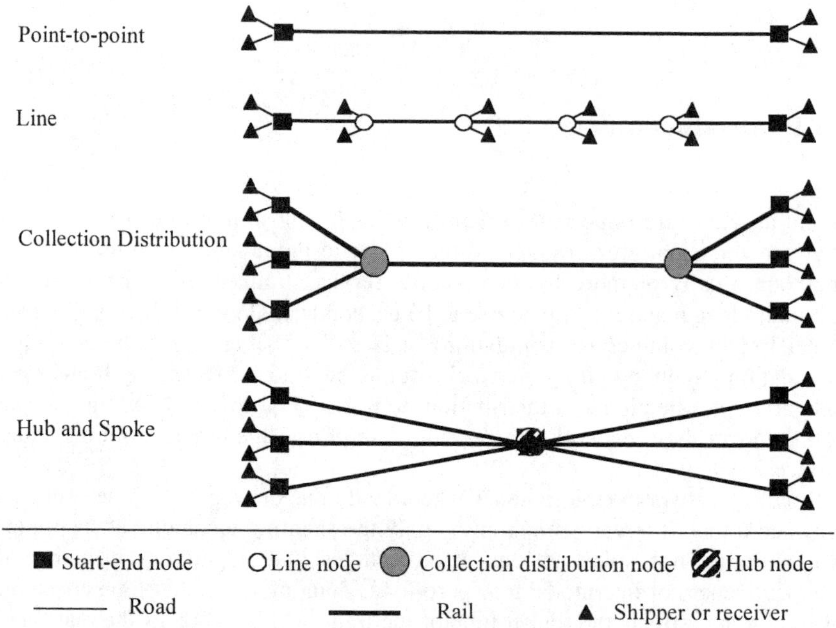

Figure 12.1 Point-to-point bundling vs. complex building networks
Source: Bontekoning and Kreutzberger (2001), adjusted.

Explorative and descriptive studies (see, e.g., Woxenius 1998; Bontekoning and Kreutzberger 1999) describe, among others, four hub-terminals designs, namely, respectively: Commutor; Fast Handling System; Compact Terminal; and Megahub. Bontekoning and Kreutzberger (2001) evaluate and compare the performance of 14 new-generation terminal designs, among which the four above-mentioned hub-terminal designs, with shunting and a conventional reference terminal (see also Bontekoning 2000). These studies indicate that innovative terminal solutions could overcome most of the disadvantages of shunting yards listed above. Consequently, intermodal networks with hub-terminals instead of shunting yards might improve the efficiency and profitability of intermodal transport. However, although hub-terminals are considered to be an alternative for shunting yards, the operations and performances of the two types of operations have not yet been thoroughly compared.

The objective of this chapter is to take a first step towards a better understanding of the characteristics of the two types of operations and a thorough performance comparison. At this stage of comparing the two types of operations, we carried out an in-depth qualitative comparative analysis to explore the operational details and structure. The scientific value of this work is twofold. First, it contributes to a better understanding of new operating strategies that hub-terminal operations bring with them and which need further optimization. Second, the comparison with shunting reveals the strengths and weaknesses of hub-terminal operations.

The outline of this chapter is as follows. In Section 12.2, the layout and operations of the shunting yard and the hub-terminal are described. The research approach and data set applied are discussed in Section 12.3. Section 12.4 contains the results of the qualitative performance comparison between shunting and hub-operations, followed by a discussion of the findings in Section 12.5 and the conclusions in Section 12.6.

12.2 Layout and Operations: Shunting Versus Hub-operations

In this section, we describe the layout and operations of the existing shunting yard Metz-Sablon (France) and of the specially designed hub-terminal that could replace shunting in Metz-Sablon.

12.2.1 Layout and Operations of the Shunting Yard

The shunting yard in Metz-Sablon is the central hub in ICF's Pan-European intermodal Qualitynet, which is depicted in Figure 12.2. Intercontainer-Interfrigo (ICF) is an association of the different European Railway Companies. The shunting yard consists of:

- an arrival siding with 10 tracks;
- a shunting hill;
- a combined sorting and departure siding of 37 tracks;
- a small siding not connected to the shunting hill but connected to the arrival siding of 8 tracks;
- two sets of switches (from Siemens, dating back to 1914).

The Metz-Sablon yard covers an area of 250,000 m^2. Nine of the 45 sorting and small siding tracks are reserved for general cargo rail operations (Terminet 2000; ICF 1999). The general layout of the shunting yard is schematically depicted in Figure 12.3. Two shunting locomotives and 130 ICF workers (in three shifts) are available to carry out the work at the shunting yard (La Vie du Rail 1999). The shunting yard is open 24 hours per day, but the majority of the shunting operations takes place between the times 22:00 and 6:00.

Figure 12.2 ICF's Pan-European network
Source: Based on ICF brochure in Terminet (2000).

Operations in Metz are similar to those at many other shunting yards (see, for instance Jung and Talke 1998a and 1998b; Huet and Vallet 1977/78). Trains arrive at the arrival siding. Locomotives are uncoupled and leave. Data about the wagons, such as weight and destination, are transmitted into the yard's control computer. The wagons (or groups of wagons) are uncoupled. The shunting locomotive arrives, picks up a train, pushes it to the shunting hill and sequentially pushes the wagons (or groups of wagons) over the hill. The shunting locomotive then returns and picks up the next train at the arrival siding, which in the meantime has been uncoupled, and takes it to the hill. All the while, wagons roll off the hill, and are braked and directed by switches into the right departure track. Wagons are not precisely braked, so that a locomotive has to push the wagons together. This assembling of wagons can be partly conducted concurrently with pushing the wagons over the hill. The faster the wagons are pushed over the hill, the more chance there is that one wagon will overtake another and move into the wrong departure track because the switches were set for the situation without overtaking. These mistakes are corrected by moving wagons in and out of the departure track, in order to remove the wrong wagon. Finally, the wagons in a departure track are coupled, the network locomotive is coupled in front, and a brake test is carried out. Some trains are assembled in two steps. First, two or more wagon-groups are formed at the departure siding. Next, these groups (blocks) are combined at the small siding (see Figure 12.3).

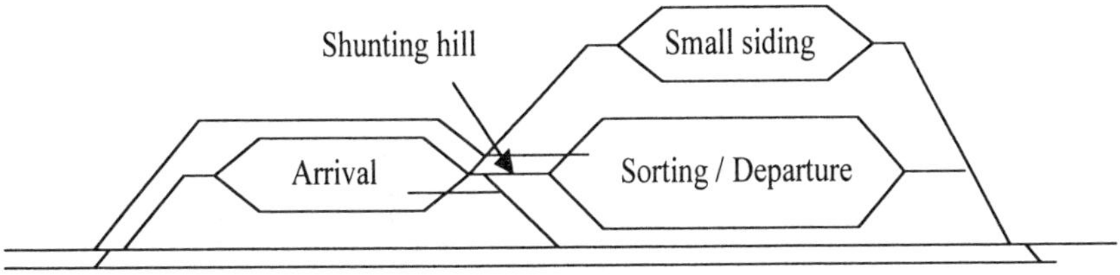

Figure 12.3 General layout of shunting yard Metz-Sablon
Source: Based on ICF (1999).

12.2.2 Layout and Operations of the Hub-terminal

In the context of the Terminet study, Vogtmann and Franke (2000) designed a hub-terminal for the location Metz-Sablon, based on data provided by ICF. The hub-terminal that could replace the shunting yard at Metz-Sablon would then consist of:

- six manually-driven cranes (which can easily be robotized), each covering six rail tracks, two road lanes, three storage lanes and a terminal transport system (three lanes); each crane has a capacity of 56 moves per hour;
- a robotized linear roller pallet transport system to sort load units and move them between crane sections; the sorting system consists of two driving lanes, one parking lane and 15 roller pallets, and can handle 75 units per hour.

The terminal layout allows direct and indirect (via storage) rail-rail transhipment, as well as direct and indirect rail-road transhipment. The surface area of the terminal is 10 ha. See Figure 12.4 for the general layout of the terminal.

The operational process at the hub-terminal, as described by Vogtmann and Franke (in Terminet 2000) would be as follows. The first train of the first batch arrives. Electrically powered trains move into the terminal, decelerate and stop at a position where the locomotive can be charged again. All following trains of the same batch arrive at intervals of 8 minutes. When the last train of the batch has arrived, the time spent simultaneously by all trains amounts to at least 16 minutes. The trains then leave the terminal at intervals of 8 minutes. After the arrival of the first train, one or a few cranes start their transhipment operations according to an optimal working order, which is calculated by the control computer. The more trains have arrived, the more cranes become active. There are six cranes; each crane takes care of the direct rail-rail transhipments within its own section. Load units that have to be exchanged between crane sections are put on the sorting system. Switching load units between crane sections implies two crane handlings, while direct transhipments within a crane section imply only one handling by the same crane. Load units for connecting trains which are not part of the same batch are stored in one of the buffer lanes, unless the (original) train remains in the terminal and is waiting for the next batch of trains to arrive. Finally, the first train of the subsequent batch enters the terminal as soon as the last train of the previous batch has left (unless a train is to remain in the terminal).

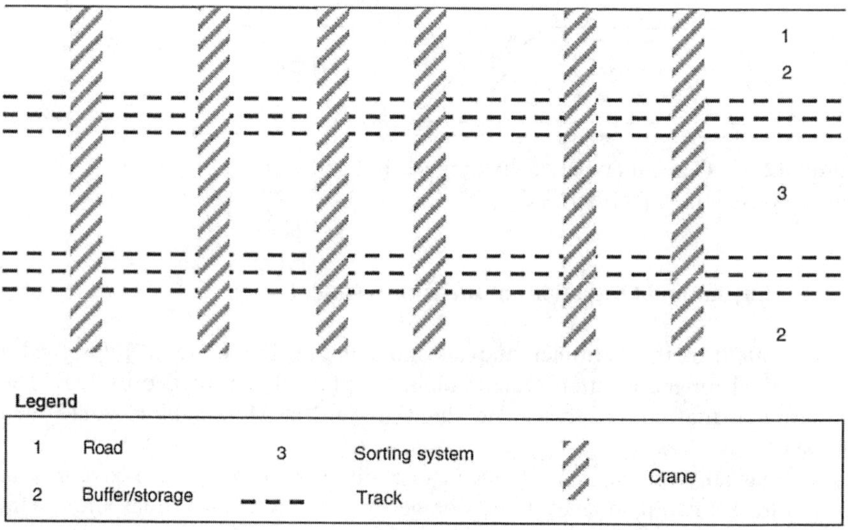

Figure 12.4 General layout of the hub-terminal designed for Metz-Sablon

12.3 Research Approach

The objective of this study is to obtain more insight into the differences in operations and performances between shunting and hub-terminal operations. The study is of an explorative and descriptive nature. The available data used in this work are rather unique, because it is seldom that commercial actors allow such openness about operating strategies and costs. Nevertheless, there are some limitations, which prevents us from carrying out a more thorough comparison of both types of operations by means of integral and quantitative methods, such as multi-criteria analysis, cost-benefit or computer simulation. However, before applying more integral evaluation methods, a thorough understanding of the operations is necessary anyway. This study contributes to such thorough understanding of the operations.

We have chosen a case study approach. Empirical data were available for the existing shunting yard Metz-Sablon (France) and for a specific hub-terminal design of the Megahub of the firm Noell (from now on called hub-terminal) for the situation in Metz-Sablon, elaborated by Vogtmann and Franke (in Terminet 2000). The advantage of this case study is that both the shunting yard and the hub-terminal can be analysed for the same daily throughput. With the available data set we have been able to calculate the daily throughput handling time and the train handling time. The first indicator is of interest for terminal operators, the second indicator for network operators.

The data we obtained consist of the daily throughput of one week in 1999, as well as a train timetable. On average 25 to 28 trains are formed daily (La Vie du Rail 1999). Unfortunately, the data provided by ICF are confidential. Due to this confidentiality we cannot provide the exact number of wagons and trains per day; however, we will discuss in this chapter how we determined performances in order to make most of our work verifiable.

There are two fundamental differences in operations between shunting and terminal rail-rail exchange. The first difference is that shunting involves wagons (or wagon groups), while at a hub-terminal load units are transhipped. Consequently, the available data on throughput in Metz in wagons per day must be transformed into load units per day. The average number of load units on a wagon was estimated by ICF to be 1.6 (Terminet 2000). The second difference is that the Metz shunting yard has enough tracks to receive and reassemble all 25 to 28 trains at the same time, while at the hub-terminal only six trains can be dealt with at the same time. Consequently, a transformation of the arrival and departure schedule into sequential batches of two to six trains is necessary. A new arrival and departure schedule was determined, based on arrival train number, departing train number and the origin and destination terminals of trains. In the available data, for each arrival train number it was indicated how many wagons were assigned to which departing train number. By plotting this information in a matrix, it was possible to identify trains that were related and trains that were not. Due to the variety in origin and destination terminals (Figure 12.2), it turned out that the exchange between the trains could be rearranged into batches of two to six trains. The exchange of load units between trains belonging to two different batches could be solved in two ways. Either such a train stayed for a period of two batches at the terminal, or the load units were exchanged sequentially, which means via the storage.

Comparing existing shunting operations and non-existing hub-operations is a delicate task. There are quite a number of differences in the underlying facts and assumptions, and in performance calculation methods. The underlying facts and assumptions in addition to those mentioned elsewhere in this chapter are:

- the shunting capacity is estimated at two wagons per minute (Terminet 2000);
- a continuous arrival and departure pattern of trains and no disturbances of the operational process are assumed. Waiting times due to irregularities in train arrival and departure patterns or shunting or hub-operations are ignored;
- the capacity needed for a hub-terminal in the situation at Metz-Sablon was calculated by Vogtmann and Franke of the firm Noell, using the data provided. The number of cranes and the number of rolling pallets were determined on the basis of peak demand, combined with the results of a previous hub-terminal study (Meyer 1998). Each crane has a maximum

capacity of 56 moves per hour, but the combined productivity of the cranes is assumed to be 70 percent (based on simulation results for the node Hannover-Lehrte). This is caused by a number of factors, including fluctuations in load volume over the hour, the uneven distribution of loads between cranes, and waiting times. The actual capacity of one crane is 39 moves per hour (56 moves * 70%). In the peak period 228 moves need to be carried out, which requires six cranes;
- it is calculated that in a worst case situation, 2/3 of the transfer flow needs to be handled by the sorting system.

12.4 Results of the Performance Comparison

Different actor groups apply different criteria to assess node performance. Consequently, the discussion of the results of the performance comparison of the hub-terminals and the shunting yard is split in two. The first part of this section deals with the daily throughput handling time, which is an indicator of interest to node operators. However, network operators judge a node's performance on another indicator: namely, the train handling time. Therefore, the train handling time in both types of operations is discussed in the second part of this section.

12.4.1 Daily Throughput Handling Time: Node Operator's Perspective

Figure 12.5 shows the daily throughput handling times for shunting and for hub-operations for the week of analysis. The throughput handling times for the hub-terminal were determined as follows. The batch size (number of trains) determines the number of interval periods of 8 minutes between the trains belonging to the same batch. For each batch the handling time was determined. Adding up all batch handling times generates the daily throughput handling time, which is always a multiple of 8 minutes. The daily throughput times for the shunting yard were calculated in a different way than those for the hub-terminal. ICF estimated the handling time for each activity that is carried out at the shunting yard. Various activities can be executed concurrently, while other activities are related to the number of trains or the number of wagons. Table 12.1 shows the estimated handling times, with an example of the calculation of the daily throughput handling time for one day's throughput.

Figure 12.5 clearly shows that, under the assumptions considered, the hub-terminal needs less time to handle the same throughput than the shunting yard (the difference amounts to 2:24 to 3:38 hours). Among other things, however, this finding is sensitive to the assumptions of the throughput capacity of the hill and the number of cranes. Although the hill capacity of two wagons per minute is an empirical figure provided by ICF, we may have to question the value of it. In an article published in 1977/78 (!), Huet and Vallet discussed the performance improvement of the French shunting yards Sotteville and Hourcade after automation was introduced. At the shunting yard Sotteville, a performance of between 4 to 6 wagons per minute was reached before the automation of the brake system, which often improved to 8 after automation. The performance figures of shunting yard Hourcade were in the same range. Performances of up to 10 to 12 wagons per minute could be reached, although this is not always considered desirable due to the increasing chances of false runners (wagons overtaking each other and moving into the wrong sorting track).

Table 12.1 Example calculation of daily throughput time at the shunting yard for one day of operations

Type of operation	min.	Handling time at shunting yard in minutes
Uncoupling of wagons (incl. uncoupling of locomotive)*	30	(only first train, because other trains are dealt with parallel to other activities)
Moving trains to hill (incl. driving shunting locomotive to arrival siding)*	280	(10 min. per train)
Hill time	248	(2 wagons per minute)
Pushing-together of wagons*	20	(partly parallel action)
Correcting improperly positioned wagons*	40	(partly parallel action)
Coupling time wagons, locomotive plus brake test*	40	(only one train is counted, can be considered as parallel action with hilling)
Approximate total time		**10 hours and 58 minutes**

*Estimation for a batch of 6 trains by ICF (Terminet, 2000), but assumed to be applicable here too.

As it would seem, therefore, that the situation in Metz is not state-of-the-art, we calculated the throughput time for a scenario for shunting in which we increased the hill capacity to 8 wagons per minute. The duration of other shunting activities is assumed to remain the same. The results are plotted in a graph (Figure 12.6), and presented together with the original results. The x-axis now shows the number of wagons, instead of the weekdays. Trend lines are indicated as well.

Figure 12.6 clearly shows the impact on the comparison of the variation in the shunting hill capacity between shunting and hub-terminal operations. For a hill capacity of 2 wagons per minute, the hub-terminal outperforms the shunting yard for any level of throughput. When the hill capacity is increased to 8 wagons per minute, however, the hub-terminal performs better up to a maximum of 525 wagons. For higher throughput volumes shunting performs better. This does not mean that the hub-terminal concept performs faster than shunting in general; however, it does perform faster on daily throughput for the Metz case study.

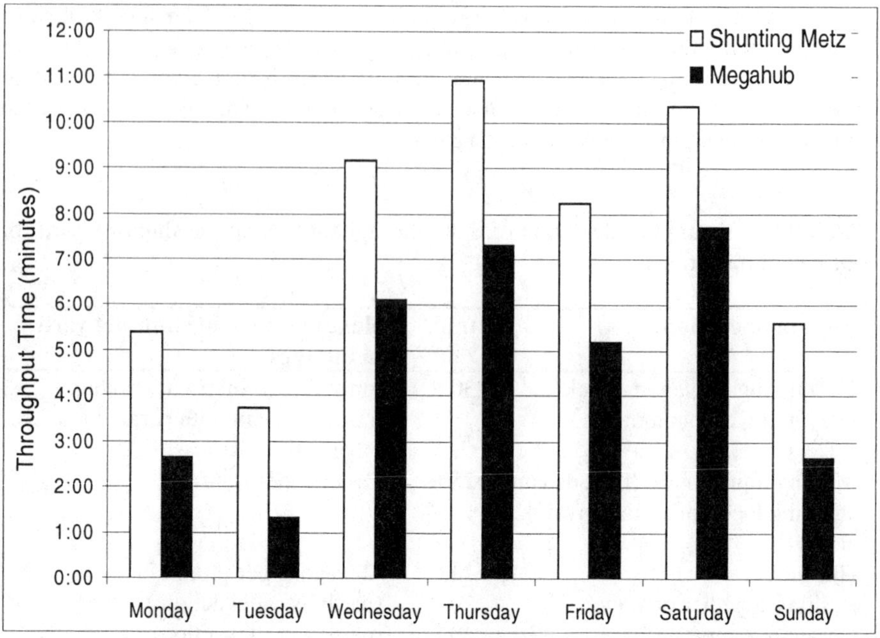

Figure 12.5 Comparison of daily throughput handling time for shunting and hub-terminal operations (wagons and load units per day)

Two issues influence the position of the curves presented in Figure 12.6, and must therefore be considered in the comparison between shunting and hub-operations. First, the curves for shunting and hub-operations contain a fixed time element, which differs for each type of operation. At the hub-terminal it takes 5 minutes for a train to enter the terminal. Only one train is incorporated in the throughput time, because other trains arrive at the same time when other operations are being carried out. Shunting in Metz has a fixed time element of 2 hours and 10 minutes, caused by four of the six shunting activities (see Table 12.1). Driving locomotives and pushing wagons over the hill are the only time elements that are not fixed. Furthermore, a number of factors determine the slope of each curve. With respect to shunting, there are three such factors. The first factor is the throughput capacity of the locomotives: the faster the locomotives or the more locomotives, the less steep the slope of the curve. The second factor is the throughput capacity of the shunting hill. As we can see in Figure 12.6, the higher the capacity, the more gentle the slope. Finally, Figure 12.6 shows the impact of the ratio of the number of trains to the number of wagons. The bend between 150 and 225 wagons in the curves for shunting is caused by the fact that the number of wagons increases, while the number of trains decreases. The number of wagons per train therefore increases. Consequently, less time is needed for coupling and uncoupling of wagons and driving the locomotive to and from the hill. Between 450 and 475 wagons, the curve for shunting also shows a bend, but in the opposite direction. This happens because the number of trains increases faster than the number of wagons; the number of wagons per train therefore decreases.

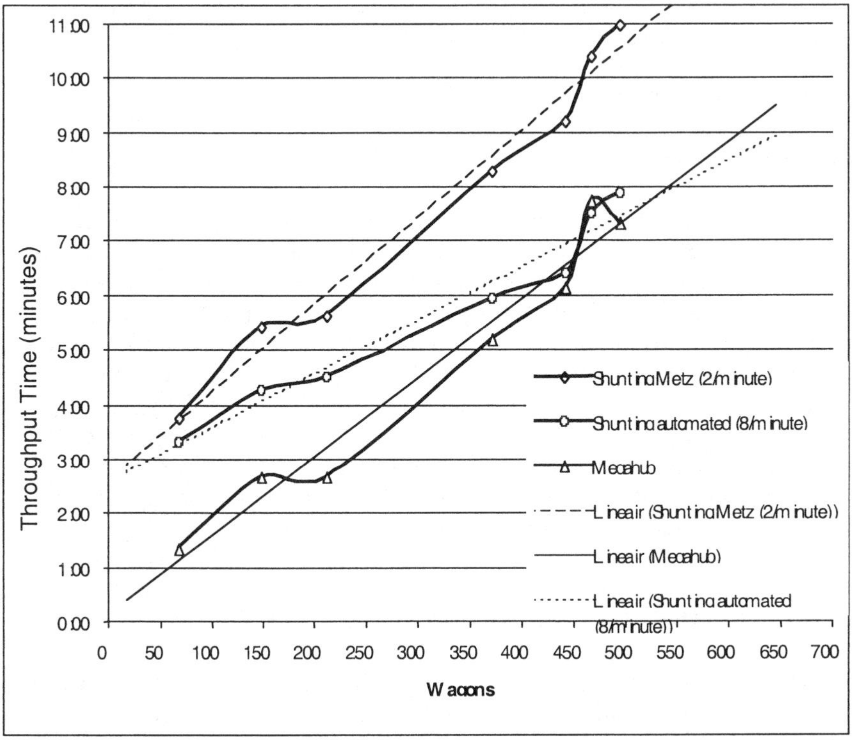

Figure 12.6 Comparison of throughput time for shunting and hub-terminal operations for increased hill capacity at the shunting yard

Something comparable occurs in the slope of the curve for hub-operations. The slope of the curve for hub-operations depends on the number of batches, the number of trains per batch and the number of load units per batch. The curve changes between 150 and 200 wagons, because the number of load units per batch increases, while the number of trains decreases. Consequently, less 8-minute intervals are needed. The increase in units can be handled without increasing the handling time, because the crane capacity is less restricting than the 8-minute arrival and departure pattern. The spare crane capacity is therefore able to absorb the increase in throughput. Of course, there is a limit to the increase in the number of load units per batch. This limit is reached when the capacity of the cranes or sorting system is not sufficient to guarantee the 8-minute interval periods. At that point cranes or roller pallets could be added. Something similar happens between 450 and 475 wagons. Here too, there is a bend, but in the reverse direction. This implies that more 8-minute intervals are needed for a comparable number of wagons (load units), due to smaller batches.

Concluding, in respect to the daily throughput handling time for the actual situation in Metz, the hub-terminal performs better in terms of throughput handling time, and could therefore be an alternative to improve the productivity of the node Metz. Productivity in Metz could also be improved by three other measures:

- increasing the hill capacity;
- reducing the time for 'fixed' activities;
- reducing the time to take trains to the shunting hill.

Generally speaking, it seems that a hub-terminal is a good solution for throughput volumes of less than 550 wagons (880 load units) per day. For higher throughput volumes shunting seems a more adequate solution. The advantage of hub-operations is that start-up and departure activities hardly take any time, while for shunting of lower volume this 'non-productive' time places a heavy burden on the throughput time. Pushing wagons over the hill is in itself a much faster operation than rail-rail terminal exchange. However, the shunting yard loses a lot of its capacity due to the activities which cause the fixed time burden of 2 hours and 10 minutes shown in Table 12.1. The hub-terminal loses a great deal of its crane capacity due to the crane hindrance factor (0.7) and the high number of moves related to load units (1.9 moves for each unit).

12.4.2 Train Handling Time: Network Operator's Perspective

In contrast to node operators, network operators judge a node's performance by train handling time, rather than daily throughput handling time. At the hub-terminal the train handling time can vary, depending on:

- the batch size (2 to 6 trains);
- the period of concurrent stay (this normally amounts to 2 * 8 minutes);
- the possibility that a train of one batch stays in the terminal waiting for the next batch. The handling time of a train in this case is influenced by:
  – the order of arrival of the train in the first batch;
  – the order of departure of the train in the second batch.

We determined train handling times for three situations. In a batch of 3 trains, the train handling time is 32 minutes (first train leaves after 2 incoming trains * 8 minutes plus a period of concurrent stay of 2 * 8 minutes). In a batch of 6 trains, the train handling time is 56 minutes (first train leaves after 5 other incoming trains * 8 minutes plus a period of concurrent stay of 2 * 8 minutes). In the situation that one train is part of two sequential batches (arriving as first train and leaving as first train in the second batch), the train handling times is 1:44 hours (5 incoming trains * 8 minutes plus 2 * 8 concurrent period plus 4 outgoing trains * 8 minutes plus 4 incoming trains second batch * 8 minutes plus second 2 * 8 concurrent period).

In the case of shunting, we can hardly speak of a train handling time, because trains arriving at the shunting yard do not continue after operations as they do at the hub-terminal. Instead, they are split into wagons, which are reassembled into new trains. An estimate of train-handling times can be obtained if we consider the first arriving train together with the first leaving train to be one train, and do the same thing with the second arriving and second leaving train, and so forth. It appears that train-handling times in shunting operations vary. The first trains in a batch show longer handling times than the later trains. Table 12.2 indicates the various handling times (see Appendix 12.1 for a detailed description of how these results were elaborated). Table 12.2 also shows that the train handling time is shorter at the hub-terminal than for shunting. We can conclude that, from the perspective of a network operator, the hub-terminal is more favourable than shunting, even if a (rather high) hill capacity of 8 wagons is considered.

Table 12.2 Comparison of train handling times for hub-terminal and shunting

	Hub-terminal	Shunting		
In a situation of		2 wagons/ minute	Hill capacity	8 wagons/minute
a batch with 3 trains	0:32	2:00 1:39 1:17	1st train 2nd 3rd	1:34 1:21 1:08
a batch with 6 trains	0:56	3:05 2:43 2:22 2:00 1:39 1:17	1st train 2nd 3rd 4th 5th 6th	2:12 2:00 1:47 1:34 1:21 1:08
one train in 2 batches	1:44	4:55	1st train	3:18

12.5 Discussion

Compared with earlier operational and performance evaluations carried out (see Bontekoning and Kreutzberger 2001), the analysis in this chapter provides new and more in-depth insights. In previous studies, the hub-performances of various general hub-terminal designs were compared with a general and average figure for existing shunting time. In addition, the throughput for the hub-operations was hypothetical. In this chapter, we considered a realistic situation with real throughput data for an existing node. Specifically for this node, a hub-terminal was designed, based on the existing throughput data. Compared with previous

results, which indicate that hub-operations were more favourable than shunting, our comparison indicates that hub-terminal operations are favourable to shunting only for volumes up to about 550 wagons per day (which is about 275,000 load units annually). However, the outcomes of the analysis in this paper are very indicative and cannot be generalized. The value of the analysis lies more in the understanding of how both types of operations are executed and how they differ from each other, than in the exact results of the performance evaluation. A few factors should be investigated further, in order to obtain more reliable outcomes:
the hill capacity of Metz-Sablon appeared not to be representative for traditional shunting yards in Europe; therefore, a sensitivity analysis was carried out in this case study with a more representative hill capacity based on the literature. This sensitivity analysis only considered the hill capacity and not other measures to improve shunting time. In addition, more shunting yards should be approached in order to determine what are representative shunting operations and average performance levels in Europe;

- as well as comparing traditional shunting versus hub-operations, hub-operations should also be compared with state-of-the art shunting;
- the hub-terminal in this study is not representative for hub-terminals in general. The fundamentals of hub-exchange operations (transhipment of load units from one train to another) are the same for all other hub-terminal designs, but layout, technology and operating strategies differ to a large extent. On the basis of the results of the evaluation study by Bontekoning and Kreutzberger (2001), a rough first indication is that other hub-terminals, although different in layout and operations, would also perform better than shunting for the lower volumes. But here too further investigation is needed;
- to obtain more insight into the operational process and the factors influencing the performance, the in-depth qualitative approach seems to be appropriate. However, a more complete and integral comparison could be obtained with a multi-criteria, cost-benefit analysis or computer simulation, under the condition that more data such as costs, input of resources such as labour, land surface, and equipment can be obtained. With such an approach, more variables and objectives, and trade-offs between them, can be considered.

Will there be substantial changes in the overall railway network operations as a consequence of replacing shunting yards by rail-rail terminals? In the calculations, we considered a continuous flow of throughput at both the rail-rail terminal and shunting yard. The purpose of this was to show the efficiency of the node operations themselves. The rail-rail terminal evaluated requires a follow up time of a maximum of 8 minutes, the shunting yard a maximum of 22 minutes (for the latter figure, see Appendix 12.1). Of course, these follow up times would become shorter when measures are taken to enlarge handling capacity. Longer follow up times are possible, but imply that the performance times for both rail-rail transhipment and shunting will increase due to waiting time. The question is: Are time slots/paths with such follow-up times possible in present railway network operations? When train follow-up times of 8 minutes cannot be realized, the value of fast and relative expensive equipment decreases. As soon as significant buffering capacity is needed, because network and infrastructure operators cannot manage to synchronize train schedules, a shunting yard is much better equipped to park trains on the large number arrival and departure tracks available.

There are two additional advantages apart from the fast transhipment of the rail-rail terminal. First, fewer rail wagons are needed, because load units for different destinations can be combined in the same rail wagon. Second, a rail-rail terminal also allows for rail-road transhipment. In this way, additional flows can be attracted and the utilization rate of equipment can be increased.

12.6 Conclusions

This chapter contributes to new and basic insights into the differences in operations and performances between traditional shunting and new-generation hub-terminal operations. Operational activities of the hub-operations show more complex relations and dependencies than shunting. There are two fundamental differences in operation between shunting and terminal operations. First, shunting involves marshalling of wagons, while at a hub-terminal load units are transhipped. Second, a shunting yard has enough tracks to receive and reassemble 25 to 28 trains at the same time, while at the Megahub only 6 trains can be dealt simultanteouly. Consequently, not only might the rail network need to be rescheduled but also the arrival and departure procedures at the node. As a whole, the preliminary results obtained by this study indicate that new-generation hub-terminals could be a feasible alternative for shunting for node and network operators. However, before such hub-terminals can be implemented, many questions still have to be investigated. Additional insights into performances and costs need to be obtained.

Acknowledgements

The authors would like to thank Mr. K.-P. Franke and Mr. M. Vogtmann of the firm Noell for their assistance. Mr. Vogtmann provided us with additional data, while he and Mr. Franke critically read earlier drafts of this article. We also wish to thank Professor H. Priemus and Professor R. Dekker for valuable comments on the earlier version of this paper. However, the content of this article is the entire responsibility of the authors.

The research was financed by the European Union and the TRAIL Research Programme Freight Transport Automation and Multimodality (FTAM) of Delft University of Technology, the Netherlands.

References

Bontekoning Y.M. 2000: The Importance of New-Generation Freight Terminals for Intermodal Transport, Journal of Advanced Transportation, Vol. 34, 3, 391–413

Bontekoning Y.M. and Kreutzberger E. (eds.) 1999: Concepts of New-Generation Terminals and Terminal Nodes, Delft University Press, Delft

Bontekoning Y.M. and Kreutzberger E. 2001: New-generation Terminals; a Performance Evaluation Study, Delft University Press, Delft

Huet J. and Vallet R. 1977/78: Die Laufzielbremsung der Guterwagen auf den französischen Rangierbahnhöfen Sotteville and Hourcade, Rangiertechnik und Gleisanschlusstechnik, Vol. 37, 31–44

ICF 1999: Map of Shunting Yard Metz-Sablon

Jung M. and Talke W. 1998a: Effizienzsteigerung in Zubbildunsanlagen – Teil 1 Analyse und Rationaliserungspotential, Eisenbahntechnische Rundschau, Vol. 47, 11, 691–695

Jung M. and Talke W. 1998b: Effizienzsteigerung in Zubbildunsanlagen – Teil 2 Innovative Lösungen, Eisenbahntechnische Rundschau, Vol. 47, 12, 768–775

Meyer P. 1998: Entwicklung eines Simulationsprogramms für Umschlagterminals des Kombinierten Verkehrs, Dissertation, University Hannover, Department of Mechanical Engineering (Maschinenbau)

La Vie du Rail, 1999: Metz-Sablon: le premier point nodal européen, 6 janvier 1999, 25–26

Oum T.H. 1979: Derived Demand for Freight Transport and Inter-Modal Competition in Canada, Journal of Transport Economics and Policy, Vol. 13, 149–168

Terminet 2000: Performance Analyses for 5 Terminal Case Studies, Deliverable D10, Delft University of Technology/OTB Research Institute, Delft

Vogtmann M. and Franke K.-P. 2000: In Terminet, Performance Analyses for 5 Terminal Case Studies, Deliverable D10, Delft University of Technology/OTB Research Institute, Delft

Woxenius J. 1998: Development of Small-Scale Intermodal Freight Transportation an a Systems Context, Department of Transportation and Logistics, Chalmers University of Technology, Göteborg

Appendix 12.1 Calculation of train handling time at the shunting yard: an example

The example consists of a batch of six trains and a hill capacity of 2 wagons per minute. The average length of a train is assumed to be 24 wagons.

Train 1 t = 0	**Min.**	**Train 2 t = 22***	**Min.**	**Train 6 t = 110***	**Min.**
Uncoupling train 1	30	Uncoupling train 2	30	Uncoupling train 6	30
Loc driving train 1 to hill	5	Loc driving train 2 to hill	5	Loc driving train 6 to hill	5
Shunting train 1	12	Shunting train 2	12	Shunting train 6	12
Loc drives back	5	Loc drives back	5	Pushing together***	0
Loc driving train 2 to hill	5	Loc drives train 3 to hill	5	Coupling	30
Shunting train 2	12	Shunting train 3	12		
Loc drives back	5	Loc drives back	5		
Loc drives train 3 to hill	5	Loc drives train 4 to hill	5		
Shunting train 3	12	Shunting train 4	12		
Loc drives back	5	Loc drives back	5		
Loc drives train 4 to hill	5	Loc drives train 5 to hill	5		
Shunting train 4	12	Shunting train 5	12		
Loc drives back	5	Loc drives back	5		
Loc drives train 5 to hill	5	Loc drives train 6 to hill	5		
Shunting train 5	12	Shunting train 6**	10		
Loc drives back	5	Pushing wagons together***	0		
Loc drives train 6 to hill	5	Coupling	30		
Shunting train 6**	9.5				
Pushing wagons together***	0				
Coupling	30				
Train handling time	**184**	**Train handling time**	**163**	**Train handling time**	**77**

Notes:* Trains 2 to 6 arrive every 22 minutes in order to achieve a continuous operation. This is determined as follows: A train needs to be uncoupled at the moment that the loc arrives at the arrival siding to pick up the train. The time needed for uncoupling is 30 minutes. Loc driving to and from the hill takes 2 times 5 minutes and shunting 12 minutes, and altogether the loc is occupied for 22 minutes. Consequently train 2 needs to arrive 8 minutes before the end of uncoupling of train 1, which is at 22 minutes, in order to be ready to be picked up when the loc arrives.

** Assuming that the wagons for the destination trains are equally spread over the arrival train, train 1 can be assembled as soon as the first of the last 6 wagons is shunted (thus 5 wagons are left for shunting: (24-5)/2 = 9.5 min.) For train 2 this implies that it can leave as soon as 4 wagons are left for shunting, etc.

*** Pushing wagons together is done at the same time with loc driving.

PART D: Scenario Perspectives

13 Light Rail: Backbone of European Urban Regions

Hugo Priemus

OTB Research Institute for Housing, Urban and Mobility Studies,
Delft University of Technology, Delft, The Netherlands

13.1 Introduction

In NW Europe, we can discern a time-honoured pattern of urbanization characterized by polynuclear urban regions in which the cities – with the exception of London and Paris – do not reach the level of global cities. Such areas as the Randstad, the Rhine-Ruhr Area, the Flemish Diamond, the British Midlands and Central Scotland incorporate networks of cities and smaller municipalities which lack any strong hierarchy.

The advantage of these polynuclear urban regions is that areas of green, water and open space abound in the vicinity of residential areas, and that an attractive intermingling of residential areas, industrial estates, green and water areas can be created. A problem that can then arise, however, is that the accessibility of the central cities falls short of what is desired and the conditions for an effective and efficient public transport service are relatively unfavourable. An awareness of this problem is derived from experience in the Netherlands.

Urban regional public transport in the Netherlands is languishing. Public transport, which accounts for about half the budget of the Ministry of Transport, Public Works and Water Management, only accounts for 14 percent of passenger kilometres travelled. Since 1986 the share of public transport in the total number of passenger kilometres travelled has certainly increased (from 12 to 14 percent), but the train accounts for almost all of this small rise (Table 13.1). The market share of bus, tram and metro within public transport has fallen to less than a third. The use of city and district transport has only risen in the Randstad.

The cost recovery level of public transport is low: 50 percent for train transport, 30-35 percent for city and district transport. The shortfall of city and district transport amounts to nearly €1 billion annually.

In this contribution, drawing heavily on Priemus and Konings (2001), we observe that the Netherlands government is on the point of introducing light rail in a number of urban regions, following the example of some successful German and French cities. But also Japanese cities could be considered as a benchmark. The central issues in this contribution are, referring to the current situation in the Netherlands: *How could light rail support the economic vitality of European cities? What should the design of the system look like, and how could public transport and real estate development be better integrated? Under what conditions could the introduction of light rail in the European urban regions be successful?*

Table 13.1 Transport performance of public transport systems, 1986-1998, selected years (in billion passenger kilometres and percentages)

	1986		1990		1994		1998	
	billion km.	%	billion km.	%	billion km.	%	billion km.	%
Train	10.6	57	13.9	65	16.7	67	19.1	69
Bus/tram/ metro	8.1	43	7.6	35	8.4	33	8.4	31
Total	18.7	100	21.5	100	25.1	100	27.5	100

Source: CBS 1999, The mobility of the Netherlands population.

In addressing these questions, we refer to experiences in French, German and Japanese cities where a strong synergy between urban public transport, urban renewal and real estate development has been realized.

This contribution is constructed as follows. In Section 13.2, we define the concept *light rail* and we present a typology. Section 13.3 gives a brief overview of successful applications of light rail in France and Germany (and which are also making an impression in the Netherlands), such as those in Strasburg, Nancy, Saarbrücken and Karlsruhe. Section 13.4 affords us a glimpse of the experiences with public transport in the large urban areas of Japan, where public transport and real estate development have been integrated to a considerable extent. Section 13.5 gives an overview of the current policy discussion in the Netherlands, where efforts are being made to strengthen the economic vitality and accessibility of central cities and where the concept of the 'urban network' has recently been introduced. In this section, we establish that urban revitalization policy and urban accessibility policy are poorly tuned to each other, so that the chances of success of the light rail plans have been affected. In Section 13.6, we draw a number of lessons from international experience. In Section 13.7, we discuss further the structure of an urban regional public transport system with feeder and interconnecting public transport. Finally, in Section 13.8, we formulate some conclusions.

13.2 Light Rail: Definitions and Typology

The Dutch Ministry of Transport, Public Works and Water Management (1997) defines light rail as follows:
'Light rail is a rail-associated public transport concept related to journey distances of 10 to 40 km between a central city and its direct area of influence, or exclusively directed to the more national region. The system technologies applied for vehicles and infrastructure are tuned to that market. The vehicles have ample pick up and set down points, rapid acceleration and short stopping times, adequate top speed and are often so adaptable that they are also capable of running on the infrastructure for existing train, express tram and metro networks. One-man operation and the relatively light weight of the carriages lead to lower operating costs and more simple rail maintenance than would be the case with train operation.'

Table 13.2 Four sorts of light rail

	Combination of own infrastructure and shared use of railway net	(Practically) exclusive use of the railway net
Urban regional main connections	A. Randstad Rail model	B. Randstadspoor model (mixing with other trains)
Regional connections around medium big cities	C. Karlsruhe model	D. Dürener model (no mixing)

Source: Ministry of Transport , Public Works and Water Management (1997).

Four types of light rail can be distinguished (see Table 13.2):

1) *Urban regional light rail on its own (new) track, with the possibility of additional shared use of railway tracks (Randstad Rail model)*
Frequency: high. The technology makes it possible to use existing railway infrastructure in addition to the metro and express tram tracks. Direct connections between important commuter areas and the heart of the central city can thereby be created. Examples: metro/express tram Rotterdam, express tram Utrecht-Nieuwegein/IJsselstein, Ring line Amsterdam, urban region Noord-Zuid line in the Amsterdam region. The light rail project Randstad Rail (Rotterdam-The Hague) will be developed according to this model.
2) *Urban regional light rail on the existing railway net (Randstadspoor model)*
Frequency: 4 to 6 times per hour. Enables direct connections between important commuter areas and the railway stations in the central city.

Examples of Dutch plans are: Randstadspoor in the Utrecht region; initiative in the KAN district.[1]

3) *Connections to (medium) big cities with shared use of the national railway net (Karlsruhe model)*
Frequency: usually, services every quarter of an hour. Transfer free connections between the residential cores and the regional city centres. Examples: Karlsruhe, Cologne/Bonn, Manchester, Newcastle, Salzburg, Bern, Kassel and Saarbrücken. Dutch plans exist for Leiden-Gouda (Rijn-Gouwe line), Zuid-Limburg (as predecessor to the Euregional MHAL[2]-network) and Groningen.

4) *Regional subsidiary lines (Dürener model)*
The particular goal here is a cheaper form of operation than the traditional train concept. As a substitute for languishing regional rail services, it is aimed at improving the service level by increasing the number of stops, the speed and frequencies. Frequency: 1 to 4 per hour per direction. Integration with bus services is essential. Examples: Düren, IGO + in the Achterhoek.

Light rail is thus indicated for various formulas. However, the types that combine seamless services on own (new) track and existing railway tracks by dual system vehicles (types 1 and 3) seem most promising in terms of correspondence with customers' need and its associated economic viability (Schaffer 1999; Jefferson and Kühn 1996). Such systems permit the avoidance of interchanging for passengers and, through the use of railway infrastructure, reduce investment costs. This also renders light rail systems economically attractive for agglomerations whose centres have less than 100,000 inhabitants. In the Randstad, different forms (alternatives 1+2+3 above) will be combined. The goals involved are: high cost coverage, many travellers and a substantial contribution to the economic revitalization of cities. These are the success criteria for light rail.

'Light rail stands with one foot in the rail transport world and the other in the urban and urban district transport world' (Ministry of Transport, Public Works and Water Management 1997). Light rail is a rail-associated transport system that can be positioned in the triangle between train, tram, and metro (see Section 13.2). Light trains are often referred to in the context of the trains which are soon to run under the authority of regional transporters on about 30 tracks shortly to be abandoned by the Netherlands Railway Company (NS) (known as the unprofitable NS lines). Express tramlines on their own tracks are often labelled as light rail, as are sometimes the city trams. The Netherlands is not yet familiar with the most integrated formula: one rail vehicle connecting urban and urban district transport (Wansbeek 1999). It is anticipated that this new formula will be introduced in the foreseeable future in such projects as RandstadRail and the Rijn-Gouwe line.

Figure 13.1 gives an overview of regional initiatives for light rail in the Netherlands.

It appears that the realization of light-rail connections is going to require considerable efforts. Investment costs will have to be higher than expected and the construction of transport networks is no sinecure, although the lines can be implemented in stages and can be combined with other rail transport lines. Light rail must not be perceived as the solution for all ills (see also Schaffer 1999). Important functions remain for the classic formulas, such as bus transport. In urban regions, however, light rail can indeed provide a large added value to the integration of city and its surrounding district.

Figure 13.1 Regional initiatives for light rail in the Netherlands
Source: Derived from Ministry of Transport, Public Works and Water Management 1997.

13.3 Successful Light Rail in Some French and German Cities

On the basis of a review of projects abroad, researchers of the Dutch Ministry of Transport conclude (Van der Maas et al. 1997a; Van der Maas et al. 1997b; Koolen et al. 1998) that high value, organization and vision are the key concepts underlying the success of urban regional public transport in certain French and German cities. On the one hand, the concern is for the comfort and convenience of the passenger, while, on the other hand, urban planning integration, image and social safety are involved. Public transport must be integrated seamlessly into the urban environment. Examples in France and Germany show that the development of public transport lines and the revitalization of central cities can go forward together (Koolen et al. 1998).

Partly as a result of its exemplary public transport, the number of visitors to the central city in Strasburg has increased. A synergy between urban vitalization and the improvement of public transport in Strasburg has brought about a cost recovery level for the tram of 110 percent. The central city has been restructured. The liveability and economic attraction of the city centre have been strengthened. Ongoing car traffic is discouraged from entering the city centre. Public transport has priority here. Park & Ride areas have been laid out at some of the larger tram stops with bus stations. Mostly, alternatives 1 (Randstad Rail model) and 3 (Karlsruhe model) apply.

Here public transport is not a purely local activity. A more agglomerative approach leads to better harmonization of service schedules, fares and infrastructure. There is a growing appreciation of the structuring effect of the infrastructure.

In France and Germany, there is evidence of a regional approach oriented to a better harmonization of both service schedules and fares. Cities join together on a regional basis (in Germany: *Verkehrsverbund*; in France: *Communauté Urbaine*).

Communautés are regional, largely autonomous government authorities. They receive 80 percent of their budget from local taxes levied on companies and residents. As a result, they are less dependent on the national government (additional taxes can be levied for large public transport investments).

In Germany, the *Bundesländer* have similar powers. The *Länder* are the direct financiers of public transport investments. The *Verkehrsverbunden* formulate the vision and the public transport companies put the public transport concept into action. The *Verkehrsverbund* system has proved to be a successful formula for public transport organization, not just in attracting more public transport travellers into central city areas, but also in integrating suburban services with city centre services to produce a truly coordinated, regional public transport system (Pucher and Kurth 1996).

13.4 Japanese Cities: Synergy between Public Transport and Real Estate Development

The close relationship between public transport and real estate in Japan, which has demonstrably gone further than in the Netherlands, could act as a benchmark for Dutch cities (Van de Velde 1999). In Japan, the public transport operators are the owners of the rail infrastructure and the stations. This situation came about together with a strong diversification of these companies. They are very active in the development of real estate along the lines of public transport and at stations, as well as in the operation of transport services (train and feeder bus services). The companies develop new residential areas, apartment complexes, department stores, shopping centres and office locations. Van de Velde (1999) reports that the old terminal stations of the private railway lines have mostly been turned into large department stores, in which the station takes up just one floor. For the relaxation of the commuter, the railway companies are also developing amusement arcades, sports centres and museums. They choose sites along their lines which generate recreation transport, preferably against the direction of peak hour travel, to generate extra income with few extra costs through a better operation of the existing public transport infrastructure.

Van de Velde (1999) stresses the involvement of the Japanese railway companies with real estate as a key factor. The quality of transport services and real estate strengthen each other: if the transporters perform poorly, the value of the real estate falls. If the development of the real estate is inadequate, there will be fewer customers in the trains. According to Van de Velde (1999), this interaction ensures a certain measure of self-regulation through the market and reduces the need for government control.

The consequence of the Japanese approach is that public transport has a market share in the total number of passenger kilometres of 35.7 percent (29.0% train and 6.7% bus): 2.5 to 3 times as great as in the Netherlands. The car has a share of 59 percent, air transport 4.9 percent, and 0.4 percent of passenger transport is carried out by ship (Van de Velde 1999). Japanese public transport is operated on the assumption that costs should be completely recovered, including the cost of the infrastructure. Subsidies are available, but on a much smaller scale than in the Netherlands. Large infrastructure investments (such as metro tunnels and the connecting up of new towns) are only partly funded by the central and provincial government authorities (Van de Velde 1999). The involvement of the Japanese Railway Companies in real estate development has resulted in substantial additional income. For most companies these sideline activities have even become the most profitable business (Cervero 1998). The railway companies benefit from the increases in value of land around the railways (compare: Cervero 1993; Cervero and Landis 1996). With land they already owned they achieved large profits from real estate development. For example, along the Tokaidoline (nearby Tokyo), the value of commercial parcels within 50 metres from the stations increased by 57 percent (Cervero 1998).

Nowadays, the railway companies cooperate with the real estate developers, who provide cheap land for the construction of rail lines, partly finance the investments and pay over a part of the increase in land values as a contribution to the railway company.

The situation in the Netherlands differs from Japan. In Japan, rail investments can be mixed with broader development by large financial conglomerates. The toll on Japanese roads promotes public transport, but in this respect the Netherlands will follow.

In the Netherlands, it is less immediately obvious that NS (abbreviation for Dutch Railways, Nederlandse Spoorwegen) Passenger and NS Real Estate should merge and immediately embark on project development and residential areas, shops and offices. In the Netherlands, the solution can be found in consortia, in which not only development companies and investors, but also the transport companies participate. In the absence of a sustainable urban regional administration, the directive role of the municipal government is indispensable in the spatial context in which the interaction between public transport and real estate development should occur. The municipality's authority with respect to the zoning plan is of crucial significance here.

The central city ought to direct the process at regional level. At the interregional level a directive role for the province is of importance: for example, in the context of policy development in the design of a corridor. Actors such as housing associations and environmental movements ought also to play a part in the decision forming.

The experiences in Japan are of particular interest, because there the emphasis is laid so strongly on the relationship between public transport operation, real estate development and real estate management. Van de Velde (1999) reports in this context that, when developing a residential neighbourhood, real estate developers draw up a contract with a bus company in which usually both companies are allied to a railway company. The house owners in the neighbourhood pay a fixed contribution for the development costs of the bus line. As soon as the line covers costs, a frequency of six buses per hour, also outside the peak hours, is achieved.

Japanese experience suggests that a closer cooperation between public transport operation and real estate development is worthy of consideration. In the Netherlands, these worlds are still largely distinct. Portheine (NS Real Estate) states: 'The real estate-public transport relationship is not yet very close, we can do very much better there' (de Vos 1999). In the Netherlands, public transport companies do not usually have much to contribute to construction plans. If only offices are built, the transporters only acquire one sort of customer, one which they already have more than enough of: peak time passengers. A well-thought-out, differentiated approach (dwellings, shops, recreation) will attract the sort of customers they desperately need: passengers who sit in empty trains in the off-peak hours, or travel in the opposite direction to the rush hour traffic (de Vos 1999).

An example of a good relationship between real estate development and public transport is the Utrecht Centre Project: 350,000 m^2 office space, 1,500 dwellings, 40,000 m^2 shopping space, an urban entertainment centre (including a casino and mega cinema), the redesign of Utrecht central station and a bus station, a 1.6 billion euro project. The redesign of the Den Bosch station and the Rotterdam Central rail station are other interesting examples. The future High Speed Trains stations are seen as opportunity-rich locations, where public transport operation and real estate development can strengthen each other.

Garvelink, (de Vos 1999) quote the Director of a private consultancy from Holland Railconsult: "It is no longer appropriate at this time first to develop a housing or office location and then to take care of the public transport connections. It is too costly and too time-consuming. So you must look for the corridor."

At present, the emphasis lies in the search for synergy between real estate development and public transport operation in the redesign of station areas, in which the highest priority is given to the High Speed Trains stations. Stations other than HST also ought to be transformed into transfer complexes with differentiated real estate development (Bertolini and Spit 1997).

13.5 Promotion of Economic Vitality and Accessibility of Central Cities in the Netherlands

To assist the establishment of priorities in investment alternatives for the strengthening of the spatial-economic structure of the Netherlands, the Dutch Cabinet has defined a number of policy packages. The largest packages are those for *accessibility* and *vital cities*. If we direct our attention to urban regional public transport, where endeavours to achieve accessibility and urban vitality come together, then we find ourselves at the focus of the State's endeavours towards strengthening the spatial-economic structure of the Netherlands. If urban and urban regional public transport is not given a powerful impulse, efforts to enhance the economic vitality of cities will be undermined, and we can only look forward to a 'city infarction'. A city infarction (Steenhuis 1999) could be avoided by casting the urban freight distribution in a new mould, making the city accessible for car traffic in a more selective manner, through improving the telecommunication infrastructure in the city, through basing both passenger and freight transport on the assumption of the interconnectivity of transport chains, and particularly through substantial improvement of urban regional public transport, to in order to modernize and to integrate it (Brand-Van Tuijn 2000).

This call for action strongly appeals to the concept of the 'Transit Metropolis' (Cervero 1998). By addressing the potential of public transport services for different urban settlement patterns, this concept could become a paradigm for sustainable urban district development.

13.5.1 Coalition Agreement of the Cabinet Kok II

In the 1998 coalition agreement of the Kok II Cabinet, an amount of €13 billion was allocated for the period 1999-2010 for the strengthening of the spatial-economic structure of the Netherlands (Kok 1998). The total investment will be allocated to several policy packages.

The accessibility package is the largest one: €5.5 billion. City and urban district transport forms part of this accessibility packag,e and has an allocation of €1.8 billion. The emphasis in the accessibility package further lies to a great extent on (inter)national connections of passenger and freight transport.

The *vitality of cities* policy package (€2.2 billion) is closely associated with the accessibility of the cities. Part of the road network expenditures (in particular, expenditures on the subsidiary road network) and the expenditures on city and urban district transport are of strategic importance in ensuring the economic vitality of the cities. The promotion of city and urban district transport must, in the years to come, be based primarily on considerations relating to the promotion of the economic vitality and accessibility of the central cities. Improvement in car accessibility of the city is also fundamental to this endeavour, but seeking a solution in this direction is subject to some physical limitations and is questionable from an environmental point of view. If, as in the Appendix to the Perspectives Memorandum of the Ministry of Transport, Public Works and Water Management (1999), the costing of (car) mobility is tackled seriously through the introduction of pay drive, paylanes and/or toll roads, the accessibility of cities is in danger of getting into difficulties unless a high value urban district transport is developed that in particular connects effectively to the centrally-situated parts of the city in a fast, reliable and comfortable manner.

The Coalition agreement of the Kok II Cabinet stands for the decentralization and competitiveness of public transport. It is worthy of comment that there is a certain tension between the investment ambitions of the Kok II Cabinet and the endeavours towards decentralization and market working in public transport. Concessions will, in principle, be granted for six years. In granting licences, particular attention will be paid to the financial aspects. These are hardly favourable conditions for large investments. If investments in light rail are to be taken seriously, much longer concession periods must be considered (12 years, for example). Contracting out would then have to relate to the question of how such investments should acquire form and substance (public-private partnership?), how investments in public transport infrastructure and real estate could strengthen each other, and what form the operation of public transport will take after the large investment.

13.5.2 Investment Budget Urban Renewal (ISV)

Meanwhile, the Kok II Cabinet has given urban renewal policy in the Netherlands further form and substance through the introduction of the Investment budget Urban Renewal (ISV), in which the Cabinet makes direct agreements with 30 cities which need substantial renewal. In their approach to big-cities policy, the Cabinet follows the advice of the Social and Economic Council 'Together for the city' (SER 1998): in the ISV a number of financial resources are bundled together in the sphere of the urban physical living environment: subsidies for the environment, housing, space, green and (physical) economic activities.

The criteria (performance areas) have been formulated so as to stimulate municipalities to arrive at a broad, cohesive approach to urban renewal. The municipalities' development programmes will be tested on the basis of 12 performance areas, the first 6 of which have a process character, and the remaining 6 are related to content. The municipalities must formulate their aims and measures in 12 performance areas in testable terms and, in the areas which refer to content, wherever possible in quantitative units.

It is beyond belief that there is no mention of traffic infrastructure or public transport in the ISV, so that as a result the theme of the accessibility of the city is conspicuous by its absence. The Ministry of Transport, Public Works and Water Management does not participate in the ISV, even though the lion's share of the ISV criteria is concerned with the policy area of this Ministry (sometimes emphatically so).

Consideration of the 12 criteria that the ISV deal with, makes it evident that a close relationship between urban revitalization and urban district transport is a necessity. That conclusion makes the participation of the Ministry of Transport, Public Works and Water Management in the ISV, or a close relationship with the ISV, devoutly to be wished. Where the Ministry is at present endeavouring to realize light rail projects in certain urban regions, that is still taking place much too independently from urban renewal.

At first sight, it seems strange that the Ministry of Transport does not participate in urban renewal policy, even though transport can contribute considerably to the accessibility and economic vitality of cities. The explanation is that this Ministry sees the province and the region as the appropriate level to adopt transport and infrastructure policies at a decentralized level, and not cities. Another reason is that the Ministry of Transport does not want to be involved in too many coordinating activities which could hamper the speed of decision making processes which the Ministry has in mind.

Nevertheless, improving the accessibility of cities is nowadays becoming a more prominent issue in the Netherlands because the economic performance and the growth of employment of central cities have been greater than the national average since the mid-1990s.

13.6 Lessons from International Experiences

Two important conclusions can be drawn from experiences in Japan, Germany and France, presented in the Sections 13.3 and 13.4. In the first place, the mutual importance of urban public transport, city development, and urban renewal through various projects abroad is emphasized. A substantial restructuring of public transport can, supported by appropriate measures, make an important contribution to the liveability and economic attraction of the city centre. On the other hand, the upward revaluation of the city centre has a stimulating effect on the number of visitors; public transport can benefit and the profitability and quality of public transport can be further improved. In other words: high value urban regional transport promotes urban renewal and urban renewal promotes public transport. Figure 13.2 illustrates the interaction between the economic revitalization of the city and the improvement of urban public transport via the increase of the numbers of visitors and tourists in the city, the increase of numbers of companies and employees in the city, the increasing purchasing power in the city, the higher cost coverage of public transport, the higher quality of public transport and the increase of the real estate value around transport stops. Due to their locational flexibility and rapid transit potential a more prominent role of light rail systems is foreseen (Campion et al. 2000).

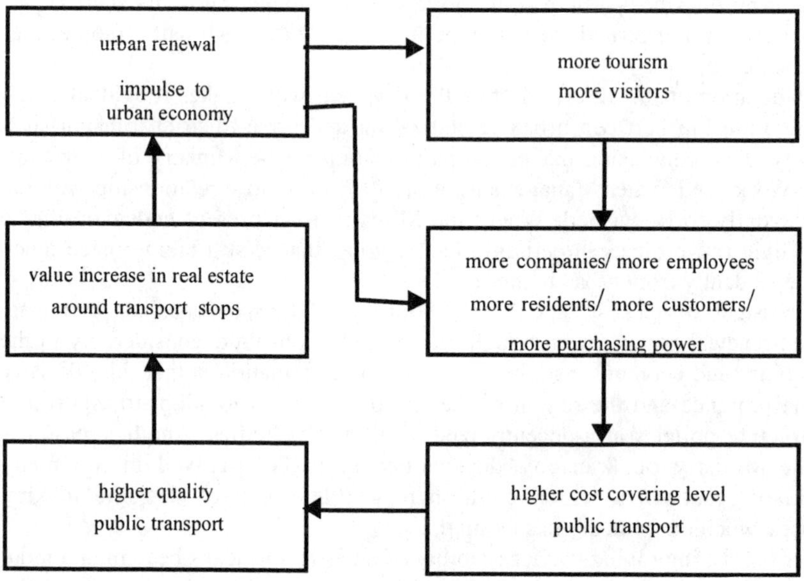

Figure 13.2 Interaction between economic revitalization of the city and improvement of urban district public transport

The first success factor is an integral approach. The international projects show that an integral approach yields synergy: investment in public transport ought to be an element of urban investment policy. The retrospective adaptation of public transport in urban planning usually leads to higher costs and poorer results.

A second success factor is formed by the associated policy (such as the planning of Park + Ride areas, discouragement of through car traffic across the city, parking policy and road pricing). Without adequate associated policy, the effectiveness of improvements in public transport will be diminished.

A third very important success factor relates to the system structure of the public transport network. German and French cases exemplify an approach in which a clear system philosophy has been chosen. Various system levels with various functions can be distinguished. There is repeated evidence of a transport system that is required to function as the supporting system, that is to say a system that is primarily designed to accommodate interconnecting transport, and feeder system levels which fulfil a collect and deliver function. The technology allows a variety of solutions, depending on the specific, local circumstances: in Cologne, the S-bahn has the interconnecting function while the express tram has the delivery function, in Strasburg, the express tram has the main support function; in Lille this function is fulfilled by the metro. and in Oberhausen by express buses and trams (Koolen et al. 1998).

A fourth success factor lies in the quality of the urban and urban regional public transport. In general, it appears that a quality leap has led not only to high value public transport, but also to a higher cost recovery level. That is precisely one of the goals the Netherlands government is striving to achieve!

An important further conclusion based on the international benchmarks is that the municipalities are usually independent, or in any event operate very closely with other local/regional government authorities; they can take the lead in the restructuring of public transport and, consequently, have also been able to set an important stamp on this process with respect to content and technical implementation. In the examples where there is evidence of integral urban renewal (as in Strasburg), this seems to be self-evident, but this assumes far-reaching municipal authority with respect to the urban public transport. Usually such authority signifies that the municipalities themselves can generate their own resources, or in a construction of close mutual cooperation (as in France in the *Communauté Urbaine*). This approach is in line with the view that problems can best be solved at the scale level at which they most strongly occur, so that those who have the responsibility must also have the authority and opportunity to exercise that authority. Simply passing on resources to lower-level government authorities is not compatible with such a view and so would not appear to be the most effective approach. On the other hand, the creation of sources of finance at the municipal level does accord with this view.

At present, the Netherlands municipalities have hardly any financial opportunities of their own for the development and implementation of their own traffic and transport policy. For the financing of infrastructure, the municipality is usually directed to subsidies from the state, or from ear-marked grants. Their own resources from municipal funds derived from local taxes (property tax) are very limited. Certainly, in most cases, a municipality can utilize the yields from parking revenues as it sees fit, but on balance the financial clout of the municipalities is rather limited. This means, among other things, that integral city projects, such as those which have been implemented with success in countries such as France and Germany, would be much more difficult to realize in the Netherlands under the present circumstances. Hopefully, the regional mobility funds formula, which has been recently announced by Dutch central government, will bring a certain decentralization of public infrastructure finance and creates more opportunities for the implementation of light rail in urban regions.

13.7 Feeder and Interconnecting Public Transport

13.7.1 Feeder Public Transport: Pruning, Differentiation and Making Dynamic

In the years to come, urban public transport policy will in many countries be a question of growing and cutting back. Particularly in the fine-meshed public feeder transport, a considerable dynamic is to be expected which will compel flexibility and selectivity. It is precisely in the capillaries of public transport that we are likely to encounter low cost-recovery levels. The relationship between the multiplicity of individual and integrated transport is of prime importance here. It concerns the complete repertoire of individual transport, demand-dependent integrated transport and traditional integrated transport. Increasingly, a distinction must be drawn between line services (running according to a service timetable, mostly in the interconnecting public transport) and the charter services (in the context of the transport management of companies and the demand of private parties).

In public feeder transport, many unprofitable line services will have to be scrapped and replaced by alternative formulas. That undertaking would be no 'once and for all' intervention, but a continuous adaptation process to changes in societal processes, mobility patterns and the built environment. The infrastructure facilities at this level will have to be far more limited. As few limitations as possible must be imposed on the desired changes in public feeder transport.

13.7.2 Interconnecting Public Transport: Growth and Connection with Urban Renewal

Interconnecting public transport at urban and urban region level demands infrastructure facilities, a thoroughly thought-through urban transport policy and considerable synergy between public transport stopping places and spatial design. The relationship between public transport and urban renewal is of crucial importance here, as Figure 13.2 makes clear. In many cities a process has been under way for years, in which the city centre becomes car shy and car traffic is concentrated on the ring roads. This attitude enhances the attraction of central cities for cyclists and pedestrians. In consultation with companies and municipalities, in a number of cities the centre management is directed towards the improvement of the residential quality and the cultural and tourist power of attraction. This assumes that the central cities are accessible. The more that car use in the city is discouraged, partly through physical measures and partly with the help of the price mechanism (paid parking), the more important it becomes for the central city to be readily accessible via public transport. This requires an adaptation of public space, usually including a redevelopment of the station environment.

The endeavour to facilitate chain mobility implies a strategic location and an adequate capacity of *transferia*, parking garages, bus stations and cycle storage, and a redesign of space to facilitate changing transport mode, staying, and 'one-stop-shopping'. In this manner, tourist and cultural facilities can be strengthened, and justice can be done to the protected cityscapes and monuments.

13.7.3 Chain Mobility in the City

The central cities provide unique opportunities for public transport. Existing, often monumental, buildings impose physical restrictions on cars which prevent their unlimited access. Because car driving is restricted here, as parking is restrained by parking fees, and because the area where parking is permitted is limited, because trains stop in the vicinity of the city centre, and because large streams of shopping public, tourists, employees, suppliers and residents are attracted, public transport in city centres finds itself in a splendid position. Here lies a multitude of opportunities for strategic city projects, transfer centres, the quality improvement of public space and the interconnection of public transport networks. Public transport is indispensable for the economic strengthening of the central city, while the upward valuation of the central city in its turn provides opportunities for public transport (see Figure 13.2).

In the city, investments in real estate, public transport and the environment ought to go hand in hand, usually in the form of public-private partnership. Furthermore, in the central city the customer must be able to choose and combine, not only public transport, but also private transport (with a price tag) and combinations of public and private transport in journey chains.

13.7.4 Increase in Scale and Differentiation in Travel Patterns

In society, a tendency towards increase in scale can be observed. At the root of this lie not only economic scale advantages, but also the far-reaching specialization of employment, products and services. Increase in scale is also letting itself be felt in the traffic and transport sector (Brand-van Tuijn and Govers 1998). Economic growth ensures that the average number of kilometres travelled per person increases. There is an increase not only in the kilometres travelled as a whole but also in the distances covered per journey. This form of scale increase occurs not only with respect to commuter and commercial traffic: the increase in the home to work distance is well known. In the recreational sphere as well, long distances are more easily covered: the orientation on recreational facilities is no longer restricted to those near one's own place of residence. This general trend is expected to continue. Besides more and longer journeys, the journey pattern has, moreover, also become more diffuse. Altogether the demand for mobility, and with it also public transport, has become more differentiated. The ability to cater for chain journeys, with fast, smooth transfer opportunities between various forms of public transport, is becoming increasingly important in this perspective.

For a competitive integrated transport supply, an integral solution needs to be developed where a variety of instruments and measures must be closely harmonized with each other: district transport, city transport, rail transport, parking and traffic policy, shop opening times, etc. Such a policy requires extensive cooperation between many regional parties (public and private), together with an intensive form of cooperation between the public transport companies in the region. The regional scale instead of the local scale has to be the focus (Cervero 1998).

13.8 Conclusions

European urban regions are more and more polynuclear. Promoting the economic vitality of the city centres in these polynuclear configurations has an increasing political priority. The transformation from traditional industries to a service economy, stimulated by information and communication technologies, gives new opportunities for urban development and urban strength. Recent statistics show that the urban economy in the Netherlands is flourishing. There is, however, an increasing danger for the development potential of cities to be frustrated through shortcomings in the accessibility of the cities. The opportunities to travel to a city by car and to find a parking place on arrival must be better exploited, but they are constrained by physical limitations. Opportunities to make cities more accessible to car traffic are thereby restricted. The accessibility of cities could be substantially improved if urban regional public transport services were also improved. The introduction of light rail networks on an urban regional scale could render this improvement in accessibility feasible, particularly if a synergy could be brought about between public transport, real estate development and spatial planning.

The quality of urban regional public transport stands or falls with the system structure. Drawing a sharp distinction between the interconnecting and feeder systems is essential. In contrast with France and Germany, in the Netherlands the potential of this twofold distinction has to date, been hardly recognized. It requires a two-pronged approach in which, in particular, the fine-meshed feeder public transport is set up with flexibility and selectivity in mind. The infrastructure investments are extremely limited here. Interconnecting public transport at the urban district level demands infrastructure facilities, sophisticated urban transport management and substantial synergy between public transport stopping places and spatial design. It is principally there where the relationship between public transport and urban revitalization must be exploited and strengthened.

Considering that chain journeys, whether with the same public transport system, whether through the use of several public transport systems (train, bus, tram, metro), or whether through a combination of public and private transport, form an important part of all journeys in urban regions, a seamless harmonization between the various networks is also of considerable importance for a well-functioning public transport system. This concerns both the harmonization between the feeder and interconnecting network (vertical interconnectivity) and the mutual connection of networks at the same scale level (horizontal interconnection), in which transfer opportunities to and from the car must also not be missed. By giving the nodes between the networks not only a traffic function, but also a meaningful residential function, transfer centres are created with potential for synergy between the urban economy and public transport.

The need for a systematic structure of public transport is apparent on all geographic scale levels, but in the Netherlands particularly, at the urban regional level it is a matter of concern. A far-reaching integration of city and district transport to urban regional transport, is then also desired. Light rail offers an opportunity rich solution for this, whereby this integration can physically be brought about. With a selective development of light rail, at agglomerative level a better quality can be achieved, and with it in due course also a higher cost-recovery level for the urban regional public transport. Concurrently, an important spatial impact of this system can be expected. Light rail seems to be an effective instrument to bring about a synergy between qualitative high value public transport, real estate development and urban vitality.

At the same time, light rail is one of the conditions for the successful development of polynuclear urban regions. With only slight exaggeration, we may consider light rail as the backbone of sustainable polynuclear urban regions: it may compete and combine with other modes, it promotes quality-of-life in urban areas by reducing noise and improving urban air quality, and it improves the accessibility of city centres.

References

Bertolini L. and Spit T. 1997: Herontwikkeling van Stationslocaties in Internationaal Perspectief (Redevelopment of Station Locations in an International Perspective), Rooilijn, Vol. 30, 8, 368–374

Brand-van Tuijn R. and Govers B. 1998: Schaalvergroting in de Randstad: Nieuwe Kansen voor Collectief Vervoer (Increase in Scale in the Randstad: New Opportunities for Integrated Transport), Verkeerskunde, May, 14–19

Brand-van Tuijn R. 2000: Stadsgewestelijk Railvervoer: Sturen op Potenties (Urban Regional Rail Transport: Guiding on the Base of Potentials, Goudappel Coffeng, Deventer

Campion D.R., Larwin T., Schumann J.W. and Wolsfeld R.P. 2000: Light Rail Transit: Future Opportunities and Changes, Transportation Research Board, Transportation in the New Millennium, State of the Art and Future Directions, Washington DC, 5

Cervero R. 1993: Rail Transit and Joint Development: Land Market Impacts in Washington DC and Atlanta, Journal of the American Planning Association, Vol. 60, 1, 83–90

Cervero R. 1998: The Transit Metropolis, A Global Inquiry, Island Press, Washington DC

Cervero R. and Landis J. 1996: Twenty Years of BART: Land Use and Development Impacts, Transportation Research, Vol. 31, 4, 309–333

Jefferson C. and Kühn A. 1996: Multimodal LRT Vehicles: A Development for the Future? in Recio J. and Sucharov L. (eds.), Urban Transport and the Environment II, Barcelona, 433–442

Kok W. 1998: Brief van de Formateur aan de Voorzitter van de Tweede Kamer der Staten-Generaal, Regeerakkoord 1998: Sdu, The Hague, (Letter from the Mediator to the Chair of the Second Chamber of the States General), Coalition agreement 1998

Koolen R., Maas C. van der and de Bruin R. 1998: Lessen in Regionaal Openbaar Vervoer (Lessons in Regional Public Transport), Verkeerskunde, March, 60–64

Maas C. van der, Koolen R. and Bruin R. de 1997a: Succesvol Regionaal Openbaar Vervoer. Ervaringen in het Buitenland (Successful Regional Public Transport. Experiences abroad), AVV, Rotterdam, July

Maas C. van der, Koolen R. and de Bruin R. 1997b: Succesvol Regionaal Openbaar Vervoer. Lessen voor Nederland (Successful Regional Public Transport. Lessons for the Netherlands), AVV, Rotterdam

Ministry of Transport, Public Works and Water Management 1997: Light Rail op een Rij (Light Rail in a Row]), Ministerie van Verkeer & Waterstaat, The Hague

Ministry of Transport, Public Works and Water Management, Adviesdienst Verkeer & Vervoer 1997: Succesvol Regionaal Openbaar Vervoer. (Successful Regional Public Transport], Ministerie van Verkeer & Waterstaat, The Hague

Ministry of Transport, Public Works and Water Management 1999: Perspectievennota Verkeer en Vervoer (Perspectives Memorandum Traffic and Transport), Ministerie van Verkeer & Waterstaat, The Hague

Priemus H. and Konings R. 2001: Light Rail in Urban Regions: What Dutch Policymakers Could Learn from Experiences in France, Germany and Japan, Journal of Transport Geography, Vol. 9, 187–198

Pucher J. and Kurth S. 1996: Verkehrsverbund: The Success of Regional Public Transport in Germany, Austria and Switzerland, Transport Policy, Vol. 2, 4, 279–291

Schaffer H. 1999: Five Rules for Successful Light Rail Systems, Tramways & Urban Transit, October, 372–374

Sociaal Economische Raad (SER) 1998: Samen voor de Stad. Advies Grotestedenbeleid, (Together for the City. Advice for Big Cities Policy), Sociaal Economische Raad SER, The Hague

Steenhuis P.H. 1999: Het Stadsinfarct. Overlevingsplannen voor de Stad (The City Infarction. Survival Plans for the city), SMO, The Hague

Velde D. van de 1999: Met Onroerend Goed Stimuleert Japan het Openbaar Vervoer (Japan Stimulates Public Transport with Real Estate), OV Magazine, Vol. 6, February, 10–13

Vos A. de 1999: 't Wil maar niet Lukken met Light Rail (Light Rail is still not Succeeding) Binnenlands Bestuur, 13 August, 22–25, 32–33

Wansbeek J. 1999: Light Rail: Sleutel tot de Binnenstad: Rijn-Gouweproject zet de Toon (Light Rail: Key to the Central City: Rijn-Gouwe Project sets the Tone), Verkeerskunde, June, 20–23

Endnotes

[1] KAN = region of Arnhem-Nijmegen.

[2] MHAL = Dutch/German/Belgian region of Maastricht-Heerlen-Aachen-Liège.

14 Scenario Building for the Future of the Tel Aviv Metropolitan Area and Its Transportation System Using the Delphi Method

Yoram Shiftan, Sigal Kaplan and Shalom Hakkert

Transportation Research Institute, Technion, Haifa, Israel

14.1 Introduction

Tel Aviv is the largest metropolitan area in Israel and serves as the country's main financial and cultural centre. The Tel Aviv Metropolitan Area[1] stretches over 1,475 square kilometres and is inhabited by 2.6 million people, of which 90 percent are living in urban areas. Data from 1999 show that the Tel Aviv Metropolitan Area produced 50 percent of the GDP and was responsible for 46.4 percent of the jobs in Israel. Because of its geographical position and for historical reasons, the Tel Aviv Metropolitan Area is a transportation focal point. All traffic between the north and the south of Israel, as well as traffic connecting the main cities (Jerusalem, Tel Aviv, Haifa and Beer Sheba) to each other, must pass through the Tel Aviv Metropolitan Area. Thus, the metropolitan transportation infrastructure must fulfil two major functions: a) supply the demand for transportation on the national level; and b) supply the demand for transportation within the metropolitan area on the regional and local levels.

Currently, Tel Aviv suffers from severe problems caused by transportation, such as air pollution, traffic congestion and open land depletion. Since the natural and economic resources in Israel are limited, sustainable development of the transportation system in Tel Aviv is extremely important in order to maintain future quality of life. In the present study, the future development of the Tel Aviv Metropolitan Area is described by means of two scenarios: an expected scenario, and a desired scenario. Based on the opinions of the experts, our research examines the potential policy measures which may contribute to a sustainable transportation system in Tel Aviv.

14.2 Background

14.2.1 The Transportation System Supply and Demand

The road system in the Tel Aviv Metropolitan Area includes about 5,800 lane-kilometres, of which 11 percent are divided highways. The public transportation system in Tel Aviv is based mostly on buses, which serve 96 percent of the transit trips. Most buses share lanes with private vehicles, and there are only 43 lane-kms of bus lanes. Most of these bus lanes are not physically separated from the regular traffic, and there is no rigorous enforcement to prevent trespassing. Bus services are operated from 5:00 am to 1:00 am, except for Friday nights and Saturdays, on which there are no bus services. Bus fares depend on the distance travelled. Discount fares are available for multi-trip tickets. Monthly passes are in widespread use. Other discounts are available for senior citizens and for youngsters under 18. The train services in Tel Aviv are operated by the national railways company, and include intercity lines as well as three suburban lines. The suburban lines operate at a frequency of 3-4 trains per hour during the peak hours.

The motorization rate in the Tel Aviv Metropolitan Area is about 350 cars per 1,000 inhabitants. This level is higher than the national level of about 280. In some cities within the metropolitan area, the motorization levels are as high as 400 private cars per 1,000 inhabitants. Trips mode-split between private and public transportation is about 70:30 in favour of the private car.

Traffic during the peak hours is directional, moving into the CBD area during the morning peak hours and out of the CBD during the evening peak hours. About 38,000 vehicles per hour enter the CBD area during the morning peak hours, between 7:00 and 9:00 am. About 35,000 vehicles per hour leave the CBD area during the evening peak hours, between 15:00 and 19:00 hours.

14.2.2 Major Transportation Problems in the Tel Aviv Metropolitan Area

1. *Lack of road infrastructure and consequently a low level of service* – the existing infrastructure is not sufficient to meet the growing traffic demand. As a result, large portions of the road system are heavily congested, especially during the peak hours.
2. *Lack of coordination between the transportation system and the spatial land use* – the growing demand for housing as well as the various local authorities' wish to create jobs inside their area of jurisdiction and attract investors are provoking a rapid decentralization of population and activities, regardless of the ability of the transportation system to serve them efficiently. This process encourages a growing dependence on cars and a worsening of traffic problems.
3. *Lack of coordination between the different authorities and interested parties in the transportation sector* – there are at least 10 different authorities and interested parties currently involved in the transportation section without any

clear status hierarchy. This often causes problems in the implementation of new projects, maintenance of infrastructure and the coordination and regulation of public transportation.

4. *Low level of service of the public transportation system* – the main problems of public transportation are lack of right-of-way and the low level of service including delays, lack of time schedule reliability, insufficient frequencies on some of the lines, and lack of coordination between the two bus cooperatives and the railways authority, which creates difficulties for the passengers to interchange between lines or modes.
5. *Negative economical, social and environmental external effects* – as a result of the growing loads of traffic and the problems stated above, the negative external effects of the transportation system are growing. These effects include economic, social and environmental problems, such as the waste of time and energy, traffic accidents, limited accessibility that aggravates social gaps, air pollution, noise, and open land depletion. Currently, only a few research projects have aimed at an estimation of the external effects of transportation in Israel. From the few articles which have been published on the subject, we can summarize the following data:

- Traffic accidents – In 1999, in the Tel Aviv Metropolitan Area, there were 10,589 traffic accidents, of which 155 were fatal accidents. This represents 47% of the national number of accidents and 38 percent of the national number of fatal accidents (statistical data from the official website of the Israeli Police Department, 2000).
- Air pollution – During the years 1995-1997, between 100 and 300 violations of the Israeli permitted level of NO_X were registered at different locations around the CBD area (Ginsberg et al. 1998). Ginsberg et. al. (1998) consider the emissions of pollutants from vehicles as the cause for 15.7 percent of the mortality from respiratory, lung and heart diseases. They also state that every year about 400 people die as a result of emissions from traffic and respiratory diseases. The method by which this estimate is reached is, however, highly speculative and need to be confirmed by additional research.
- Lack of open lands – The open land quotas per person are less then $1m^2$ in the CBD area, 30 m^2 around the core and 60 m^2 in the countryside. These quotas are much lower than the recommended quotas according to Israeli and European standards. In addition, the open lands in the metropolitan area are extremely sensitive because they serve as infiltration areas of ground water resources. They constitute some of Israel's unique landscapes and some of them are of archeological importance as well (information derived from a survey of national and regional plans, submitted by the postgraduate class of urban and regional planning, June 2000).

14.3 Sustainable Development and Transportation

14.3.1 The Need for Sustainable Development

The need for a sustainable society has already been expressed in the early 1970s with the publication of several articles such as 'A blueprint for survival' (Haq 1997). Since the 1970s, many articles (Newman and Kenworthy 1999; Nijkamp et al. 1998; Cervero 1998; Haq 1997) have discussed sustainable development, and there have been a number of international and world conferences (such as the establishment of the Intergovernmental Panel on Climate Change in 1988, The UN Rio 1992 Earth Summit, The 1995 European Conference of Ministers of Transport, and The 1997 Kyoto Convention on Climate Change).

One of the often-cited definitions of sustainable development is the definition stated in the report of the World Commission on Environment and Development, also called the Bruntland Report after its Chairman. This report defines sustainable development as: 'development which meets present needs without compromising the ability of future generations to achieve their own needs and aspirations'. This definition contains the understanding that natural resources are limited, along with the concern for the well-being of human society. Thus, it combines economic, social and environmental goals, and in so doing it acknowledges their mutual relations and the possibility of conflicts included in the accomplishment of all or some of them. An extremely important aspect of the definition, as we see it, is the time dimension. The time dimension constitutes both advantages and limitations. The advantages are: a) the possibility of technological innovations and social changes that might help to solve current and future problems, b) the recognition of uncertainty in the decision-making process, which improves the ability to cope with it, and c) the possibility of a multiple-step implementation, which can be flexible and adaptive.

On the other hand, the time dimension becomes a limitation when there is a need to consider limited natural resources that should be saved for the well-being of future generations. Also, the wide range of uncertainties involved in the definition of sustainable development over time opens it to a variety of interpretations, and thus creates conflicts around decision making and implementation. Finally, a multiple-step implementation may also have the disadvantage of being sensitive to political changes.

14.3.2 Key Factors in Transportation

According to the literature (Nijkamp et. al. 1998; Masser et al. 1992, 1993; Nijkamp 1999), there are five main factors that influence the development of transportation: spatial and land use planning, economic forces, technology, government policy, and social and behavioural trends. Each of these factors evolves over time, and affects both the other factors and the transportation system.

More specifically, these factors influence mobile technology, infrastructure design, modal split, the level of motorization and policy measures. Thus, these factors will determine whether the progress in transportation will be sustainable. The five individual factors discussed in the present study are outlined below.

1. *Spatial and land use pattern* – The demand for travel and thus the transportation system are influenced by the size of the urban area, building density, level of specialization and the spreading of activity centres within the urban area. Because of the long lifespan of buildings and infrastructure, the influence of spatial patterns on transportation is a long-run one.
2. *Economic forces* – Large-scale transportation projects are highly sensitive to economic changes, because of the huge financial investment involved, the primary capital needed, the long time span of the project and the slow rate of returns.
3. *Technology* – Technology can contribute to sustainable transportation as it promotes improved waste treatments and a reduction of further pollution, as well as creating substitutes for physical travel through communication. Technological development in transportation is, however, relatively slow for reasons such as the scale and cost of the projects, the long periods of research and development, and the long life expectancy of infrastructure and mobile equipment.
4. *Government policy* – The various authorities are strongly involved in transportation development for several reasons, e.g. the need for long range planning to reserve rights-of-way for future development, the huge financial investment and the consideration of negative external effects. Thus, transportation development is sensitive to political priorities.
5. *Social and behavioural trends* – Individual behaviour is a combination of habits, practical and emotional considerations. Thus, social values and norms may greatly affect the transportation choice of individuals.

14.3.3 The Sustainable Transportation Development Goals and Policy Measures

For this aspect of the research, we defined the following nine sustainable transportation development goals grouped in three main categories:

(i) *Environmental goals*: 1) the reduction of air pollution and noise from road vehicles; 2) the preservation of open land; and 3) protection of wildlife and natural habitats.

(ii) *Economic goals*: 1) energy saving; 2) minimizing transportation infrastructure cost; and 3) travel time saving.

(iii) *Social goals*: 1) the improvement of accessibility to employment, cultural activities and open land areas; 2) the maximization of the availability of public transportation to the population; and 3) the increase of road safety by decreasing the rate of road accidents.

We also defined 26 policy measures of sustainable transportation development classified into five main categories (a-e):

(a) *Spatial policy measures*: 1) the definition of car-restricted, pedestrian-friendly zones in city centres; 2) the designation of high density land uses along main public transportation corridors; 3) the construction of high density development near major public transportation stations; 4) the encouragement of mixed land use development; and 5) the promotion of high density development in the CBD area.

(b) *Economic policy measures:* 1) the imposition of heavy taxes on more than one vehicle per household; 2) high parking fees in the CBD area; 3) the introduction of congestion pricing around the CBD area; 4) the heavy subsidization of public transportation in order to decrease fares; 5) the privatization of public transport; and 6) the consideration of external costs in the evaluation of new projects.

(c) *Technological policy measures:* 1) incentives for buying zero-emission vehicles; 2) the intensive development of ITS systems; 3) a high quality public transport system based on buses and Light Rail Transit (LRT); 4) a high quality public transport system based on a subway in the core area; 5) the development of communication infrastructure and local centres for telecommuting; and 6) the increased provision of parking spaces in the core area by automated parking.

(d) *Governmental policy measures:* 1) the daily operation of a 24-hour public transport service; 2) reducing the development of new roads; 3) the extensive development of new roads; 4) the limitation of parking spaces in zones that are well served by public transport; and 5) the granting of business licenses on the basis of the provision of adequate public transportation to employees.

(e) *Social and behavioural policy measures:* 1) the provision of public information as an aid for trip planning (e.g. timetables, different means of travel, the fastest path from an origin to a destination); 2) information about the negative external effects of transportation on the environment; 3) information about how telecommuting can reduce the number of trips; and 4) educational programs to increase public transportation use and car pooling.

14.4 Methodology

In this study, we construct a likely and a desired expert-based scenario, using the Delphi method to evaluate the possibilities for sustainable development in the Tel Aviv Metropolitan area.

14.4.1 The Expert-based Scenario

According to the definition of a scenario by Kahn and Weiner (1967) and later by Ayers (Pearman 1988), a scenario is a hypothetical sequence of logical and plausible (but not necessarily probable) events, constructed in order to focus attention on causal processes and decision points. Scenario-based planning falls within the class of conjectural forecasting methods most commonly associated with future research (Pearman 1988).

Scenario construction is a useful tool in long-range planning, especially in complex policy issues such as transportation planning (Reibnitz 1987; Ringland 1998). Scenarios may help to identify conflicts, problems and decision points and explore various directions of future development. Scenario making enables us to internalize the component of uncertainty in the decision-making process. Examples of the use of scenarios in transportation planning in recent years are Nijkamp et al. (1998), Nijkamp (1999), and Masser et. al. (1992, 1993).

Scenarios can be of a descriptive or normative nature; be built from a forecasting or a backcasting point of view; be based on a common opinion or the knowledge of experts; and be constructed as menu-driven or policy packages. In this research, we chose to describe the development of sustainable transportation by means of an expert-based backcasting scenario, constructed first as a menu-driven and then as a policy package scenario. This type of scenario was chosen for a number of reasons. First, the backcasting scenario is a normative approach; therefore, scenarios are concerned with the desirability of future images, and are based on a certain vision about the future, thus enabling value judgement. Secondly, in the process of backcasting scenarios, the future image is first defined independently, and then connected to the present situation, thus encouraging creative and unconventional ideas. Thirdly, backcasting derives from teleology, an approach that states that human behaviour is explained by causal or functional relations as well as desires and beliefs, and therefore is purpose-oriented (Dreborg 1996). Thus, a backcasting scenario can help to identify future goals and conflicting interests, and explore different paths of development that may change existing functional relations. Fourthly, experts may be the best source for designing creative yet realistic images of the future, since they have the best knowledge of scientific progress, technological innovations, political change and economic and social changes in their fields of interest. And, finally, first constructing menu-driven scenarios and then policy package scenarios, we benefit from both the creativity of the experts in the menu-driven scenario, and the ability to consider synergetic effects of measures in the policy package scenario.

The time span we chose for our research is 30 years, which in the field of transportation is reasonable, considering the long R&D periods, the scope of the projects, the large investments and the long lifespan of the infrastructure.

The expert-based method of scenario construction, chosen for the present research, is the Delphi method (Linstone and Turrof 1975; Nijkamp et al. 1998). This method is based on a series of several questionnaires with a controlled feedback, with the purpose of reaching a relatively narrow range of future images, by comparing opinions in an iterative way. The process may result in either a consensus or several different opinions, and a single solution is not mandatory. Two examples of recent use of the Delphi method in scenario construction in long-range transport planning are presented by Masser et al. (1993) and by Nijkamp et al. (1998).

The Delphi method has two benefits concerning flexibility. First, it offers a degree of freedom in answering the questionnaire – allowing time for careful consideration of the subject and for convenience in answering. Secondly, it provides scope for the presentation of the results – the method is not limited to a single solution, but gives the possibility of several opinions.

In this research, we conduct a two-round Delphi survey. This chapter discusses only the results of the first round. The questionnaires were sent to 63 leading Israeli experts in the transportation sector and other relevant fields such as regional planning, economy, environment and geography. The list of experts included academics, consultants and decision makers.

The experts were given a menu-driven questionnaire with the 26 policy measures described above. The experts were asked about the implementation probability of each measure and about the ability of each measure to achieve each of the nine goals of sustainable development defined above (see 14.3.3).

14.4.2 Scenario Construction

From the results of the survey we constructed two scenarios: an expected scenario, and a desired scenario. The *expected scenario* was constructed from the policy measures that received the highest probability scores. This scenario is basically a projected descriptive scenario, although the experts did not explicitly define the path leading from the present situation to the expected image of the future. The *desired scenario* was constructed from the policy measures that received the best scores according to the sustainable development criteria. Thus, it is a backcasting normative scenario.

14.4.3 Data Analysis

Data analysis included frequency analysis and non-parametric statistical tests (Sheskin 1997). These descriptive statistical methods were used because of the relatively small sample size and the relatively detailed questionnaire.

The answers of the experts were analysed as described below.

Method 1. A frequency analysis of the probabilities of the 26 policy measures. The most probable policy measures were chosen to create the most probable scenario, designated as the 'Expected Scenario'.

Method 2. A frequency analysis of the desirability of the 26 policy measures, in order to create the desired scenario. This analysis was performed on three types of measure aggregates:

- Achieving all three groups of goals by each of the five groups of measures (spatial, economic, etc.), in order to find the most efficient group.
- Achieving each goal by each of the five groups of measures, in order to find the most efficient group.
- Achieving each goal by each of the measures.

Method 3. An analysis based on the Wilcoxson Matched-Pairs Signed-Ranks Test and the method of quantile-quantile (Q-Q) plot was conducted in order to identify differences among three pairs of specific similar measures. The Q-Q plot is a plot of the quantiles of one data set against the quantiles of another data set. If the two sets come from a population with the same distribution, the points should fall approximately along a 45 percent reference line. The greater the departure from this reference line, the greater the evidence for the conclusion that the two data sets have come from populations with different distributions. Any such difference can be important, since these measures are currently a pending issue among Israeli experts. The three pairs of measures are:

- Congestion pricing around the CBD area versus high parking fees in the CBD area.
- A public transportation system that is based on a Metro, versus a public transportation system that is based on Light Rail Transit and buses.
- The spatial development of high density areas around major nodes of public transportation versus spatial development along the major transportation corridors.

A further analysis was performed based on a Kruskal-Wallis One Way Analysis of Variance by Ranks, in order to identify differences among experts in different aspects of transportation (consultants, academy members and decision makers).

14.5 Results and Discussion

14.5.1 Rate of Response

32 responses were received out of the 63 questionnaires distributed, of which 28 were found to be valid and were analyzed as described above. This rate of response is considered to be high, given the time (typically several hours, by personal communication with 10 of the experts) and effort required to answer our detailed questionnaire.

The highest rate of response to the questionnaire, 55 percent, was that of the consultants. The rate of response from academic members and decision makers was 45 percent and 36 percent, respectively. The high rate of response indicates that there is a great interest in sustainable development in transportation among Israeli experts. The high rate of response by the consultants may indicate that this group will probably lead the efforts towards sustainable development.

14.5.2 Measure Probabilities

The probability of a given measure, as indicated by the completed questionnaires, determines its importance in the expected scenario. By the probability analysis, the scores of the economic measures were found to be similar to the scores of the governmental policy measures, and were lower than the scores of the spatial technological, and social measures, which in turn were similar to each other (see Table 14.1). In each of the groups, however, individual measures received high probability scores (median values between 60 percent and 72 percent and average values between 55 percent and 68 percent). These policy measures are described later in the expected scenario.

14.5.3 Expected Scenario for the Year 2030

In the Tel Aviv Metropolitan Area in the year 2030, there will be a public transportation system with a high level of service based on Light Rail Transit and buses. High density zones of population and employment will be located along the main corridors of this system. The highest density zones will be located near major public transportation stations. The use of private cars will be restrained through parking limitations (high prices and limited parking space) in the CBD and in the inner bound. Also, some areas will be pedestrian-friendly restricted car zones, and will be served by high quality public transport.

The road network will be highly developed. The infrastructure road kilometres will be much greater than the current 2000 figure. Intelligent Transportation Systems will be used heavily, and will include information systems, urban transportation control systems and incident detection and response. A massive expansion of the infrastructure of electronic communication systems at home, or at local centres for telecommuting, is expected. Educational programs will explain the benefits of public transportation and the advantages and possibilities of trip planning. These programs will attempt to increase public awareness of the possible negative effects of transportation on the environment.

14.5.4 Ability of Measures to Achieve the Goals

From the analysis of the ability of the measures to achieve all three types of sustainable development goals, the following results were obtained. 25 percent of the experts stated that sustainable development must be achieved through a combination of policy measures, for two important reasons: a) the complexity of the concept of sustainable development; and b) some policy measures are interdependent, and their combination may have a synergetic effect. A policy measure that was not mentioned in the questionnaire, but which was added by 25 percent of the experts as a vital element for sustainable development, is the establishment of a metropolitan transportation authority that would be responsible for strategic transportation planning in the metropolitan area, and would coordinate the various transportation elements in the metropolitan area. Currently, such a metropolitan authority does not exist in Israel.

The group of spatial and land use measures received the highest scores for the achievement of all goals (see Table 14.1). The group of technological measures is rated second-best. The group of economic policy measures and the group of

advantage over the governmental measures. The group of social and behavioural measures (e.g education and information) received the lowest scores. Figure 14.1 describes the frequency of the scores for the ability of the different groups of measures to reduce air and noise pollution.

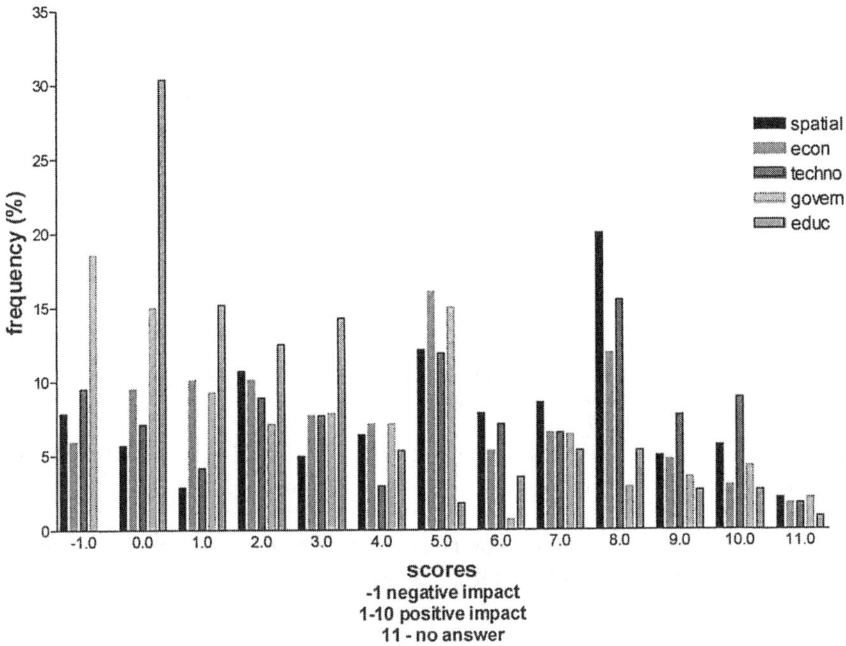

Figure 14.1 Frequency of scores for the ability of measure groups to reduce air and noise pollution

Among spatial policy measures, the best one is the development of high density areas of population and major activities near main public transportation stations. The second-best measure is the development of high density areas along major public transportation corridors. According to the Wilcoxson test (at a confidence level of 95%) and Q-Q plot analysis (as illustrated in Figure 14.2), the difference between the best and the second-best measures is significant. Among the technological policy measures, the best one is the development of a mass transportation system based on a subway in the CBD area. The second-best measure is a public transportation system based on Light Rail Transit (LRT). The results of the Wilcoxson test showed no significant difference between these two (at a confidence level of 95%). The Q-Q plot showed a slight advantage of the subway in terms of time saving (see Figure 14.3A), the preservation of open land and the reduction of air pollution, and a slight advantage of LRT in terms of cost saving (see Figure 14.3B) and accessibility. With respect to the rest of the goals, there was little difference between the two measures.

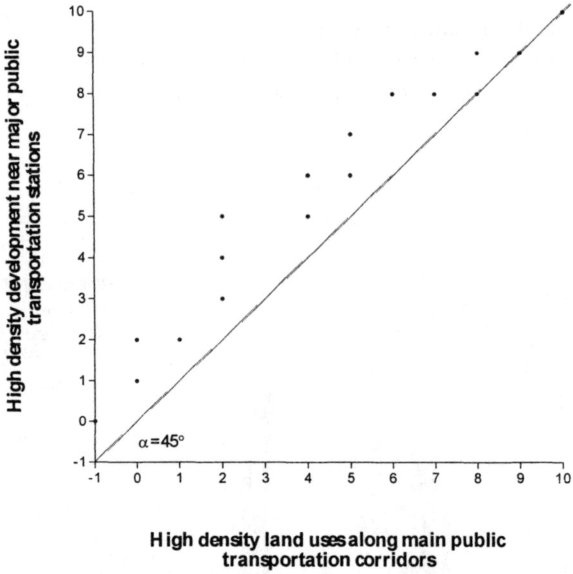

Figure 14.2 Q-Q plot of the scores of the best and second-best spatial measures for reducing air pollution and noise

None of the economic measures can be considered better than others, since most of the scores are medium or low and there is a wide dispersion of the scores.

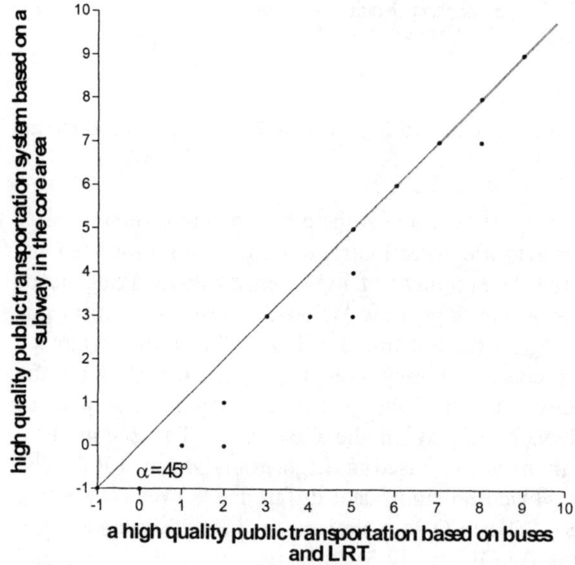

Figure 14.3a Q–Q plot of the scores of a subway system versus LRT and buses for minimizing infrastructure costs

However, congestion pricing, high parking fees and the consideration of external costs of transportation in the primary stages of project evaluation have the best potential to contribute to sustainable development. In a comparison of high parking fees and congestion pricing, the results of the Wilcoxson test showed no significant difference between these two measures (at a confidence level of 95 percent). The results of the Q-Q Plot points to a slight advantage of congestion pricing for achieving the goals.

None of the governmental policy measures is significantly better than the others, since most of the scores are low or widely dispersed. All of the educational measures received very low scores. Furthermore, the scores were quite homogeneous. The median of the scores of all the policy measures in this group was 0-2, and more than 80 percent of the scores were smaller than 3.

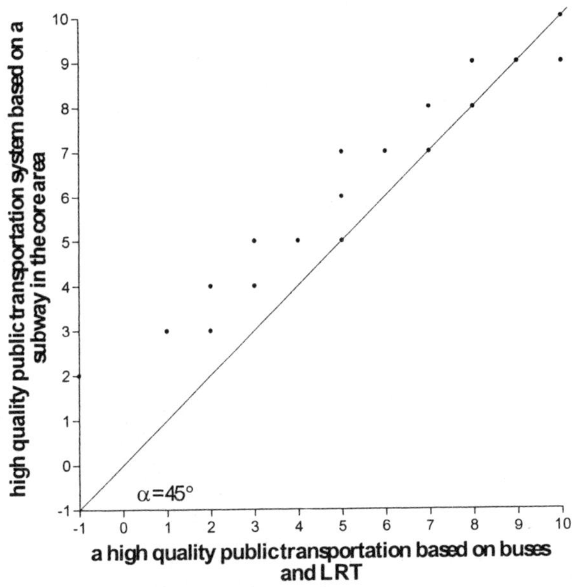

Figure 14.3b Q–Q plot of the scores of a subway system versus LRT and buses for travel time saving

A comparison among the three groups of experts (consultant engineers, academics and decision makers), in which all measures and goals were analyzed by the Kruskal-Wallis test, did not show a significant difference in the scores given to the different measures among experts from different groups.

Finally, there is a similarity between the expected scenario and the desired scenarios, as can be seen in Figure 14.4. The main elements in the desired scenario can also be found in the expected scenario, which is favourable. Therefore, sustainable development of the transportation system in Tel Aviv is a possible and realistic vision of the future.

Table 14.1 The probability and desirability of the different groups of policy measures

Policy measures	Desirability									Probability (%)
	Increasing road safety	Increasing public Transporation availability	Improved Accessibility	Travel time saving	Minimizing the costs of infrastructure	Energy saving	Protection of bio-diversity	Preservation of open land	Reducing air & noise pollution	
Spatial and land use	4.21 (5.00) Sd 3.14	5.12 (5.00) Sd 3.35	5.39 (5.00) Sd 3.11	4.37 (5.00) Sd 3.14	4.37 (5.00) Sd 3.40	4.78 (5.00) Sd 2.88	4.27 (5.00) Sd 3.32	4.46 (5.00) Sd 3.33	4.98 (5.00) Sd 3.28	54.64 (60.00) Sd 26.20
Economic	2.49 (2.00) 2.91	2.88 (2.00) Sd 3.05	2.04 (1.00) Sd 2.90	1.93 (1.00) Sd 2.97	2.80 (2.00) Sd 3.05	3.33 (3.00) Sd 2.96	2.33 (1.00) Sd 2.86	2.51 (2.00) Sd 2.92	4.10 (4.00) Sd 3.12	40.95 (40.00) Sd 29.49
Technological	3.49 (3.00) Sd 3.30	3.15 (2.00) Sd 3.52	3.79 (4.00) Sd 3.29	3.97 (4.00) Sd 3.16	2.38 (1.00) Sd 3.22	4.25 (4.00) Sd 3.23	3.10 (2.00) Sd 3.28	2.68 (2.00) Sd 3.02	4.87 (5.00) Sd 3.54	55.65 (60.00) Sd 28.45
Governmental	2.30 (1.00) Sd 2.99	2.54 (1.00) Sd 3.40	2.98 (3.00) Sd 3.21	2.47 (2.00) Sd 3.00	2.64 (1.50) Sd 3.50	2.61 (2.0) Sd 3.07	2.04 (0.00) Sd 3.23	1.93 (0.00) Sd 3.16	2.85 (2.00) Sd 3.30	40.71 (50.00) Sd 30.09
Education and Information	1.82 (1.00) Sd 2.73	1.12 (0.00) Sd 2.73	1.67 (0.00) Sd 2.66	2.09 (1.00) Sd 2.90	2.22 (1.00) Sd 3.02	2.41 (1.00) Sd 2.92	1.69 (1.00) Sd 2.70	1.54 (0.00) Sd 2.64	2.68 (2.00) Sd 2.90	53.79 (60.00) Sd 29.30

Average......................56.25
(Median) ....(70.00)
Standard Deviation.........Sd 29.74

14.5.5 Desired Scenario for the Year 2030

Three basic and vital elements determine the desired scenario for the Tel Aviv Metropolitan Area in the year 2030. These elements are:

1. A high quality public transportation system in the CBD and the inner bound area that will be based either on a Light Rail System and buses or on a Subway system.
2. The development of high-density land uses (population, employment, commercial and cultural activities) near major stations of the public transportation system.
3. The development of high-density land use, and a stronger specialization of the CBD in the core of the metropolitan area, to maintain and enhance its dominant functional role in the metropolitan area.

Other elements that may contribute to the desired scenario are:

- Creating car-restricted, pedestrian-friendly zones in central areas.
- Either high parking fees in the CBD area, or congestion pricing around the core.
- Limited parking spaces in zones well served by public transportation.
- Intensive use of ITS systems.
- Intensive use of zero-emission vehicles.

These elements are expected to have an intermediate effect on achieving the goals of sustainable development, and may become supplementary elements, which enhance the effect of the main elements.

In order to achieve sustainable transportation, a metropolitan transportation authority should be established. This authority should be responsible for strategic transportation planning in the metropolitan area, and should coordinate the various transportation agencies in the metropolitan area.

Further extensive spreading and development of the road system is not desirable, although limited road development is considered necessary.

14.6 Conclusions

From a policy perspective, this chapter showed that, for the Tel Aviv Metropolitan Area, a combination of policy measures is necessary in order to achieve a sustainable transportation system on account of the complexity of the problem and the interdependency between measures and possible synergetic effects. In the first Delphi round, the elements in each scenario were evaluated separately. In the second Delphi round (not discussed in this chapter), each scenario will be evaluated as a whole policy package. The group of spatial and land use measures received the highest scores for the achievement of the goals. The group of technological measures was rated second-best. This is probably because these types of elements affect large parts of the metropolitan area. Thus, land use and technological measures are vital elements in the desired scenario for the Tel-Aviv Metropolitan area. The groups of economic and governmental measures received medium or low scores, with a slight advantage for the economic measures for the achievement of the goals. The scores for these two groups were highly dispersed, possibly because some of the elements examined were extreme. The consequences of other elements may be implementation-dependent. The low scores for the educational measures for the achievement of the goals indicate that, in Israel, the development of sustainable transportation will come not as a result of public pressure, but rather as a result of transportation planning.

The main elements in the desired scenario are a highly developed public transport system and better coordination between the spatial development and the transportation system. For the next 30 years, it seems that a public transportation system based on Light Rail Transit (LRT) and buses will be sufficient. Secondary measures in the desired scenario are traffic-restraint policy measures (high parking fees, congestion pricing, restricted car zones), and maintaining the functional role of the CBD area.

The expected scenario is also based on a highly developed public transport system and a better coordination between the spatial development and the transportation system. For the next 30 years, a public transport system based on LRT and buses is much more probable than a subway system. High parking fees and restricted car zones are also part of the expected scenario.

A similarity was found between the expected and the desired scenarios (see Figure 14.4). The main elements in the desired scenario can also be found in the expected scenario, which is favourable. Therefore, a sustainable development of the transportation system in Tel Aviv is a possible and realistic vision of the future.

From a methodological point of view, the Delphi method for data collection was found to be an efficient tool in our research with regard to the following aspects: a) there was a fairly high response rate to the questionnaire, b) a high percentage of reasonable answers were given by the experts, c) the data were collected in a relatively short time, and the required financial resources for the research were modest. However, the number of goals and measures were too large, resulting in a long and cumbersome questionnaire. While the response rate was high and the results agree with the literature on sustainable transportation, it seems that a more focused questionnaire would be more appropriate for future studies.

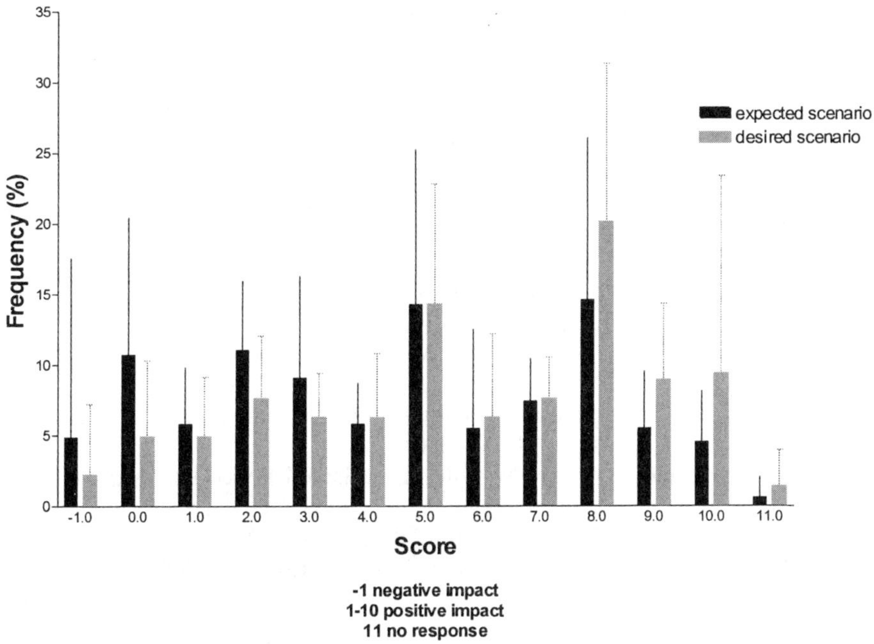

Figure 14.4 Comparison between the expected and the desired scenarios

References

Cervero R. 1998: The Transit Metropolis – A Global Inquiry, Island Press, Washington DC

Dreborg K.H. 1996: Essence of Backcasting, Futures, Vol. 28, 9, 813–828

Ginsberg G., Serri A., Elaine F., Shemer J., Koutik D. and Karsenty E. 1998: Mortality from Vehicular Particulate Emissions in Tel-Aviv Jafo, World Transport Policy & Practice

Haq G. 1997: Towards Sustainable Transport Planning, Ashgate, Aldershot England (reprint 1999)

Kahn H. and Wiener A.J. 1967: The Year 2000: A Framework for Speculation on the Next Thirty-Three Years, New York, Macmillan Co.

Linstone H.A. and Turrof M. 1975: The Delphi Method – Techniques and Applications, Addison-Wesley Publishing Company, Inc.

Masser I., Sviden O. and Wegener M. 1992: From Growth to Equity and Sustainability – Paradigm Shift in Transport Planning, Futures, Vol 24, July-August

Masser I., Sviden O. and Wegener M. 1993: Transport Planning for Equity and Sustainability, Transportation Planning and Technology, Vol. 17, 319–330

Nijkamp P. 1999: Sustainable Transport: New Research and Policy Challenge for the Next Millennium, European Review, Vol. 7, 4, 551–677

Nijkamp P., Rienstra S. and Vleugel J. 1998: Transportation Planning and the Future, John Wiley & Sons Ltd, Chichester, England

Newman P. and Kenworthy J. 1999: Sustainabilty and Cities – Overcoming Automobile Dependence, Island Press, Washington, DC

Pearman A.D. 1988: Scenario Construction for Transport Planning, Transportation Planning and Technology, Vol. 12, 1

Reibnitz U.V. 1987: Scenario Techniques, McGraw-Hill Book Company Gmbh, Hamburg

Ringland G. 1998: Scenario Planning – Managing for the Future, John Wiley & Sons, Chichester

Sheskin D.J. 1997: Handbook of Parametric and Nonparametric Statistical Procedures, CRC Press, 83–94, 291–302

Statistical data from the official website of the Israeli Police Department, 2000. Internet website: www.police.gov.il

Stead D. and Banister D. 1999: Transport Policy Scenario Building: Findings from the POSSUM Project, Paper presented at the NECTAR Conference, Delft, October

Endnote

[1] In order to simplify, the Tel Aviv Metropolitan Area will be referred to as Tel Aviv – its main city and core.

15 Assessing Spatial Planning Policy with Accessibility Indicators: The Case of Lille's Metropolis Scenario

Alain L'Hostis*, Philippe Menerault*, and
Christophe Decoupigny**

*INRETS-TRACES, Villeneuve d'Ascq, France
** CESA, Tours, France

15.1 Introduction

In the French context of the regional authorities' responsibility for the rail passenger transport system, this contribution focuses on the relations between collective transport networks and regional planning in the Nord-Pas-de-Calais region. Beyond the analysis of the existing situation, we want to build a prospective approach, in order to propose some modifications to the transport system which can respond to identified regional planning stakes. Through a modelling of the transport system it is possible to simulate these modifications of the supply and to measure their potential consequences.

At the intersection of supply and demand, the measure of actual transport services enables us to question and to go beyond the classical accessibility indicators based on average travel times and frequencies.

Many models, stemming mostly from operations research, tackle the problems of mobility and transport, but mainly from an economic perspective (McFadden 1997). The general purpose of these researches is to model and analyze the equilibrium between supply and demand. As the regional planning criteria are not reducible to an economic optimization problem, we propose to build a complementary tool dedicated to measurement and simulation. Even though our aim is different, many methodological advancements developed in the flow models may be mobilized to enrich our approach.

In order to build this assessment of the transport system in its contribution to regional planning objectives, the chapter is divided in five sections:

- The first section (15.2) identifies the major regional planning stakes through the creation of spatial organization scenarios;
- The second section (15.3) confronts the scenarios with the spatial planning orientations set at the state and the regional levels, concerning the Nord-Pas-de-Calais territory;

- In a third section (15.4), the transport service assessment method is detailed and its principles are explained;
- The fourth section (15.5) shows an application of the method to study the position of the regional capital in the urban system;
- In the last section (15.6) a prospective approach is developed, through the measurement of the gap between the existing and the desired situation; it enables us to identify the modalities of intervention.

15.2 The Scenarios

We first describe the scenarios of possible futures of the Nord-Pas-de-Calais region which have been developed by geographers and spatial planners (Bruyelle 1996; Damette 1994; Pumain et al. 1990; Giblin Delvallet 1990). The scenario "0" refers to the existing state which is in fact largely the heritage of the introduction of the high speed rail in the region and its accompanying measures.

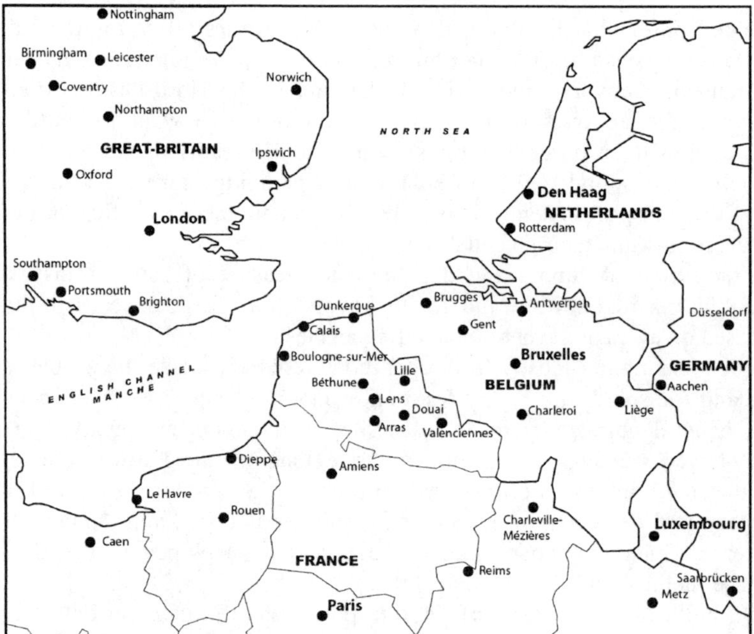

Figure 15.1 Location of the Nord-Pas-de-Calais 'region'

15.2.1 Present Scenario of Paris Attraction: High Speed Rail and Accompanying Measures

The organization of the Nord-Pas-de-Calais rail network is based on the principle of its articulation on the high speed rail system. In this perspective, the relations to Paris are highlighted.

Such a scenario has been discussed by Bruyelle (1994) and called 'the Great North (*le Grand Nord*) in the Parisian orbit'. It is confirmed by a series of rail infrastructure investments made either by the national SNCF or the Regional Council

The modifications of the network related to the high speed rail line (Menerault 1997, 1998) confirm this scenario with:

- the organization of the crossing of the TGV lines within Lille that makes it the main regional access node for the high speed rail line;
- the Arras junction that makes it the second regional entry point on the high speed rail network;
- the completion of the electrification of the Calais-Lille line;
- the electrification of the Lille-Tournai line that enables another direct link to Belgium;
- a high speed line junction to Dunkerque;[1]
- the electrification of the Boulogne-Calais line;
- the electrification of the Douai-Cambrai line.

The function of all these investments, except the electrification of the Lille-Tournai line, is to facilitate the relations between the major cities of the Nord-Pas-de-Calais and Paris.

Nevertheless, the system offers some potentialities for the development of intra-regional connections centred on Lille, or going through Lille. It is only very recently that this potential has begun to be realized with the introduction of regional high speed connections since the year 2000.

The major trends of this phase of the regional planning process correspond to the opening of Lille to the European network, and to an improvement of the accessibility to Paris. This is a planning model in which the reinforcement of the Parisian attraction dominates.

A future scenario directly inspired by this trend would further strengthen the centrality of Lille in the entire region with a 'hegemonic metropolis', as shown on Figure 15.2.

15.2.2 Hegemonic Metropolis

The 'hegemonic metropolis' scenario was developed by Damette (Damette 1994) on the basis of the following statement about the predominance of Lille:

The Nord-Pas-de-Calais region is the model of a rudimentary network: between the conurbation of Lille and the second agglomeration of the region, Valenciennes, there is a true abyss (Damette 1994).

The results of the last census in 1999 confirm the demographic dynamism of Lille. The relations analysis – through the study of the phone flows – and the functional analysis – based on the rail passenger flows– conducted by Damette shows the importance of the relations with Lille for the entire regional territory. Within this framework, the regional planning strategy – in the spirit of the 'equilibrium metropolis' policy set up by the state in the mid-1960s – considers the reinforcement of the relations with Lille as essential.

The spatial model associated with this scenario is based on centre-periphery connections between Lille and all the cities of the region. To consolidate this scenario based on an extension of the metropolitan space to the totality of the 'Région' territory, the action on the regional transport system would consist in giving the priority to:

- direct rail connections between Lille and the cities of the former mining basin;
- direct and fast rail connections from pole to pole towards Lille with the regional high speed rail line called TER-GV from the littoral area and Arras, and improved connections between Lille, Valenciennes, Maubeuge and Cambrai.

In addition, the metropolis is extended by better links with its Belgian side (Tournai and Kortrijk). In this spatial scenario, there is no reinforcement of the sub-regional identities around Valenciennes or the littoral area.

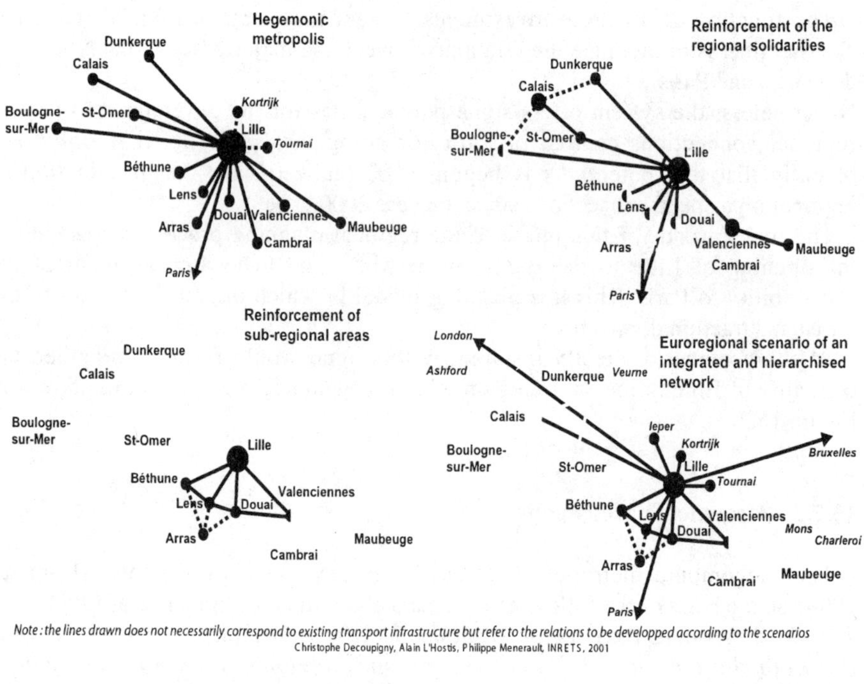

Note : the lines drawn does not necessarily correspond to existing transport infrastructure but refer to the relations to be developped according to the scenarios
Christophe Decoupigny, Alain L'Hostis, Philippe Menerault, INRETS, 2001

Figure 15.2 Four regional planning scenarios

15.2.3 Reinforcement of the Regional Solidarities

A different kind of network organization could be based on the idea of a 'reinforcement of the regional solidarities'. By relying on Arras, the second administrative centre of the region, and a gate to the high speed rail system (Barré 2000), it is possible to develop relations which are not centred on Lille.

The main idea behind this scenario is a bipolarization of the regional space around Lille and Arras, supported by relay cities – Valenciennes and Calais. But the objective is also to bring the peripheries of the region closer together, particularly by developing a railway bypass of Lille.

15.2.4 Reinforcement of Sub-regional Areas

In place of a scenario with a polarization around Lille (and Arras), one could consider as an alternative the consolidation of a set of coherent areas (Direction Régionale de l'Equipement Nord-Pas-de-Calais 1999). In such a schema, three major urban systems take shape: the central urban area around Lille; the littoral area; and, a network of cities around Valenciennes.

This scenario raises the question of the role of Arras in the regional urban system with a choice between its insulation, the Parisian influence, or its connection to an enlarged metropolitan area. In a regional scenario which privileges territorial coherence, one will choose the last option which corresponds to an inclusion of Arras into a second belt of the metropolitan area.

15.2.5 Euroregional Scenario of an Integrated and Hierarchical Network

In a Euroregional perspective, the previous scenario of reinforcement of the regional subspaces can be extended by developing cross-border relations towards Belgium and the South of England. The objective is then to balance the Parisian pressure by means of open relations with the neighbouring countries of Northern Europe.

The Belgian cross-border area can be considered through a polarization on Lille and/or a reinforcement of the Franco-Belgian relations: Belgian littoral-Dunkerque; the reopening of the Valenciennes-Mons line; the development of the Sambre-Charleroi and Sambre-Mons lines, as well as, inside the metropolis, the Comines line towards Ypres, the relations Lille-Courtrai and Lille-Tournai; and the prolongation of the urban lines.

The four scenarios discussed above develop variants between the two extreme spatial patterns of uni-polarization and multi-polarization. To that extent, they can be considered as non-exhaustive of all the possible futures but representative of the major trends of the spatial planning in the region (Brunet 1986).

15.3 Comparison of the Scenarios with the Regional Council and State Orientations

Table 15.1 Comparison of the State and Regional Council orientations with the regional planning scenarios

	State orientations	Regional Council orientations
Scenario 1		• Programme of mobilization of the high speed rail network at the regional level • Opening up the Sambre and Cambrai areas through fast and frequent connections towards Lille (Speed of 200 km/h: Maubeuge in 50 min) • Objective that no major agglomeration of the Region is at more than one hour from Lille
Scenario 2	Find solutions to the saturation of the rail traffic node of Lille for the crossing of the metropolis	• The metropolis as the 'head of a regional network of cities' with an 'equilibrated distribution of the urban functions' • The Boulogne-Etaples-Arras line • Modernization of the Lille-Maubeuge line • According to the principle of intra-regional cooperation, 'Cambrai must cooperate with Valenciennes, but also with Arras and Douai' • The Arras area – 'Southern gate of the Nord-Pas-de-Calais' – as a relay pole of the metropolis organizing a network of cities in the South of the region (St-Pol, Bapaume, Cambrai)
Scenario 3	Develop or create an integrated collective transport network at the scale of the urban area of Lille from: • the existing railway network and • a set of interconnections with the urban networks	• The Region wishes to encourage the development dynamics of the littoral area around the University and to strengthen the association of local authorities (the cross-Channel relation is seen here as an economic activity but not as a spatial relation) • The East of the Region must open itself to the Valenciennes area • Avesnes must cooperate with Valenciennes and Cambrai • Doubling of the track on the Don-Sainghin-Béthune line • Electrification of the Calais-Dunkerque line Good connection of the rail network to Calais-Fréthun (Orientations TER 2010)
Scenario 4	A hierarchical spatial organization at the regional scale that: • facilitates the access to the high speed networks • facilitates the access to the functions and services only available in the regional metropolis • reinforces the links around the sub-regional poles Cross-border relations referring to the RAFHAEL (Réseau des Agglomérations de Flandres, du Hainaut, d'Artois et du Littoral) network : Valenciennes Mons, Lille, Kortrijk and Tournai (Stratégie de l'Etat 2000-2006)	• A development of the metropolis according to the model of the Central Urban Area developed by the planners in the early 1970s (OREAM Nord) • The cross-border development of Lille • Cooperation of the East of the Region with the Belgian side • Strengthen the interregional links with 'our Belgian and British neighbours' • Make a real international station of Calais Fréthun (Orientations TER 2010)

The analysis of the orientation documents of the Regional Council[2] and of the State[3] enables us to sort out their recommendations according to each scenario, as shown in Table 15.1.

From a general point of view, according to the planning orientations, the regional rail transport system has first to ensure the intercity connections; the model of the Parisian railway network called RER ('Réseau Express Régional'), with almost no hierarchy of services, is not supported at all.

As one can see in the table, the State orientations express a clear preference for the Scenario 4: a 'Euroregional hierarchical and integrated network'.

On the other hand, the orientations of the Regional Council are much more ambiguous. It is not a unique scenario which emerges, but rather a superposition of the principles and actions which define the scenarios of a 'hegemonic metropolis', of 'regional solidarity reinforcement' and part of a 'Euro-regional hierarchical and integrated network'.

The main priorities to be pursued by the Regional Council consist in connecting the major agglomerations to the metropolis in less than one hour by high speed regional lines to the cities of the littoral, and in the modernization of the Lille-Maubeuge axis for the Eastern areas. These priorities are clearly associated with the scenario of the 'hegemonic metropolis'; nevertheless, the attention given to the connections with the littoral and Maubeuge areas also corresponds to Scenario 2: 'regional solidarity reinforcement'.

We will now evaluate this specific objective through an analysis of the quality of the transport service.

15.4 Method Principles

The method developed here aims at assessing transport systems with regard to their contribution to spatial planning objectives. The focus is set here not on the internal performance of the system itself but rather on the service produced by the system as an answer to the mobility needs. The transport service is analyzed from three points of view:

- the segmentation of demand;
- the identification of the major attracting locations;
- the urban rhythms.

The transport services can be studied with two different methodologies: the transport flow models, and the accessibility models.

The flow models rely on a detailed description of transport demand and supply (Trépanier and Chapleau 1996; McFadden 1997), particularly when public transport is involved (Orfeuil 2000). Nevertheless, one should notice that the integration of the timetables in the description of transport supply is very recent (Nuzzolo et al. 2000), despite the fact that the constraints induced by the schedules can play an important role. In effect, missing a transfer because of bad timetable coordination – when using a single mode, or between two modes – can generate a substantial increase of the total duration of a trip.

Another way to assess the contribution of transport networks to spatial planning is through accessibility indicators (Rietveld and Bruinsma 1998). In spatial analysis, the accessibility of an object can be defined as the relative easiness to reach that object, and can be applied to a unit of population, information or production (Huriot and Perreur 1994). According to that definition, the concept of accessibility contains two different and complementary ideas (Martellato and Nijkamp 1998): the idea of opportunities – i.e. the possibility for two economic agents to meet – and the idea of the distance which has to be crossed for reaching such an interaction. Consequently, two dimensions have to be considered: on the one hand, the places to be reached at the origin and at the destination, and on the other hand, the distances to be travelled over the transport system. The places to connect are selected according to the demand segments and the traffic generators to be considered. As to the measure of distances, the classical solutions based on averaged or minimum time paths are not appropriate, because of the gap effect induced by the schedules' degree of coordination. A more complex type of measurement has to be developed.

In addition, the promotion of intermodality – which has been brought to the fore in France by the law on the sustainable development of the territories (*Loi d'Orientation sur le Développement Durable des Territoires*) – requires that more attention should be given to transfers between modes, i.e. to the coordination of timetables and the connection between different transport modes. These elements must be taken into account when distances are computed.

Thus, the accessibility indicators introduced here are based on exact travel time, including transfer time. The necessary information is provided by the railway timetables for the stations located in the Nord-Pas-de-Calais. The relevant data are introduced in a graph representing the passenger rail network. It is made of approximately 30,000 arcs and nearly 300 nodes. The measurements of accessibility result from the computation of minimum time paths which integrate transfer times. This full time-table information approach is relatively new (Nuzzolo et al. 2000), and the application to strategic analysis has seldom been made before (Mathis et al. 1997)

This approach enables us to go beyond the simple notion of transport supply by explicitly including some elements of the demand that are spatially and temporally disaggregated (Rietveld and Bruinsma 1998). It applies at a regional scale a methodology that is more usually employed at an urban scale (Trépanier and Chapleau 1996). Indeed, at a regional scale, we will focus on the daily mobility that constitutes by far the most important part of the overall regional mobility (CERTU and SNCF 1998).

15.5 Is the 'Hegemonic Metropolis' a Reality?

The 'hegemonic metropolis' scenario considers that the spatial relations to develop as a priority are the bilateral links between Lille and each regional urban pole. This scenario is clearly supported by the objective of the Regional Council of ensuring that the cities are at less than one hour from Lille by collective transport. This is the policy objective that must be assessed for its contribution to the scenario of 'hegemonic metropolis'.

15.5.1 Classical Accessibility: Potential Relations

In a first step, a classical accessibility indicator with no reference to the notion of transport service will be applied (Figure 15.3). For each origin-destination pair, the set of possible rail trips is computed, from station to station, and including the transfer waiting times inside the railway network. Then, the shortest one is selected regardless of the period of time, in order to determine which agglomerations cannot meet the objective. The map on Figure 15.3 indicates the best travel times on the railway network to Lille-Flandres or Lille-Europe, the two central stations of Lille. The best return trips from Lille take similar times. It is seen that all the great agglomerations – except Maubeuge – have at least one railway service that enables Lille to be reached in less than one hour.

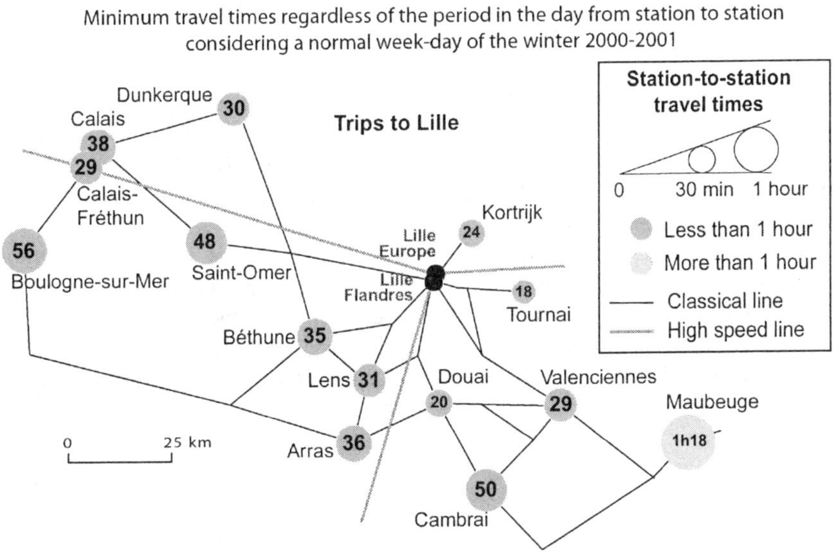

Figure 15.3 Classical accessibility
Source: Decoupigny and L'Hostis, INRETS 2001.

The shortest travel time from Maubeuge takes 1h18. This is significantly longer than the stated one-hour goal. It has been shown (Menerault and L'Hostis 2000) that the connection to Lille is possible in one hour, but only at the price of a major reorganization of the architecture of the railway service.

15.5.2 Peak Period Accessibility: Useful Relations

The problem raised by this first result in terms of transport service is the lack of references, to the rhythms of the daily mobility. Indeed, at the regional scale, 61 percent of the morning commuting trips during the week arrive between 7h00 and 9h00 (CERTU and SNCF 1998), while 67 percent of the evening commuting trips reach their destination between 17h00 and 19h30, the evening peak period being somewhat more spread out.

Hence, this first measure of travelling time, ignoring the particular period of time during the day, is not really relevant. The second indicator proposed here, that of peak period accessibility, goes one step further in the description of transport service by taking into account the urban rhythms and a first segmentation of the demand (Figure 15.4).

From the point of view of the regional polarization on Lille, only the trips during the above peak hours between Lille and the major urban centres were analyzed. The map in Figure 15.4 shows the best travel times from Lille between 17h00 and 19h30. Concerning the morning trips, only the journey from Maubeuge does not satisfy the objective. In the evening, the situation is more difficult: the returns to Calais (1h23), Boulogne (1h53) are not satisfactory. The reason is that the regional service of high speed trains does not play the same role as in the morning, as the return trips arrive only at about 20h00 which is after the peak period.

Moreover, one observes much similarity between the best travel times during the day and the shortest travel times during the peak periods for the whole of Nord-Pas-de-Calais, with the exception of the return trip to the littoral between 17h00 and 19h30. Indeed, return trip travel times are practically doubled for Dunkerque, Calais and Boulogne.

Thus, the objective of a trip time of less than one hour in the 'hegemonic metropolis' scenario is already almost a reality for a large majority of the Nord-Pas-de-Calais agglomerations. Only two areas remain beyond the limit of one hour: (1) the East with Maubeuge, for which a considerable modification of the rail service would be needed; and (2) the littoral, as far as the evening return is concerned.

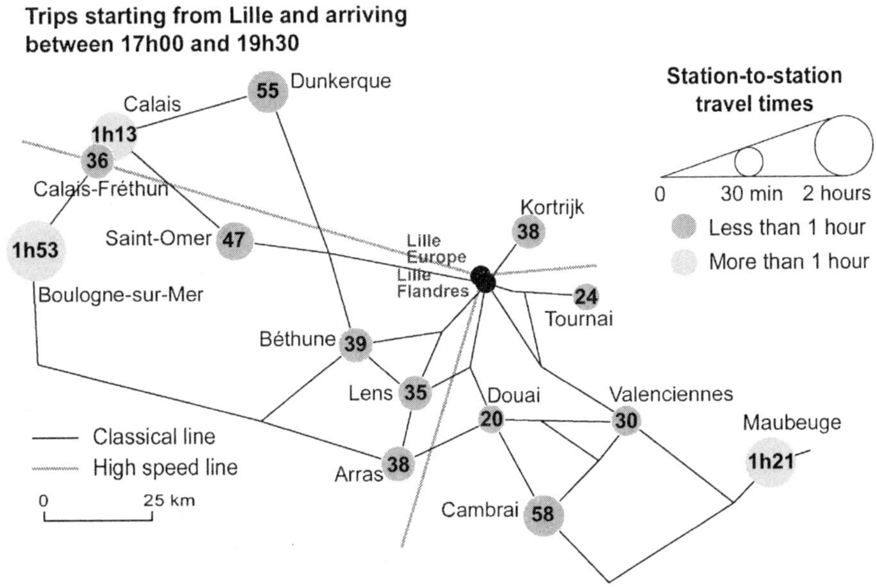

Figure 15.4 Peak period accessibility
Source: Decoupigny and L'Hostis, INRETS 2001.

15.5.3 Scheduled Accessibility: Real Relations

In order to fully integrate all aspects of transport services, the peak period indicator has to be extended in two directions: on the one hand, the rhythms of urban activities impose time constraints that are often more precise than the indication of a two-hour period and, on the other hand, the trips do not end at the destination railway station.

Focussing on the commuting segment towards Lille as a destination, it can be seen that, within a walking radius of 10 minutes around the central stations, one can find a set of traffic attracting locations with a very developed commercial and tertiary centre (Bertolini and Spit 1998). This additional time must be integrated into the analysis. Furthermore, commuting mobility is strongly constrained by the starting time of activity. This aspect of mobility must also be taken into account by an indicator aiming at the quality of service for this important part of the demand. From this point of view, the transport service can be defined by the conditions of access to the town centre of Lille for the purpose of work starting at a well defined time. This new indicator differs strongly from the two previous ones: it is not the simple trip duration which is considered, but the duration computed from the *departure time* from each node considered. It may include a waiting time at the destination, if the time of arrival is not equal to the starting time of activity. In doing so, we focus more on the transport service quality than on the performance of the transport system itself.

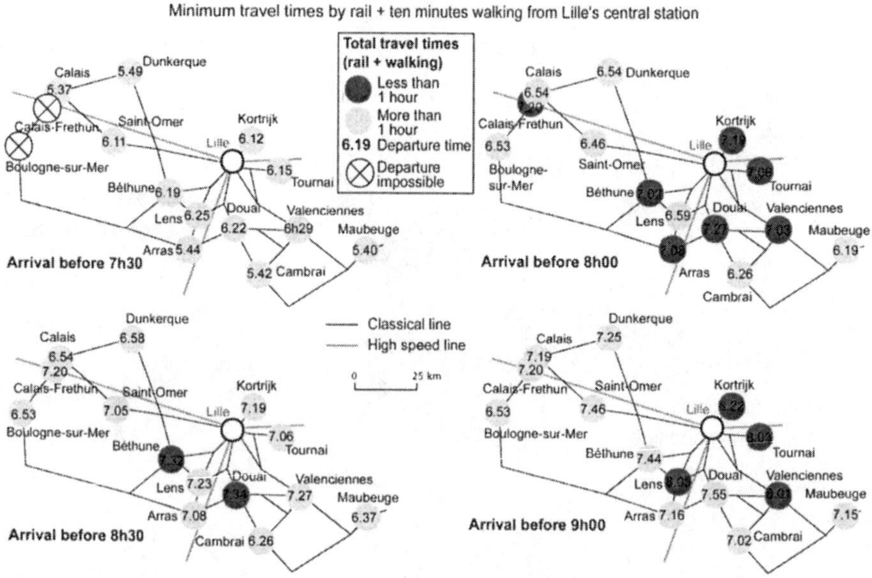

Figure 15.5 Scheduled accessibility to Lille
Source: Decoupigny and L'Hostis, INRETS 2001.

The maps in Figure 15.5 show the departure times from the major urban centres corresponding to a time constraint of arrival in Lille which takes into account the 10 minutes that are necessary to reach the final destination. Under these conditions, the objective of a trip time of less than one hour for four time constraints during the morning peak period is not achieved by any agglomeration. In the best cases, this objective is achieved for two time constraints only: namely, in the case of Douai, Béthune, and Valenciennes, and in that of Courtrai and Tournai.

The most favourable of the four constraints is that of arrivals before 8h00, where 6 agglomerations out of 13 are located at less than one hour of the metropolis, and where the regional high speed rail places the littoral at 1h10. In contrast, for an arrival before 7h30, no agglomeration is at less than one hour, and from Boulogne it is not possible to reach the metropolis.

The regional valorization of the high speed line produces an obvious effect on the Lille-bound access from the littoral. However, the railway supply could still be improved to meet these constraints. In particular, Calais could be better linked if the connection with the Fréthun station was optimized: for arrivals before 8h00 and 8h30, the connection between Calais and Calais-Fréthun takes as much as 26 minutes, which means that it takes hardly less time than the high speed trip between Fréthun and Lille-Europe.

15.5.4 'Hegemonic Metropolis' Scenario Assessment

On the basis of all these indicators an assessment of the objective to set the major agglomerations of the Region at less than one hour of Lille can be conducted. Two main observations can certainly be made.

First, the measurement of best rail travel times, regardless of the time of day, shows that the direct connections Maubeuge-Lille must be considerably modified to meet the objective. Concerning all the other agglomerations, the railway supply makes it possible – at least potentially – to establish connections in less than one hour. In this first approach, the accessibility indicator reveals a regional situation that is not very far from that of Scenario 1: 'hegemonic metropolis'.

However, the approach emphasizing the quality of transport services identifies the problem of evening returns to the littoral and raises the question of the connection between Calais and Calais-Fréthun. The improvement of this connection would certainly contribute to the role of Lille as a hegemonic metropolis, despite the fact that it is seen by the Regional Council as an objective for supporting Scenario 2: 'integrated and hierarchic network'.

The stated objective by the Regional Council to link the major cities of the Region to the metropolis in less than one hour appears to be more a transport *supply* objective than a transport *service* objective. Actually, the classical and peak period accessibility indicators have shown a situation that is not very far from that objective, with the exception of some peripheral parts of the territory. On the other hand, the transport service analysis – expressing the user's point of view – raises the question of the internal coherence of the metropolitan area.

15.5.5 Methodological Lessons

The transport service, and the related planning strategy objectives, can be assessed through a new family of accessibility indicators based on a very detailed analysis of the transport supply.

In order to adopt a service perspective, and to go beyond a strict supply analysis, elements of demand have to be included. Demand segmentation, spatial disaggregation and time-schedule constraints, associated with full timetable information, enables us to build these new accessibility measurements. These elements permit us to focus on the real services produced by a transport system. Nevertheless, the lesser degree of aggregation in space and time, results in a reduction of the range of validity of these measurements, which are relevant only for specific relations and demand. The above analysis focused on a very important segment of transport demand, commuting traffic. Obviously, different but similar indicators would be relevant for shopping trips, for instance. In any case, this is the type of indicators, defined in view of set objectives, that are relevant for developing a spatial planning strategy.

15.6 Intermodality for Reinforcing the Internal Coherence of the Metropolis

In order to obtain a more comprehensive understanding of the transport services supplied to the population, it is necessary to extend the analysis to another important transport demand segment, i.e. that linked to education activities. Actually, education-related mobility constitutes about half of the regional clientele of the rail network

To begin with, it is necessary to identify the major points of attraction corresponding to this demand segment. Figure 15.6 shows Lille's five major university campuses and the urban transport system that can be used to reach them, starting from the regional rail transport network.

Initially, the heavy urban transport system was designed to link the major campuses to the central station of Lille-Flandres by means of an automatic metro called VAL (Véhicule Automatique Léger). The regional users can access the campuses by the metro line number 1, with an interconnection at the central station. However, the regional rail network includes a set of stations located within the metropolis, some of them well-positioned near the campuses. In order to *better exploit the existing network*, our approach is based on the principle of the *virtual network* developed by Dupuy (Dupuy 1991). By exploring all the possible ways to reach the destinations – the virtual network – one can build a 'real network' that is made of credible alternatives which enhance the connectivity of the network. The virtual network appears in Figure 15.6 with interconnections between rail and metro: at the hospital station named 'Halte CHR', at the 'Porte-de-Douai' station, and at the University Human Sciences campus Pont-de-Bois station, as well as between rail and buses at Ronchin, Porte-de-Douai, Baisieux, Tressin, and Lesquin stations.

The normal way to reach the University Sciences campus from Lille-Flandres and with the metro is taken as the base accessibility measure. The students must be able to reach the University campus before 8h00, the normal starting time of the lectures. The map in Figure 15.7 shows the best departure times from each railway station to arrive in good time at the 'Ecole Centrale' on this campus. This does not show a very favourable situation since, with the exception of Douai, no major urban centre of the region is located at less than one hour of the campus. The accessibility to the Sciences Campus is even less favourable than the accessibility to the town centre of Lille, as was shown in Figure 15.5.

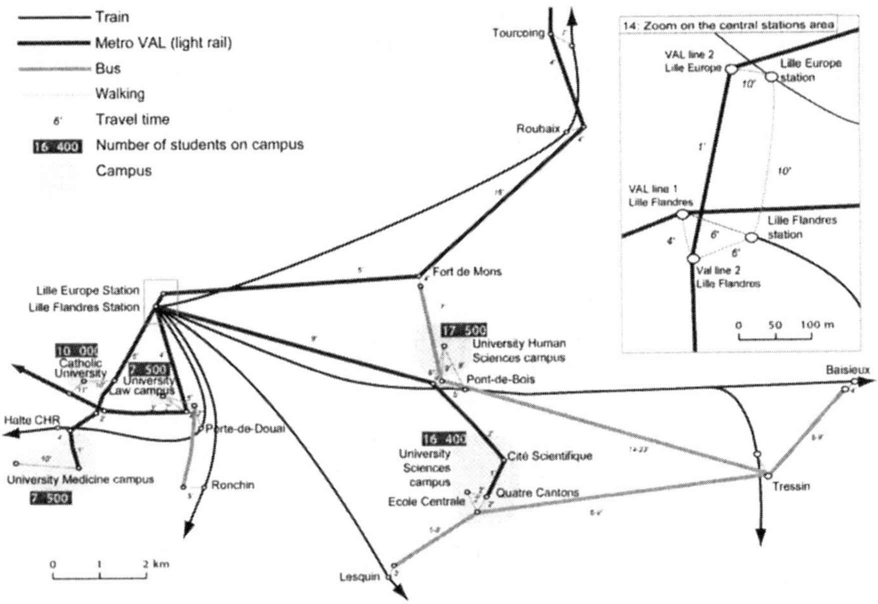

Figure 15.6 Train-bus-metro connections to the campuses of Lille
Source: Decoupigny and L'Hostis, INRETS 2001.

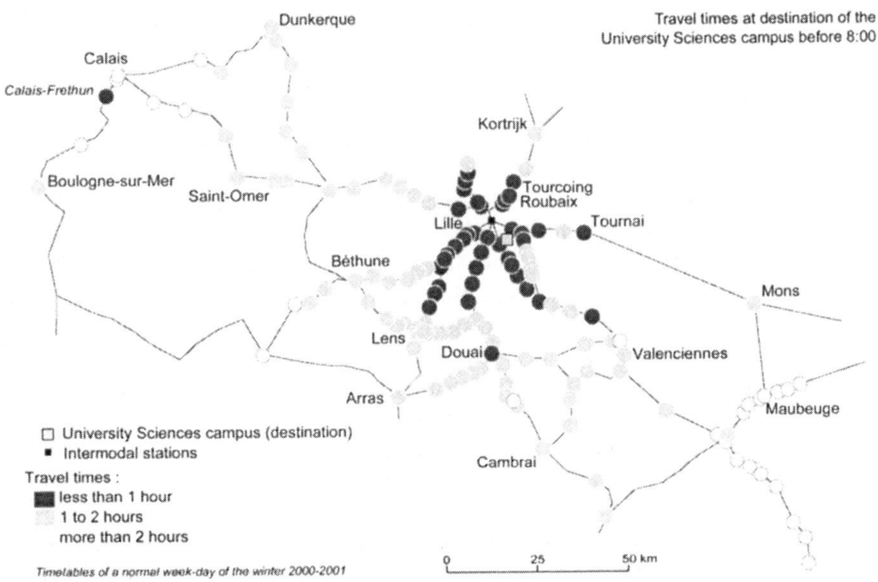

Figure 15.7 Scheduled accessibility to the University Sciences campus
Source: Decoupigny and L'Hostis, INRETS 2001.

We now consider the access to the University Medicine campus, located in the East of the metropolis, it seems difficult to take into account the central stations only, and to ignore the Hospital railway station ('Halte CHR'). Actually, its introduction into the network analysis provides a better level of accessibility from some towns. In particular, Béthune enters the set of stations at less than one hour, while the accessibility from Douai is slightly reduced.

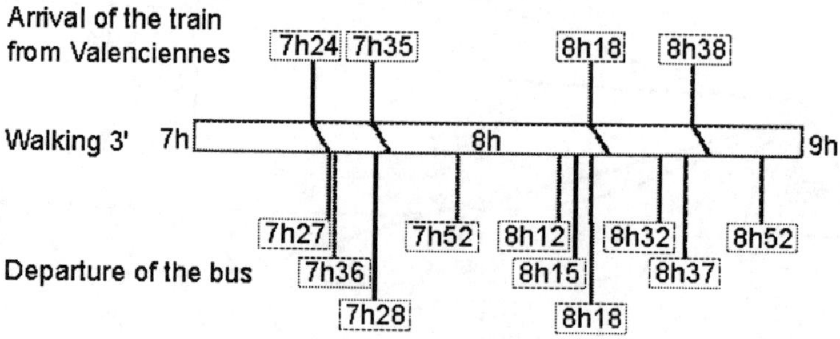

Figure 15.8 Existing train-bus connections at Lesquin in the direction of the University Sciences campus

At this point, it is worth considering whether the use of this new station would also improve the accessibility of some other campuses. Figure 15.6 shows that the University Law campus can also be reached that way using the two metro lines. Indeed, this alternative solution can be considered as credible, because of the very high frequency of the metro system during the peak period – 1'30'' between two metros – and the quality of the pedestrian link between the two platforms facing each other. Moreover, the arrival times at that station are likely to be more reliable than at Lille-Flandres, given the railway congestion at this central station. The effect of the use of the new station on accessibility is measured by the difference of departure times when this solution is included. It reveals a strong effect, resulting from the fact that the few minutes saved by going through the Hospital station means that it is possible to take a later train in Béthune and the surrounding stations. A gain of 45 minutes can thus be obtained for some stations in the Western part of the region, and the University Law campus can be reached in less than one hour from Béthune.

The *virtual network* approach can be applied to analyze access to the Eastern campuses using some bus lines. Among these, the bus connection at the Lesquin railway station can be taken as a credible alternative for the students coming from Valenciennes. A close examination of the connection conditions is presented in Figure 15.8.

Arrival times of trains from Valenciennes are given along the top of Figure 15.8, while departure times of the buses to the University Sciences campus are indicated along the bottom. Only one of the four trains during the morning peak period is connected to a bus, if we count three minutes walking from the railway station to the bus station. Clearly, the interconnection of trains and buses is not currently a criterion in the design of the bus timetables. Nevertheless, marginal modification of the bus timetables could clearly improve the quality of the rail-bus connections, as shown in Figure 15.9.

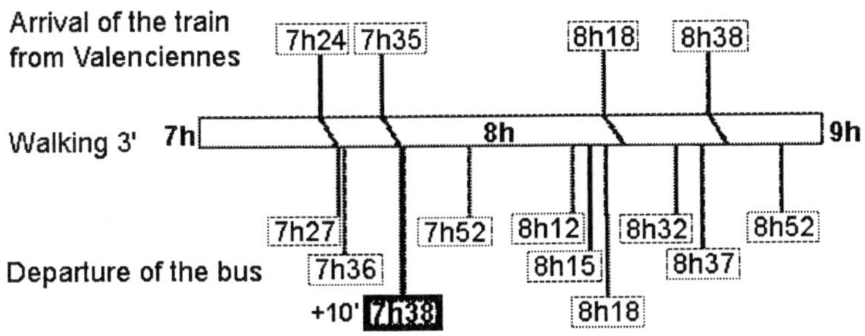

Figure 15.9 Proposed train-bus connections at Lesquin in the direction of the Sciences campus

Figure 15.9 shows that a simple 10 minutes delay for the arrival time of one bus at Lesquin would suffice to realize a good connection for the7h35 train which is the one predominantly used by the students. The introduction of a new bus service could also be considered. It is worth mentioning that the use of this second alternative solution would have no effect on the fares paid.

The improvement of accessibility brought by this modification is measured in Figure 15.10 by the difference between departure times when these two alternative connections – via the Lesquin and the Hospital stations – are activated. 15 minutes is gained at the departure from Valenciennes, which brings the city into the one-hour trip category. A similar travel time saving is observed at St-Amand, Orchies and Templeuve. Beyond Valenciennes, the savings are very high: half an hour from Le Quesnoy and 45 minutes from Maubeuge.

There is also an improvement in access to the University Sciences campus. Indeed, the metro line between Lille-Flandres central station and the Eastern campuses begins in the hospitals area. Hence, the students coming from the Lille-Béthune area and beyond may use the Hospital station instead of the Lille-Flandres station to take the metro. By doing, so they avoid the congestion of the central station and they save a few minutes that enables them to take the next train.

There is also some improvement in the accessibility of the University Human Sciences campus. Therefore, these modifications would provide much better accessibility to the major University campuses, which attract about one-third of the total regional student population.

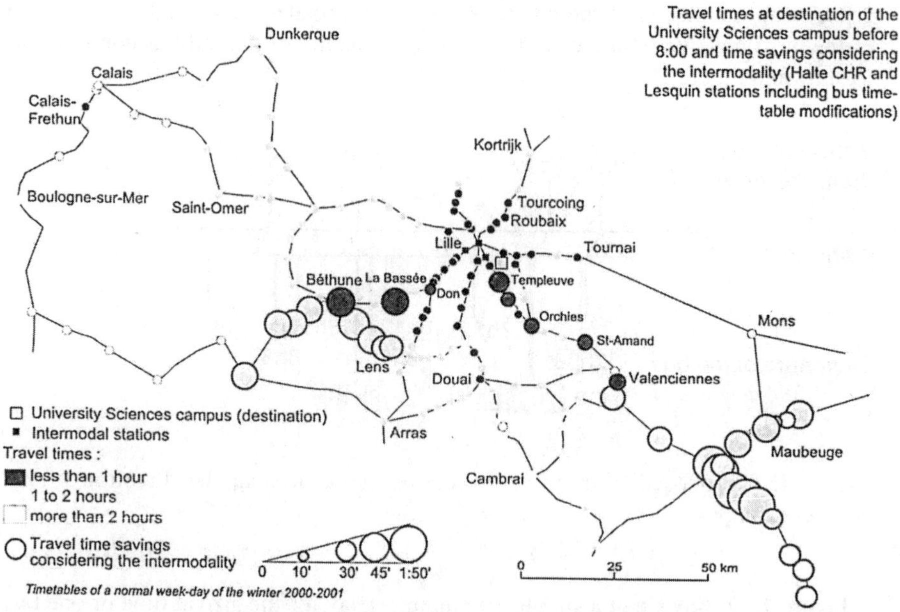

Figure 15.10 Accessibility improvement by the development of intermodality
Source: Decoupigny and L'Hostis, INRETS 2001.

15.7 Conclusion

This chapter proposes a methodology to assess of the transport system's contribution to the regional spatial organization. To illustrate this approach, the focus was placed on the role of Lille, the region's metropolis, in the regional urban system.

This methodology belongs to the field of strategic analysis. Nevertheless, it is shown that, in order to develop a correct analysis of transport services, it is necessary to proceed to a very detailed modelling of the transport supply that includes the full timetables of the different transport means and the specific needs of the different demand segments at different times of the day. A new family of accessibility indicators was proposed, which incorporates the timetable information as well as the timing and location of activities. As tools for the regional planning of transport services, they enable us to better measure the adequacy of the transport supply compared with the transport needs, and the quality of transport services.

The objective of connecting the cities of the Region to Lille in one hour – the emblematic measure of the 'hegemonic metropolis' spatial scenario – corresponds to a transport supply objective. Traditional accessibility indicators can be used in its assessment, but the proposed approach is better able to identify the real problems of accessibility. Applied to the case of the Lille region, it suggests, for instance, that the development of the scenario 'hegemonic metropolis' would benefit from improved connections to and from Maubeuge, and from a better frequency of evening return trips towards the littoral area. Combined with an analysis of the regional 'virtual network', this approach is also able to identify concrete alternatives like a new scheduling of buses or an improved connection of existing railway stations to the urban network. The proposed modifications would also contribute to a more successful transport implementation of the planning objectives.

References

Barré A 2000: Réseaux de transport et Réseaux Urbains dans de Nord-Pas-de-Calais, Revue du Nord, Vol. 82 (335–836), 527–535

Bertolini L. and Spit T. 1998: Cities on Rails: The Redevelopment of Railway Station Areas, E & FN Spon, London

Brunet R. 1986: L'Espace Règles du Jeu, in Auriac F. and Brunet R., Espaces, Jeux et Enjeux, Fayard, Paris, 297–315

Bruyelle P. 1994: Quels Avenirs Possibles? Le Chantier Grand Nord, Les Cahiers de l'Office Régional d'Habitat et d'Aménagement, Vol. 11, 22–25.

Bruyelle P. 1996: Le Grand Nord, in Leclerc R., Paris Y. and Wachter S. (eds.), Les Régions au Futur, DATAR/Editions de l'Aube, Paris, 67–71

CERTU and SNCF 1998: La Mobilité Régionale: Le Train et les Autres Modes de Transport, CERTU, Paris

Damette F. 1994: La France en Villes, La Documentation Française/DATAR, Paris

Direction Régionale de l'Equipement Nord-Pas-de-Calais 1999: Stratégie de l'Etat dans la Région Nord-Pas-de-Calais (2000/2006), Préfecture de la Région Nord-Pas-de-Calais, 15

Dupuy G. 1991: L'Urbanisme des Réseaux, Armand Colin, Paris

Giblin-Delvallet B. 1990: La Région, Territoires Politiques: Le Nord-Pas-de-Calais, Fayard, Paris

Huriot J.-M. and Perreure J. 1994: L'Accessibilité, in Auray J.-P., Derycke P.-H. and Huriot J.-M. (eds), Encyclopédie d'Economie Spatiale, Economica, Paris, 55–60

Martellato D. and Nijkamp P. 1998: The Concept of Accessibility Revisited, in Reggiani A. (ed.), Accessibility, Trade and Locational Behaviour, Ashgate, Aldershot, 17–40

Mathis P. and Baptiste H. et al. 1997: Grille de Niveau de Service, CESA

McFadden D.L. 1997: The Theory and Practise of Disaggregate Demand Forcasting for Various Modes of Urban Transportation, in Oum Th., Dodgson J.S. and Henscher D.A. et al., Transport Economics: Selected Readings, Harwood Academic Publishers, Amsterdam, 51–79

Menerault P. 1997: Dynamiques et Politiques Régionales Autour du Tunnel sous la Manche et du TGV Nord, Les Annales de Géographie, (593–594), 5–33

Menerault P. 1998: Processus de Terrotialisation des Réseaux: Analyse de la Grand Vitesse Ferroviaire à l'Échelle Régionale, NETCOM, Vol. 12, 1-2-3, 161–184

Menerault P. and L'Hostis A. 2000: Analyse des Relations Réseaux/Territoires Restructuration de l'Offre Ferroviaire de l'Axe Lille-Valenciennes-Jeumont, GRRT, Vol. 84

Nuzzolo A and Crisalli U. et al. 2000: A Behavioural Choice Model for the Evaluation of Railway Supply and Pricing Policies, Transport Research Part A, Vol. 5, 395–404

Orfeuil J.-P. 2000: L'Evolution de la Mobilité Quotidienne, Comprendre et Éclairer les Dynamiques, Éclairer les Controverses, INRETS, Arceuil

Pumain D. and Saint-Julien T. et al. 1990: La Géographie Universelle: France, Europe du Sud, Hachette/RECLUS, Paris

Rietveld P. and Bruinsma F. 1998: Is Transport Infrastructure Effective? Transport Infrastructure and Accessibility: Impacts on the Space Economy, Springer, Berlin

Trépanier M and Chapleau R. 1996: Un Modèle d'Analyse Désagrégré des Générateurs de Déplacements: Vers une Connaissance Détaillée de l'Utilisation des Lieux Urbains, 31ème Congrès de l'Association Québécoise du transport et des Routes, Mont Saint-Anne, Québec

---

Endnotes

[1] The French spelling of Dunkerque (not the English spelling 'Dunkirk') is used throughout this chapter, so that it matches the spelling on the maps and diagrams of the locatity.

[2] 'Document d'Orientations Régionales' (Document of Regional Orientations) juin 1999, and 'Rapport d'orientation sur le schéma régional ferroviaire des transports de voyageurs' (Orientation Report on the Passenger Rail Transport Regional Scheme), 1995.

[3] 'Schémas multimodaux de services de transport de voyageurs et de transport de marchandises', automne 2000 (Multimodal Freight and Passenger Transport Service Schemas) and 'Stratégie de l'Etat dans la Région Nord-Pas-de-Calais 2000-2006', 1999 (Strategy of the State in the Nord-Pas-de-Calais Region 2000-2006).

16 European Innovations in Multi-modal Assessment: How Will They Support and Enhance the Public Policy Process?

John Nellthorp

Institute for Transport Studies, University of Leeds, Leeds, United Kingdom

16.1 Background

This chapter is written as a comment on current practice in the multi-modal evaluation of transport projects. In particular, it attempts to trace the recent development of ideas underlying practical transport evaluation, during a period which has seen changes in methodology in many countries.

The chapter will highlight four specific areas of development which appear at first to be diverse in nature, even unconnected. By identifying the linkages between the four, the chapter will attempt to present them as part of an ongoing process of renewal – mostly of practice but also of theory – in this area of decision support.

This chapter will argue that the science of multi-modal transport evaluation is developing fast, and that at a time like this, a key challenge for policy makers is to achieve a balanced overall approach in multi-modal evaluation, making the most of the new and the established techniques.

In setting the scene, it is worth making clear what we mean by multi-modal evaluation and what are the sources of evidence on which this commentary is based.

16.2 Multi-modal Transport Appraisal

The subject matter of this chapter is current practice in multi-modal transport appraisal. That means that methods in use such as the UK Multi-Modal Studies appraisal (DETR 2000), the Netherlands OEEI framework (OEEI 2000), the German FTIP and EWS (PLANCO et al. 1993; FSV 1997), the French Secretaire d'Etat circulars (1995) and so on, are of central interest.

For the reader who is not familiar with these methods, one way of clarifying what we mean is to imagine a spectrum of evaluation methods. At one end of the spectrum are some theoretically more interesting evaluation techniques and methods which are less commonly used in practice. The methods at this end of the spectrum are diverse, but include sophisticated forms of multi-criteria analysis (MCA) which are capable of a much more advanced treatment of uncertainty than anything in common practice (e.g. Beuthe et al. 1998; Scanella and Beuthe 1999). The work of Nijkamp, Rietveld and others on spatial analysis and MCA also lays the theoretical foundations for much more rigorous forms of decision analysis in spatial planning (Nijkamp, Rietveld and Voogd 1990).

At the other end of the spectrum are simple planning aids which have little analytical content in themselves, but are useful in structuring the decision problem. Examples could include a simple statement of objectives, or a goals achievement matrix used without any supporting analysis. These sorts of tools consume few resources, and can be useful and appropriate for low cost projects, projects in the early stages of development, or where there are limited resources available for evaluation. Unsurprisingly, these are widely used in practice.

This chapter is concerned with what lies between these two ends of the spectrum – methods which:

- are in widespread use, to evaluate fairly substantial and complex changes to the transport system;

and

- have substantial analytical content, including the ability to estimate the benefits and costs of transport initiatives based on quantified predictions of the effects on users, operators and other agents within the economic-environmental system.

A key feature of these methods is that transport is analysed as a network service, with demand arising between origins *i* and destinations *j*. In multi-modal appraisal, the theoretical framework also recognizes that there are certain functional relationships between modes of transport *m* within the transport system. For example, where modes (or routes or services) are in competition, reductions in generalized cost on one mode can impact negatively on the demand for the competing mode. Where modes (or routes or services) are complementary, generalized cost reductions on one mode can impact positively on demand for the complementary mode. There are also relationships on the supply side: for example, provision of infrastructure for one mode *M* between i and j can improve supply conditions for other modes through the relief of congestion.

In principle, there is no reason why modes *m* cannot be extended to include non-motorized (and cleaner) modes such as walking and cycling. The fact that this is not often done in practice is a reflection of the scarcity of empirical data and models of pedestrian and cycle transport – and that situation is improving (e.g. Wardman, Hatfield and Page 1997).

Once a multi-modal *ijm* framework has been established, with appropriate functional relationships between modes, it becomes possible to give a comprehensive social evaluation of a range of initiatives:

unimodal[1] initiatives – such as rail infrastructure investment or inter-urban HGV[2] charging;

- multi-modal initiatives or 'packages' – such as combined transport terminals/rail infrastructure, or public transport interchanges, or complete urban transport strategies.

Predicting outcomes (equilibria of some kind) on a multi-modal network is a non-trivial task attended to by a range of forecasting models. Yet for evaluation, more is required than a set of matrices representing changes in flows and costs across the network:

- in the cost-benefit analysis (CBA) framework, estimates are required of changes in consumer and producer surpluses, some way into the future;
- in any framework, some key information has to be extracted and provided to the decision maker. What information is provided is influenced by value judgements about what is important. For example, measures of total travel distance on the network; total emissions of CO_2; measures of journey time reliability; numbers of households experiencing a reduction in transport noise. Which of these should be provided, and in what form, are key issues in multi-modal appraisal. We return to this point in Section 16.4 below.

In fact, most of the European multi-modal appraisal frameworks cited above do adopt CBA as the basis for the 'economic efficiency' aspect of the evaluation. Some go further, monetizing environmental effects such as noise and air pollution damage. Most, however, have a broader organizing framework related to a set of transport policy objectives – the organizing framework may be based on MCA principles, although the use of explicit scoring and weighting is surprisingly limited (Nellthorp, Mackie and Bristow 1998).

The eagle-eyed reader will have noticed that we have already used three different terms for assessment – apparently interchangeably. We will continue to do this throughout the chapter, whilst recognizing that in some circumstances it can be useful to draw distinctions between the three meanings. For example, 'appraisal' is often, but not always, used to refer to *ex ante* assessment; 'evaluation' is sometimes used to refer specifically to *ex post* assessment. Other authors use 'evaluation' as the general term for this field of study rather than 'assessment'. Fortunately, these semantics are not of great importance for this chapter. We will continue to use all three terms interchangeably. But it is perhaps worth noting that *ex ante* assessment – as an aid to forward planning – is the main application being discussed here.

It was also promised to say something about the evidence underlying the views expressed in this chapter. The author has been fortunate to be part of two studies involving comparisons of practice in different countries – the research study EUNET and the thematic network TRANS-TALK.[3] Comparative evidence in this chapter is drawn from these. The discussion of multi-modal cost-benefit analysis in the chapter draws heavily on the experience with EUNET, and on the author's involvement in the development of methodology for the current Multi-Modal Studies in the UK. It also relates to work just beginning on the IASON 5th

Framework project, which is concerned with integrating the spatial output and employment effects into evaluation. This background may also help the reader to understand any bias in the author's viewpoint.

In Sections 16.3-16.6, we go on to the main content of the chapter – four areas of development in multi-modal appraisal which reflect changes in thinking:

- firstly, alternative methodologies for forecasting and evaluating secondary economic effects, including spatial and network effects. The key developments here relate to spatial computable general equilibrium models (SCGE) and land-use and transport (LUTI) models, and the way in which their outputs can be brought into the assessment.
- secondly, refinements in the theory and practice of multi-modal evaluation in the UK – in particular, overcoming outstanding problems with the treatment of taxation, the management of information, and outputs to satisfy multiple stakeholders.
- thirdly, participative evaluation – where stakeholders are not only recognized but brought actively into the evaluation process, with various consequences for depth and flexibility of the assessment.
- fourthly, a substantial shift in 'official' evaluation methods from an efficiency perspective to a strategy-led perspective, with implications for the breadth and scope of the assessment (e.g., the 'New Approach to Appraisal' in the UK (DETR 1998)).

In bringing the four together in Section 16.7, the focus will be on the role of evaluation information in supporting decision makers and contributing to the wider debate on longer-term strategy.

16.3 Spatial Output and Employment Effects

The UK SACTRA and Dutch OEEI reports (SACTRA 1999; OEEI 2000) mark a subtle change in official thinking about wider economic impacts in transport appraisal. The old official 'line' in both countries was that the total economic benefits of new transport infrastructure were well represented by the transport sector benefits from a conventional CBA. This view was often backed-up by references to the theoretical and empirical literature, including the articles by Mohring and Williamson (1969) and Dodgson (1973). In practice, this line meant that the key appraisal information relevant to economic impacts was a cost-benefit analysis focused solely on the transport sector (we will abbreviate this to TCBA (Transport Cost-Benefit Analysis)).

16.3.1 Inadequacy of Conventional Approaches

The old 'line' was sometimes unsatisfactory from a policy viewpoint. This was evident from the use of various types of 'economic impact assessment' alongside

TCBA (for examples, see SACTRA 1999, Chapter 9). These assessments displayed great variety in methodology and assumptions, but what they had in common was the aim of identifying changes in employment and output in specific regional areas, as a consequence of the transport investment.

Some studies were concerned with the very localized area, or corridor, in which the project was located. This made sense where the appraising organization had a purely local remit or when most of the effects would be felt locally. The analysis usually proceeded by examining the potential for bringing unemployed resources into employment – e.g. empty commercial property; vacant sites for business and housing; and unemployed labour. This sort of analysis was most often used for local projects in localized areas of high unemployment. As exercises in evaluation, these types of analysis are notable for:

(i) their reliance on very detailed local information;
(ii) a high degree of input from stakeholders in the local economy who are in a position to understand the potential market responses to the project; and
(iii) the professional judgement of the analyst in filtering and assimilating the objective data and the judgements of others in reaching a conclusion – in place of a formal analytical model.

SACTRA (1999) pointedly remarked of these local evaluations that 'little of what is done can be described as standard practice'. Indeed, the variation in methodology from study to study has prompted calls for something more robust, consistent and transparent.

At the regional scale, other types of 'economic impact assessment' have been used. Certain types of models appear frequently: input-output analysis; regional growth/multiplier type models; and (sometimes crude) business location models. Again, however, SACTRA were critical:

- there is concern about partiality – the assessments tend to focus on impacts in the geographical area of interest to the sponsoring organization, so effects elsewhere (which may be positive or negative) are given little consideration. Such a partial picture is problematic from a country-level or an EU perspective.
- there is also a concern about double-counting – since the direct benefits in terms of travel time and cost have already been calculated in money terms in the transport sector, it is unclear to what extent any changes in output (or employment) are manifestations of the same benefits in the final goods market, or to what extent they are additional.
- perhaps most fundamentally, doubt exists as to whether the models used are capable of representing the complex network of linkages between the various markets involved: the labour market; the property market; passenger transport; freight transport; product markets; etc.

In view of these problems, part of the TRANS-TALK discussion focused on policy makers' needs in relation to the 'wider economic impacts' (Pearman, Mackie and Nellthorp 2001). It emerged that the needs of policy makers both at the EU and country level appear to be:

- for practical methods which can predict, *ex ante*, the spatial employment and output impacts of projects;
- for a clearer understanding of the relationship between this information and the conventional CBA results – in particular: Are there novel methods which can not only predict these wider economic impacts but also provide a comprehensive and robust measure of the net benefits of projects?

16.3.2 New Approaches to the Problem

In fact, there are two substantial streams of research addressing these questions, based on two model types;

- spatial computable general equilibrium (SCGE) models, and
- land-use transport interaction (LUTI) models.

There are substantive and fundamental differences between the two types, which we will come back to shortly, but at a conceptual level they have much in common. Both attempt to link spatial models of trade in the 'wider economy' with models of the (multi-modal) transport network. Transport is an input to production and to various consumption activities, so changes in transport costs have an influence in these models over business and residential location decisions, and over the level of activity at different locations. This means that at least some quantitative predictions of employment and output change can be produced, by spatial area.

The power of these models to predict spatial output and employment impacts derives from:

- representation of the main markets in the spatial economic system: product markets; labour market; land and property market; and the transport market;
- the way in which the forces which drive trade are captured – especially representation of economies of scale and economies of agglomeration, and productivity effects;
- the inclusion of appropriate feedbacks and equilibration.

Spatial CGE models, which relate to the new economic geography (Fujita, Krugman and Venables 1999), are beginning to be built to address transport policy issues (Oosterhaven and Knaap 2001; Venables and Gasiorek 1999). Spatial differentiation (into regions or sub-regions) sets these apart from country-level CGE models. When transport costs change, the system moves to a new equilibrium, which may include changes in the prices of goods and services, production, consumption and trade (in the short run), changes in the number of firms per region in each sector (in the medium term) and changes in the number of workers per region (in the long run) – as the production sector and the labour market adjust.

SCGE models are at present comparative-static models, emphasizing the comparison of alternative states of the world – this is ideal for use in project appraisal and evaluation. Another attractive feature is that SCGE models have a strong microeconomic basis, incorporating imperfect competition, heterogeneous products and economies/ diseconomies scale. These are believed to be significant in influencing trade patterns, and their omission from other families of models is a source of doubt about those models' predictive ability (SACTRA 1999).

Experience from testing an SCGE model (called RAEM – see Oosterhaven and Knaap 2001) with an inter-urban rail improvement project in the Netherlands has highlighted both the potential of these models and some problems. Some clear spatial patterns emerged from the analysis, including a regional redistribution of employment and some additional employment at the national level, as well as changes in output, consumer prices and consumer welfare.

However, problems with RAEM included the need to rely on 'guesstimates' for certain input data because not all the data required for this type of model were yet readily available. Also the analysis showed that the specification of transport costs is critical, and the 'iceberg type' cost specification used could be improved upon. Oosterhaven's view (SACTRA 1999; Oosterhaven and Knaap 2001) is that some difficulties are natural, since SCGE is in its first decade of development. In deciding whether to invest further resources in development of SCGE, he emphasizes the potential of this approach compared with the alternatives.

LUTI models are relatively well known, having evolved in the context of urban planning since the 1960s (Lowry 1964; Mackett 1990; Wilson 1998; DSC/ME&P 1999). What makes them novel, and relevant to the current policy needs, is that recent models within the LUTI family (see DSC/ME&P 1999) address the product market and labour market, not only land use. State-of-the-art LUTI models also have highly developed[4] transport sub-models, that are capable of providing the inputs for a conventional transport CBA.

Criticisms of LUTI models tend to focus on the 'trade' sub-models, which sometimes assume constant returns to scale and relatively crude pricing rules, compared with SCGE. This undermines their ability to model production and business location on an inter-urban scale (Oosterhaven and Knaap, 2001).

16.3.3 Potential to Meet the Needs of Strategic Planning

In the light of this discussion, it is interesting to reflect on whether the 'unresolved issue' is primarily a modelling problem or an evaluation problem. The modelling problem is that we lack credible and robust (and standardized) methods to predict spatial employment and output effects. Getting these quantities right clearly has a role to play in informing the decision maker and generating data for the evaluation. SCGE and LUTI appear to offer ways forward on this issue. But for evaluation, the outputs that are of interest are not only predictions of behaviour, but predictions of welfare – or changes in consumer and producer surplus – beyond the transport sector.

Oosterhaven and Knaap argue that SCGE models are inherently capable of providing outputs to evaluation, in the form of consumer and producer surplus change. The questions remain: Would these measures of welfare change supplant measures taken from the transport sector? Would the difference between the

SCGE benefits and the transport CBA benefits be treated as the 'wider economic impact' and reported separately in welfare terms? Or should the reporting of spatial output and employment effects remain as separate quantities in their own units – placing the onus on the decision maker to avoid double counting with the transport CBA?

Benefit estimation from LUTI models has not been a neglected field either (see Martinez and Araya 1998/2000; Jara-Diaz 1986). Key issues here are: in a framework containing linked land-use and transport models, Martinez and Araya (2000) show that the benefits measured in each market will be equal to the total social benefit only under certain restricted conditions; in the light, of this how should the transport sector benefits and land-use benefits be related to one another in estimating the total benefits? And how do these relate to benefit measures taken in the product or labour markets?

Further productive debate between the protagonists can be expected. The challenge for SCGE models seems to be to demonstrate their empirical validity by increasing the objectivity of the input data and testing more models in real-world applications. They also have some way to go before they can match the level of spatial detail in LUTI models, although that is less important at the EU/country/regional level than at the local urban level. Among the biggest challenges facing LUTI appear to be: representation of imperfect competition and non-constant returns to scale; and clarification of the position on benefit measures.

Given the possibilities which are created by both model types, further research and development of them seems important, including attention to the benefit measures arising. The European project IASON will take forward this range of issues.

16.4 Refinements in Theory and Practice of Multi-modal Appraisal

Multi-modality itself has been the impetus for a number of refinements in network-based appraisal.

16.4.1 The Unit of Account and the Treatment of Taxation

In multi-modal CBA, as in other CBA, there is a choice of unit of account: factor cost or market prices.[5] In the UK, for example:

$$B_{MP} = 1.209 * B_{FC}$$

where B_{MP} are benefits at market prices;
B_{FC} are benefits at factor cost; and
1.209 is the value of (1+t), where t is the average rate of indirect taxation in the economy.

Put simply, consumers think in terms of market prices, producers think of factor cost. For example, we have been aware for a long time that values of time derived from Stated Preference survey evidence among consumers will be at market prices. If the unit of account in CBA is FC, values of time at MP have to be adjusted by dividing by (1+t).

On the other hand, construction costs have often been naturally at factor cost because VAT is zero or reclaimable.[6] This implies that if the appraisal is at market prices, costs must be multiplied by (1+t) to be consistent with the unit of account.

The choice of factor cost or market prices as the unit of account does not affect the BCR or IRR measures of project performance, but it does affect NPV, and the size of the effect on NPV is obvious from the indirect tax factor, 1.209.

The recent focus on multi-modal CBA development (DETR 2000; OEEI 2000) has revived interest in the factor cost/market prices issue, not least because the tax treatment of transport differs between modes. Therefore, if transport policy is successful in securing modal shift towards public transport, governments can anticipate a fall in tax revenue. Robert Sudgen has produced two seminal papers for the UK DETR on this subject (Sugden 1999a,b); the findings have now been assimilated into standard appraisal practice. Essentially, every benefit item is expressed rigorously in market prices by making appropriate tax adjustments. For example:

Change in indirect tax revenue to government (work trips):

$$\left(F^1 - F^0\right)t'_F \frac{(1+t)}{(1+t'_F)} + \left(M^1 - M^0\right)t'\,M \frac{(1+t)}{(1+t'_M)} + \left(N^1 - N^0\right)t'\,N \frac{(1+t)}{(1+t'_N)}$$

where F are fuel costs (to private vehicle users only);
M are user charges (i.e. money costs);
N are non-fuel vehicle operating costs (to private vehicle users only);
t is the average rate of indirect taxation in the economy;
t'_F is the rate of taxation on fuel as an intermediate good;
t'_M is the rate of taxation on user charges as an intermediate good;
t'_N is the rate on non-fuel VOCs as an intermediate good.

One effect of this renewal of interest has been to bring to light some rather unclear thinking on safety values in current practice. The issue of the indirect tax correction appears to have been neglected when laying down appraisal values 10 years ago, with the result that the value used may have been too high by a factor of 1.209.

Secondly, Sugden's work prompted a review of the use of factor cost or market prices in transport appraisal. EUNET had found in 1997 that 14 out of 15 EU Member States used factor cost, although the odd one out was Sweden – arguably one of the most advanced countries in its practical CBA methodologies. The outcome of the review following Sugden's contributions was that the UK switched over to market prices, from March 2000. Many in the UK agreed with Sugden's point that the unit of account is immaterial as long as it is consistently applied.

Nevertheless, the switch over went ahead and is now embodied in the standard appraisal software (see below).

16.4.2 Management of Information

Having extended the mathematical framework for transport CBA to all modes (DETR 2000, Appendix F), it was also clear that the UK multi-modal studies would require careful management of the large volume of information which would be generated. Among the key ideas were:

- that standard software could be prepared for use by all those conducting multi-modal studies (roughly 25 studies are in progress at present) to carry out the user benefit calculations – this would greatly reduce the burden of auditing the appraisals, and would simplify the process of producing summary information from each study, since the software would have in-built the capability to summarize the outputs in various ways. This has been now been implemented (Mott Macdonald 2001).
- that the summary information could be reported in any number of formats, but at least in one standard format specified by DETR (the Transport Economic Efficiency table – Table 16.1). Behind every cell in this table, there is a formula defining exactly what quantity should appear there. This table notably gives a breakdown by group: transport users, transport operators and Government, reflecting the increasing interest in the institutional breakdown of costs and benefits in a multi-stakeholder appraisal situation.

16.4.3 Large Cost Changes

It is standard practice, in many of the countries surveyed by EUNET, to use the rule-of-a-half (RoH) as an approximation to the true consumer surplus measure of user benefits:

$$\Delta \mathrm{CS} \sim \mathrm{RoH} = \frac{1}{2}\sum_{ijm}\left(T_{ijm}{}^{0} + T_{ijm}{}^{1}\right)\left(GC_{ijm}{}^{0} - GC_{ijm}{}^{1}\right),$$

where: T^0_{ijm} are trips between i and j by mode m in the do-minimum scenario;
T^1_{ijm} are trips between i and j by mode m in the do-something scenario;
GC^0_{ijm} is the generalized cost of travel between i and j by mode m in the do-minimum scenario;
GC^1_{ijm} is the generalized cost of travel between i and j by mode m in the do-something scenario.

As the change in cost (GC^0–GC^1) increases, the more vulnerable the RoH estimate becomes to inaccuracy if the demand curve is non-linear. In situations where cost changes are large – for example new river/estuary crossings – an alternative approach to benefit estimation is needed. One has now been proposed, exploiting the RoH to its fullest extent rather than starting again from scratch (Nellthorp and Hyman 2001). This work also has some implications for the problematic subject of new modes, where GC^0 is undefined, and for the joint evaluation of transport and land-use changes.

Two consequences of the various developments discussed in this section should be to make multi-modal CBA practice more consistent with the theory which underlies it, and to reinforce the consistency of treatment between modes within multi-modal CBA. Both of these are essential to ensure the acceptability of multi-modal CBA results to decision makers.

16.5 Participation by Stakeholders

Network-based appraisal of transport projects has for a long time been complemented by 'consultation' with interested groups, such as local residents, representative agencies and pressure groups. 'Participation' by stakeholders represents a rather higher level of involvement. A recent example of road network strategy development from Finland (Linkama 2000) involved a group of invitees in setting the goals and developing the strategy, through to a final evaluation against the (agreed) goals. The invitees included government transport organizations, national and regional interest groups, district road authorities and other partners. The national road administration, FINNRA, was generally satisfied with the result, noting that the participants also gave very positive feedback. The assessment was, however, qualitative in nature, FINNRA noting that introducing the 'technical approach' of cost-benefit analysis into the process may have been counterproductive. The decision being evaluated was a choice of national strategy: broadly how to allocate resources between maintenance, construction, telematics and services, and in favour of which categories of user – car; pedestrians and cyclists; freight. Specific projects were not under discussion.

In another sector, the UK nuclear waste disposal agency, NIREX, has used participative evaluation methods to support decisions on where to locate its repositories – in that case, the participants were actively involved in a quantitative assessment. Over the course of five meetings, quantitative impacts on safety, the environment and costs were proposed by experts, debated, revised and agreed. A corresponding system of weights was also developed through discussion, leading to a ranking of projects and sensitivity testing. A linear additive MCA model was used, which is from a theoretical viewpoint equivalent to a simple cost-benefit analysis (Dodgson et al. 2001).

Participative appraisal methods are in their infancy in the evaluation of major transport initiatives. Their potential advantages include:

- greater ownership of the appraisal results by the decision makers, encouraging more use of appraisal evidence in the decision making process;

- a degree of consent to the findings of the appraisal, and to any recommendations emerging, on the part of the participants, who may be key stakeholders in the project or in potential opposition to the project;
- opening the project development process out to comment at an earlier stage, allowing the developer of the project, not only the evaluator, to benefit from the inputs of interested parties.

On the other hand, there is no question of the 'participants' becoming fully involved in every detail of a multi-modal network-based evaluation, yet such evaluations are – we have argued – important to provide an evidence-based assessment of many transport initiatives. Another view aired in the TRANS-TALK workshops was that participation will require careful study and control, if game-playing and the use of 'influence' are to be kept to a minimum.

The recent rise in experimentation with this type of appraisal has some parallels in CBA practice. In particular, we observed in Section 16.4 the trend towards presenting different perspectives in the appraisal, by giving an institutional breakdown (users, operators, government), rather than an all-encompassing 'social' NPV. In both cases, the aim appears to be to recognize the interests of distinguishable groups individually. We speculate that this could also be intimately related to governments shifting towards a more focused, objectives-led approach to transport policy: our last change in appraisal thinking, below.

16.6 Shift in Governments' Perspectives on Appraisal

Finally, we draw attention to a longer-term trend in transport appraisal methodology. Reading a transport appraisal text from the 1970s or early 1980s (see, e.g., Harrison 1974; Pearce and Nash 1981) it is difficult not to come away with the impression that social welfare maximization was the accepted goal of public intervention, and that cost-benefit analysis was the convenient tool with which contributions to this could be assessed. There are still some organizations which give this level of prominence to net social benefit, one of which is London Underground, whose appraisal methods are explicitly designed to ensure that social welfare is maximized, subject to the various constraints.

In recent transport appraisal methodology, however – including most of the examples listed in the Section 16.2 – the starting point has changed: transport projects should contribute to achievement of a *set* of transport policy objectives. A subset of these objectives may then be assessed in terms of net social benefit, and others not.

This shift towards an objectives-led approach to appraisal has had two important effects:

- it has broadened the scope of the appraisal, reintroducing a range of impacts which the cost-benefit analyst would in principle have wanted to include (for completeness), but which proved too challenging to monetize in a meaningful way;

- it has also brought home sharply the message that, for some key objectives, analytical methods in general, including evaluation methods, are poorly developed. Perhaps the best example is reliability, which consistently appears at or near the top of passengers' lists of priorities in attitudinal surveys, and has also been shown to be a key issue for logistics (Halcrow Fox 1995). Yet there is a lack of consistent measures of reliability across transport networks, a lack of data with which to calibrate models, and only the beginnings of an analytical approach in evaluation (DETR 2000). Many environmental impacts also fall into the category of 'included but not monetized' in modern multi-modal evaluation methods.

It has also been queried whether decision makers are willing to take into account non-monetized items when a major part of the benefits is given in monetary terms.

With these issues in mind, Nellthorp and Mackie (2000) probed the relationship between appraisal information and decision making. Their method was to search for a statistical model which 'explained' the decisions on 68 road schemes in the UK, decisions which were taken during a one-off Roads Review in 1997/8. *Ex ante* appraisals had been carried out using the CBA+framework type of approach, and the summary tables published (these are a good example of the management of information – see Annex 16.1). There were monetary costs and benefits for: project cost, journey time savings, safety and vehicle operating costs. There were also a mixture of quantities and scores for the other 12 sub-criteria. By regressing the variables from the appraisal (independent variables) on the decision variable (yes/no), systematic attempts were made to find a model[7] with good statistical properties. After several iterations, a model was found which gave a surprisingly good statistical fit. This model included 8 out of 15 evaluation criteria. Variables which dropped out included all the 'Accessibility' and 'Integration' criteria – it was suggested that the crudeness of the (new) evaluation methods on these criteria, and/or decision makers' lack of familiarity with them may have contributed to this. 'Air quality' also dropped out, potentially because it was so closely correlated with 'noise', which remained in.

Overall, the exercise gave some encouragement that decisions appeared to be related to most, if not all, of the appraisal information. Both cost-benefit and other forms of appraisal data appeared to have influenced the decision. In particular, the environmental and economic non-CBA variables appeared to play a role. We interpret this as evidence in support of the thesis that hybrid CBA+framework appraisals can be understood by decision makers, and that the objectives-led approach is not necessarily the end of the road for CBA.

It would be revealing to conduct further tests to explore how decision makers respond to appraisal information when a further layer of complexity is added, in the form of a quantitative risk assessment – for example using spreadsheet add-in software which is widely commercially available, and increasingly used for privately-financed projects. But this is in the future.

Table 16.1 Summarizing transport CBA results – an example from UK practice, 'Transport Economic Efficiency table'

Impact	TOTAL Present Value 1994 Prices and Values		Disaggregation by mode:			
User benefits & disbenefits						
Personal travel			Car	Bus and coach	Rail	Other
Travel time						
Vehicle operating costs						
User charges						
NET IMPACT						
Freight			Road freight	Rail freight	Other	
Travel time						
Vehicle operating costs						
User charges						
NET IMPACT						
NET IMPACT		(1)				
Private sector provider impacts			Road infrastructure	Bus and coach	Rail	Other
Revenues						
Operating costs		(a)				
Investment costs		(b)				
Grant/subsidy						
NET IMPACT		(2)				
Public sector provider impacts			Road infrastructure	Rail	Other	
Revenues						
Operating costs		(c)				
Investment costs		(d)				
NET IMPACT		(3)				
Other Government impacts			Road infrastructure	Bus and coach	Rail	Other
Grant/subsidy payments						
Indirect tax revenue						
NET IMPACT		(4)				
TOTAL						
Net Present Value, NPV		(5) = (1) + (2) + (3) + (4)				
Present Value of Costs, PVC		(6) = (a) + (b) + (c) + (d)				
Benefit:Cost Ratio, BCR		(7) = [(5) - (6)] / (6)				

Source: DETR (2000).

16.7 Links and Conclusions

We now return to the four areas of development in multi-modal evaluation, to consider the links between them, what appear to have been the driving forces of change, and what conclusions can be drawn for strategy formulation in transport and related sectors.

- We began by looking at tools with which to assess the spatial output and employment impacts of transport initiatives. We observed that the demand for this information is coming from policy makers, and that transport policy objectives such as 'Regeneration', 'Regional development', 'Employment' and so on reflect this demand. In an objectives-led appraisal (see Section 16.6), the existence of the objective creates a place in the appraisal for this information. Both SCGE and LUTI models are promising as predictive tools to generate the information.

  In relation to current multi-modal appraisal methods, SCGE and LUTI raise some interesting issues: both are capable of producing benefit measures outside the transport sector, so should these replace the multi-modal appraisal as the measure of total benefits? Are 'wider economic benefits' then the difference between the two? How robust are the benefit estimates from SCGE/LUTI models compared with conventional transport benefits?

  For strategic planning, LUTI models have already been applied at the European level, with external zones representing world trade. The continued development of these and SCGE models could be invaluable in forecasting the effect of European transport strategies on global competitiveness.
- The broadening of appraisal, following the move to an objectives-led approach, has increased the importance of the management of information. If decision makers are to be expected to weigh up projects on 10 or more criteria simultaneously, then effective summarizing and presentation will increasingly be a key task. Software has a role to play in delivering customized summaries to meet particular decision making needs.
- It has become more and more important to recognize multiple stakeholders – this has spilled over into CBA, in the form of results disaggregated by incidence group, and into an increasing interest in participative evaluation. Participative evaluation is both a potentially useful technique for strategy development, and a potential complement to analytical methods in assessing transport initiatives of all types on the network – including, for example, road pricing and network infrastructure measures.
- Multi-modality is having a profound effect over network-based evaluation. A range of theoretical and practical issues have been reconsidered, including the treatment of taxation and the appraisal of large cost changes and new modes.

In summary, then, perhaps the main driving forces behind recent appraisal developments have been:

- transport policy itself, acting through the objectives-led appraisal approach which is dominant at present;

- multi-modality, which has helped to rejuvenate transport cost benefit analysis; and
- changes in political structures, particularly increased public-private partnership and the role of interest groups as 'stakeholders' – multi-modal appraisal methods have responded with evaluation from different perspectives, based on the same underlying data, whilst interest has been stimulated in participative appraisal methods.

Looking ahead, we conclude that an important theme will be to reintegrate the various new methods described above. For example, we have argued that spatial economic models (LUTI and SCGE) have the potential to address employment, regional equity and related issues – issues which are the 'Achilles heel' of current assessment methods. We have also pointed to recent examples of stakeholder participation in assessment. Whilst we see both of these as 'progressive' moves, we recognize that one is taking appraisal in a more deeply analytical direction, with more data and more computing required; whilst the other is taking appraisal into an arena where discussion dominates. The two will somehow be reconciled in practice – the challenge for social scientists is to anticipate this and find positive roles for each within the overall appraisal process.

Another significant challenge arises in integrating spatial economic models (LUTI and SCGE) with the current state of the art in multi-modal assessment. As we have already seen, there are interesting issues here for modellers and economists to grapple with.

Overall, it seems possible that, as appraisal moves towards a broader, more policy-relevant basis, the *management* of the appraisal process – and the information which emerges from it – will become a more and more important aspect of the task.

Appendix 16.1 'Appraisal summary table

A6 Clapham, Bedford (GOER)	1996 scheme - 5km D2 bypass			Cost £30.9m
PROBLEMS	Poor safety and environment within Clapham (pop 3,200) where A6 carries up to 21,000 vpd (7% HGV). 300 residential properties + 2 schools front on to the road. Peak hour queuing occurs on length between village and northern outskirts of Bedford.			
OTHER OPTIONS	2 Pelican crossings already provided in village. Large scale traffic calming scheme would have unacceptable noise, air pollution and severance effects. Other options considered include reduced standard single carriageway bypass on proposed line; an eastern bypass and improved rail services to new Bedford North station with park and ride. All have inferior benefits.			

CRITERIA	SUB-CRITERIA	QUALITATIVE IMPACTS	QUANTITATIVE MEASURE	ASSESSMENT
ENVIRONMENTAL IMPACT	Noise	Properties within Clapham benefit from removal of through traffic.	No. properties experiencing: - Increase in noise 15 - Decrease in noise 316	301 properties experience net decrease in noise
CO_2 tonnes added 0 - 2000	Local air quality	Removal of through traffic by bypass will improve air quality within Clapham.	No. properties experiencing: - improved air quality 400 - worse air quality 0	-409 PM_{10} -2548 NO_2
	Landscape	Bypass partially within local area of Great Landscape Value and would result in loss of pasture land..	-	Slight -ve
	Biodiversity	County Wildlife Site affected.	-	Slight -ve
	Heritage	No significant impacts.	-	Neutral
	Water	Even with mitigation, there may still be: a significant risk of polluting a sensitive watercourse and an aquifer used for public water supply during both construction and operation; and an impact on flood risk as the scheme is within a floodplain and bridges a river.	-	Moderate -ve
SAFETY	-	Bypass reduces pedestrian/vehicle conflict in village and replaces a section of poor standard single carriageway.	Accidents Deaths Serious Slight 311 9 94 359	PVB £10.5m 54% of PVC
ECONOMY	Journey times & VOCs	Faster journey times on new bypass.	peak 4.6 mins inter-peak 3.5 mins	PVB £27.6m 141% of PVC
	Cost		-	PVC £19.6m
	Reliability		Route stress Before 104% After 38%	Slight Low rel to PVC
	Regeneration		Serves regeneration priority area?	No
ACCESSIBILITY	Public transport	Will help to reduce peak journey times of existing local bus services	-	Slight +ve
	Severance	Removes 80% of traffic from village.	-	Moderate +ve
	Pedestrians and others	Will improve accessibility for residents to local services.	-	Moderate +ve
INTEGRATION	-	Complements Bedford/Kempston package proposals and facilitates proposed Bedford North rail station with park and ride. Assists proposed local residential and commercial developments.	-	Positive
COBA			PVB £38.2m PVC £19.6m NPV £18.5m BCR 1.95	

Source: DETR (1988).

Acknowledgement

The author is grateful to the sponsors of the research on which this chapter draws: namely, the European Commission (Contracts ST-96-SC037 & 1999-TN.10869) and the UK Department of Transport, Local Government and the Regions, and to colleagues Peter Mackie and Alan Pearman in particular for a great many discussions. Needless to say, the author takes full responsibility for the views expressed.

References

Beuthe M., Grant-Muller S.M., Leleur S., Nellthorp J., Panou K., Pearman A.D., Rehfeld C. and Tsamboulas D. 1998: Innovations in Decision Analysis, Deliverable D10, EUNET Project, Marcial Echenique & Partners, Cambridge, UK

David Simmonds Consultancy and Marcial Echenique & Partners (DSC/ME&P) 1999: Review of Land Use and Transport Interaction Models, Report to SACTRA Committee, DETR, London

Department of the Environment, Transport and the Regions (DETR) 1998: Understanding the New Approach to Appraisal, DETR, London

Department of the Environment, Transport and the Regions (DETR) 2000: Guidance on the Methodology for Multi-Modal Studies, DETR, London

Dodgson J.S. 1973: External Effects and Secondary Benefits in Road Investment Appraisal, Journal of Transport Economics and Policy, Vol. 7, 2

Dodgson J.S., Spackman M., Pearman A.D. and Phillips L. 2001: Multi-Criteria Analysis: A Manual, DETR, London

Forschungsgesellschaft für Strassen- und Verkehrswesen (FSV) 1997: Empfehlungen für Wirtschaftlichkeitsuntersuchungen an Strassen (EWS): Aktualisierung der RAS-W 86, Vienna

Fujita M., Krugman P. and Venables A.J. 1999: The Spatial Economy: Cities, Regions and International Trade, MIT Press, Cambridge, MA

Halcrow Fox 1995: Business Benefits of Intelligent Transport Systems in Scotland, Report prepared for Scottish Enterprise, Halcrow Fox, Edinburgh

Harrison A.J. 1974: The Economics of Transport Appraisal, Croom Helm, London

Inter-Secretariat Working Group on National Accounts (CEC/IMF/OECD/UN/World Bank) 1993: System of National Accounts 1993, United Nations, New York

Jara-Diaz S.R. 1986: On the Relation between Users' Benefits and the Economic Effects of Transportation Activities, Journal of Regional Science, Vol. 26, 379–391

Linkama E. 2000: Impact Assessment of Strategic Road Management and Development Plan of Finnish National Road Administration. Paper presented at the 2nd TRANS-TALK Workshop, Hotel Metropole – Brussels, 6-8 November 2000, ICCR, Vienna

Lowry I.S. 1964: A Model of Metropolis, Rand Corporation, Santa Monica

Mackett R.L. 1990: The systematic application of the LILT model to Dortmund, Leeds and Tokyo, Transport Reviews, Vol. 10, 4, 323–338

Mackie P.J., Nellthorp J. and Pearman A.D. (eds) 2003: Transport Projects, Programmes and Policies: Evaluation Needs and Capabilities, Ashgate, Aldershot

Martinez F.J. and Araya C.A. 1998: Land Use Impacts of Transport Projects: User Benefits, Rents and Externalities, Paper presented at the 8th World Conference on Transport Research, Antwerp

Martinez F.J. and Araya C.A. 2000: Transport and Land-use Benefits under Location Externalities, Environment and Planning A, Vol. 32, 9, 1611–1624

Mohring H. and Williamson H.F. 1969: Scale and 'Industrial Reorganisation' Economies of Transport Improvements, Journal of Transport Economics and Policy, Vol. 3, 251–271

Mott MacDonald 2001: *TUBA* User Manual (Version 1.2), Mott MacDonald, Winchester

Nellthorp J. and Hyman G. 2001: Alternatives to The Rule of a Half in Matrix-based Appraisal, Paper presented at the European Transport Conference, Homerton College, Cambridge, 10–12 September 2001, PTRC, London

Nellthorp J. and Mackie P.J. 2000: The UK Roads Review: a Hedonic Model of decision-making, Transport Policy, Vol. 7, 2, 127–138

Nellthorp J., Mackie P.J. and Bristow A.L. 1998: Measurement and Valuation of the Impacts of Transport Initiatives, Deliverable D9, EUNET Project, Marcial Echenique & Partners, Cambridge, UK

Nijkamp P., Rietveld P. and Voogd H. 1990: Multicriteria Evaluation in Physical Planning, Elsevier, Amsterdam

Onderzoeksprogramma Economische Effecten Infrastructuur (OEEI) 2000: Appraisal of Infrastructural Projects: Guide for Cost-Benefit Analysis, RWS-AVV, Rotterdam

Oosterhaven J. and Knaap T. 2001: Spatial Economic Impacts of Transport Infrastructure Investments, in Pearman A.D., Mackie P.J. and Nellthorp J. (eds.), Projects, Programmes, Policies: Evaluation Needs and Capabilities, TRANS-TALK Deliverable D5, ICCR, Vienna

Pearce D.W. and Nash C.A. 1981: The Social Appraisal of Projects, Macmillan, London

PLANCO Consulting, BVU and Ingenieurbüro Heusch/Boesefeldt 1993: Macro-Economic Evaluation of Transport Infrastructure Investments: Evaluation Guidelines for the Federal Transport Investment Plan 1992, Final Report to the Federal Minister of Transport, PLANCO Consulting, Essen

Scanella G. and Beuthe M. 1999: MUSTARD User Guide, FUCAM, Mons

Secrétaire d'Etat aux Transports 1995. Instruction Cadre Relative aux Méthodes D'évaluation Économique des Grands Projets D'infrastructure de Transport, Jointe a la Circulaire du 3 Octobre 1995, Secrétaire d'Etat aux Transports, Paris

Standing Advisory Committee on Trunk Road Assessment (SACTRA) 1999: Transport and the Economy, The Stationery Office, London
Sugden R. 1999a: The Treatment of Taxation, Report to DETR, University of East Anglia, Norwich
Sugden R. 1999b: Developing a Consistent Multi-Modal Cost Benefit Analysis Framework, Report to DETR., University of East Anglia, Norwich
Wardman M.R., Hatfield R. and Page M. 1997: The UK National Cycling Strategy: Can Improved Facilities Meet the Targets?, Transport Policy, Vol. 4, 2, 123–133
Williams H.C.W.L. 1977: On the Formation of Travel Demand Models and Economic Evaluation Measures of User Benefit, Environment and Planning A, Vol. 9, 285
Williams I.N., Mackie P.J., Tsamboulas D. and Larkinson J. 1998: Assessing the Socio-Economic and Spatial Impacts of Transport Initiatives: The EUNET Project, Paper presented at the 8th World Conference on Transport Research, Antwerp
Wilson A. 1998: Land-use/Transport Interaction Models – Past and Future, Journal of Transport Economics and Policy, Vol. 32, 3

Endnotes

[1] Initiatives involving changes to one mode.
[2] Heavy Goods Vehicle.
[3] Funded by the European Commission under the 4th and 5th Framework Programmes, respectively. Both studies involved much more than a review. EUNET developed a suite of multi-modal modelling and evaluation tools and methods (Williams et al. 1998). TRANS-TALK is a Thematic Network, focusing on transport policy, programme and project evaluation methodologies. Its objective is to bring together transport researchers, professionals and policy makers who have worked on the development of these three topics, in different settings, to discuss commonalities and differences, and with a view to developing a framework for their integration (Pearman, Mackie and Nellthorp 2001).
[4] E.g. using random utility discrete choice modelling techniques, on a detailed network.
[5] Work on the System of National Accounts (SNA) (CEC et al. 1993) suggests there are theoretically more choices than these two, depending on precisely which taxes and subsidies are included and excluded, but these two relate most closely to established practice.
[6] Strictly, fuel tax on inputs should be considered also.
[7] This was a binomial logit model, with the yes/no decision represented as a 1/0 binary variable.

List of Figures

List of Tables

List of Contributors

Dr. Michel Beuthe
Groupe Transport et Mobilité (GTM)
Facultés Universitaires Catholiques de Mons
151, Ch. De Binche
7000 Mons
Belgium
Beuthe@message.fucam.ac.be

Dr. Yvonne Bontekoning
Delft University of Technology
OTB Research Institute
Thijsseweg 11
2629 AJ Delft
The Netherlands
Bontekoning@otb.tudelft.nl

Dr. Christophe Decoupigny
Université de Tours, CESA
Parc Grandmont
37200 Tours
France
Cdecoupigny@univ-tours.fr

Professor Martin Dijst
Urban Research Centre Utrecht (URU)
Faculty of Geographical Sciences
Utrecht University
PO Box 80115
3508 TC Utrecht
The Netherlands
m.dijst@geog.uu.nl

Dr. Job van Exel
Erasmus University Rotterdam (iMTA)
Institute for Medical Technology Assessment
PO Box 1738
3000 DR Rotterdam
The Netherlands
Vanexel@bmg.eur.nl

Professor Eran Feitelson
Department of Geography
The Hebrew University of Jerusalem
Mount Scopus
Jerusalem 91905
Israel
Msfeitel@mscc.huji.ac.il

Professor Marina van Geenhuizen
Faculty of Technology, Policy and Mangement
Delft University of Technology
PO Box 5015
2600 GA Delft
The Netherlands
m.s.vangeenhuizen@tbm.tudelft.nl

Professor Shalom Hakkert
Transportation Research Institute
Technion City
Haifa 3200
Israel
Hakkert@techunix.technion.ac.il

Dr. Hans van Ham
Delft University of Technology, Department of Transport
Policy and Logistics' Organisation
PO Box 5015
2600 GA Delft
The Netherlands
j.c.vanham@tbm.tudelft.nl

Dr. Leif Gjesing Hansen
Department of Environment Technology and Social Studies House 11.1
Roskilde University
PO Box 260
4000 Roskilde
Denmark
Gjesing@teksam.ruc.dk

Dr. Veli Himanen
JP-Transplan Ltd
P.O. Box 500
FIN-01621 Vantaa
Finland
Veli.himanen@poyry.fi

Dr. Alain L'Hostis
INRETS – TRACES
20 rue Elisée Reclus
09650 Villeneuve d'Ascq
France
Lhostis@inrets.fr

Dr. Sigal Kaplan
Transportation Research Institute
Technion City
Haifa 3200
Israel
Kaplan@techunix.technion.ac.il

Dr. Bart Kuipers
TNO Inro, section Logistics and Transport
PO Box 6041
2600 JA Delft
The Netherlands
Bku@inro.tno.nl

Dr. Jari Kurri
Department of Civil and Environmental Engineering
Laboratory of Transportation Engineering
Helsinki University of Technology
PO Box 2100
02015 Hut
Finland
Jari.kurri@hut.fi

Dr. Martin Lanzendorf
Utrecht University
Urban Research Centre
PO Box 80115
3508 TC Utrecht
The Netherlands
m.lazendorf@geog.uu.nl

Professor Cathy Macharis
Vrije Universiteit Brussels
Department of Business Economics and Strategic Management
Pleinlaan 2
1050 Brussels
Belgium
Cjmachar@vub.ac.be

Dr. Philippe Menerault
INRETS – TRACES
20 rue Elisée Reclus
09650 Villeneuve d'Ascq
France
Pmenerault@inrets.fr

Dr. John Nellthorp
Institute for Transport Studies
University of Leeds
38 University Road
Leeds LS2 9JT
United Kingdom
Jnelltho@its.leeds.ac.uk

Professor Peter Nijkamp
Department of Spatial Economics
Free University
De Boelelaan 1105
1081 HV Amsterdam
The Netherlands
Pnijkamp@feweb.vu.nl

Professor Hugo Priemus
OTB Research Institute for Housing, Urban and Mobility Studies
Delft University of Technology
Thijsseweg 11
2629 JA Delft
The Netherlands
Priemus@otb.tudelft.nl

Professor Aura Reggiani
Department of Economics
University of Bologna
Piazza Scaravilli 2
40126 Bologne (BO)
Italy
Reggiani@economia.unibo.it

Professor Piet Rietveld
Free University
Department of Spatial Economics
De Boelelaan 1105
1081 HV Amsterdam
The Netherlands
Prietveld@feweb.vu.nl

Professor Ilan Salomon
Department of Geography
The Hebrew University of Jerusalem
Mount Scopus
Jerusalem 91905
Israel
Msilans@mscc.huji.ac.il

Dr. Jens Schade
Traffic and Transportation Psychology
Dresden University of Technology
Hettmerstraße 1
01069 Dresden
Germany
Schade@verkehrspsychologie-dresden.de

Dr. Yoram Shiftan
Transportation Research Institute
Technion City
Haifa 3200
Israel
Shiftan@techunix.technion.ac.il

Dr. Jan Jacob Trip
Delft University of Technology
OTB Research Institute
Thijsseweg 11
2629 JA Delft
The Netherlands
Trip@otb.tudelft.nl

Dr. Luca Zamparini
Faculty of Law
University of Leece
Via per Monteroni snc
73100 Lecce
Italy
Luca.zamparini@ateneo.unile.it

Titles in the Series (continued)

M. M. Fischer and *J. Fröhlich* (Eds.)
Knowledge, Complexity and Innovation Systems
XII, 477 pages. 2001. ISBN 3-540-41969-1

M. M. Fischer, J. Revilla Diez and *F. Snickars*
Metropolitan Innovation Systems
VIII, 270 pages. 2001. ISBN 3-540-41967-5

L. Lundqvist and *L.-G. Mattsson* (Eds.)
National Transport Models
VIII, 202 pages. 2002. ISBN 3-540-42426-1

J. R. Cuadrado-Roura and *M. Parellada* (Eds.)
Regional Convergence in the European Union
VIII, 368 pages. 2002. ISBN 3-540-43242-6

G. J. D. Hewings, M. Sonis and *D. Boyce* (Eds.)
Trade, Networks and Hierarchies
XI, 467 pages. 2002. ISBN 3-540-43087-3

G. Atalik and *M. M. Fischer* (Eds.)
Regional Development Reconsidered
X, 220 pages. 2002. ISBN 3-540-43610-3

Z. J. Acs, H. L. F. de Groot and *P. Nijkamp* (Eds.)
The Emergence of the Knowledge Economy
VII, 388 pages. 2002. ISBN 3-540-43722-3

R. J. Stimson, R. R. Stough and *B. H. Roberts*
Regional Economic Development
X, 397 pages. 2002. ISBN 3-540-43731-2

S. Geertman and *J. Stillwell* (Eds.)
Planning Support Systems in Practice
XII, 578 pages. 2003. ISBN 3-540-43719-3

B. Fingleton (Ed.)
European Regional Growth
VIII, 435 pages. 2003. ISBN 3-540-00366-5

T. Puu
Mathematical Location and Land Use Theory, 2nd Edition
X, 362 pages. 2003. ISBN 3-540-00931-0

J. Bröcker, D. Dohse and *R. Soltwedel* (Eds.)
Innovation Clusters and Interregional Competition
VIII, 409 pages. 2003. ISBN 3-540-00999-X

D. A. Griffith
Spatial Autocorrelation and Spatial Filtering
XIV, 247 pages. 2003. ISBN 3-540-00932-9

J. R. Roy
Spatial Interaction Modelling
X, 239 pages. 2004. ISBN 3-540-20528-4